CURRENT
INTERRUPTION
IN HIGH-VOLTAGE
NETWORKS

Earlier Brown Boveri Symposia

Flow Research on Blading • 1969
Edited by L. S. Dzung

Real-Time Control of Electric Power Systems • 1971
Edited by E. Handschin

High-Temperature Materials in Gas Turbines • 1973
Edited by P. R. Sahm and M. O. Speidel

Nonemissive Electrooptic Displays • 1975
Edited by A. R. Kmetz and F. K. von Willisen

CURRENT INTERRUPTION IN HIGH-VOLTAGE NETWORKS

Edited by
Klaus Ragaller
BBC Brown, Boveri & Company Limited

PLENUM PRESS • NEW YORK AND LONDON

Library of Congress Cataloging in Publication Data

Brown Boveri Symposium on Current Interruption in High-Voltage Networks, Baden, Switzerland, 1977.
Current interruption in high-voltage networks.

Includes indexes.
1. Electric circuit-breakers – Congresses. 2. High voltages – Congresses. 3. Electric networks – Congresses. I. Ragaller, Klaus. II. Brown Boveri Research Center. III. Title.
TK2842.B765 1977   621.31'7   78-6057
ISBN 0-306-40007-3

Proceedings of the Brown Boveri Symposium on Current
Interruption in High-Voltage Networks held at the Brown Boveri
Research Center, Baden, Switzerland, September 29–30, 1977

© 1978 Plenum Press, New York
A Division of Plenum Publishing Corporation
227 West 17th Street, New York, N.Y. 10011

All rights reserved

No part of this book may be reproduced, stored in a retrieval system, or transmitted, in any form or by any means, electronic, mechanical, photocopying, microfilming, recording, or otherwise, without written permission from the Publisher

Printed in the United States of Amercia

FOREWORD

Shortly after the establishment of the Brown Boveri Research Center in 1967, plans were developed to organize a series of Brown Boveri Scientific Symposia, each having a different topic, to be held every other year in Baden. We choose the subject for a symposium with the following requirements in mind:

- It should characterize a part of a scientific discipline; in other words it should concern an area of scholarly study and research.
- It should be of current interest in the sense that important results have recently been obtained and considerable research is under way in the world's scientific community.
- It should bear some relation to the scientific and technological activity of the Company.

These symposia are intimately related to one of the very basic concepts which have governed the work of many modern manufacturing companies: close coupling between science and engineering. It is to this coupling that we owe the technical standard of our products, and it is this coupling which we hope to be furthered by our symposia.

It is often said that the most important technological innovations come from the basic sciences, and the transistor is taken as an outstanding example. Indeed, the invention of the transistor was the result of research into quantum mechanics, a fundamental scientific discipline which predicts the behaviour of electrons in a crystal lattice and explains the existence of energy bands. Without a knowledge of quantum mechanics, the invention of the transistor would have been impossible. This is a model case for the following order of events: scientific discovery; then followed by engineering invention.

It is important to note that this course, although frequent, is not the only one along which technological progress can take place. In fact, the opposite sequence is quite common: first, engineering invention; then discovery of the underlying scientific principles. The subject of our symposium is almost a model for this latter process. The first high-power circuit breakers were designed and successfully operated at a time when very little was known about the physics of electric arcs. A theoretical treatment was not feasible even for the steady state, and the engineers used very simple arc models to describe current interruption. These models had a more or less intuitive basis and allowed many qualitative aspects to be simulated. However, the limitations in predicting the complex phenomena shortly before and after current zero became more and more obvious as development progressed. Not only was the theory deficient, but also experimental tools could give only a crude picture of what actually went on in a circuit breaker. That, despite these weaknesses, designers were able to build circuit breakers with a breaking capacity of several GVA per chamber is an impressive demonstration of the deep intuitive insight of these men. For further progress, however, application of basic scientific principles is essential: we have come as far as intuition will take us. The inventor still plays an important part in this process, but he must interact closely with the scientist and frequently ask himself: How far from the optimum is my invention? Am I expecting something that violates the fundamental laws of nature? What can be achieved with my idea? Without answers to these questions, technical progress would be almost impossible beyond a certain point, as is the case, not only for circuit breakers, but for all fields of technology.

As we have said, the coupling of science and engineering is essential in today's technological world. Yet the scientists on one hand and the inventors and engineers on the other are two very different breeds of people with very different modes of work. Inventors and engineers, by the very nature of their activity, have to perform most of their work behind closed doors. They are barred from excessive publicity, and it is doubtful whether they would seek it even if it were permitted. Scientists on the other hand form one large, worldwide community with many ties and links. To keep scientific results secret is useless. The body of scientific knowledge is a single large pool to which many sources contribute. It is unwise for anyone to draw from this pool without making a contribution towards replenishing it. Throughout the years of its existence our company has both drawn liberally from the pool, and has also contributed its share of replenishment. In 1891, the year in which Brown Boveri was founded, Charles E. L. Brown, one of the co-founders, published a paper entitled 'High voltages, their generation, transmission and distribution' in the Proceedings of the German Electrotechnical Society. This fundamental paper described a systems aspect which was to have far-reaching effects. Brown points out that with voltages of 2 - 4 kV, then generally considered to be the upper limit of what was technologically feasible, power transmission on a larger scale would not be possible, and he proceeds to explain the systems components necessary for a power network with 30 or even 40 kV. This of course was a major step forward. The paper gives a detailed description of transformers, breakers, and transmission lines. It is impressive to see how the author places the systems aspect, not the components aspect, at the beginning of his reasoning, and designs the components to suit the system.

FOREWORD

Brown immediately proceeded to implement his ideas. He pioneered the design of transformers for higher voltages and thus opened the door to a new area in power networks. Charles Brown's paper undoubtedly was a milestone in electrical engineering. Scientific and technical papers published by Brown's successors in Switzerland, Germany and France in the following eight decades have been numerous and significant. One of the visible proofs of the Company's management determination to continue this tradition was the dedication of the building at Dättwil four years ago, where research groups are at work to help assure the technological future of the Corporation.

The symposium 'Current Interruption in High-Voltage Networks' was attended by 101 participants from 16 countries. It was both an honour and a pleasure to welcome these scientists and engineers from so many parts of the world. Their willingness to travel to Baden and spend two full days with us was a challenge as well as an obligation to us as organizers, and we sincerely hope that the expectations which prompted them to attend were fulfilled.

We sincerely hope that the papers and discussions presented during the two days have inspired further investigations which will advance our knowledge in this sphere. The future trends in current interruption in high-voltage networks were clearly indicated during the symposium.

To conclude, we should like to take this opportunity to express our sincere gratitude to every participant in the Symposium. We hope they consider the time spent with us to have been worthwhile. Thanks are due primarily to the authors for having spared no effort in preparing their papers: the contents of this volume reflect the high quality of their work. We thank also the participants in the discussions, both official and unofficial, and the editor of the proceedings. - Our sincere thanks are extended to Dr. K. Ragaller, scientific secretary and Mr. E. Arn, administrative secretary who, by their considerable efforts and close attention to detail, contributed much to the material content and smooth running of the Symposium.

Owing to our limited space and the desire to avoid an unwieldily large congress, it was not possible to extend invitations to a larger circle; and it was with much regret that we were forced to disappoint many people who wished to participate. We hope that the publication of the present volume is partial consolation for those whom we did not have the pleasure of greeting as our guests.

A. P. Speiser
Director of Research

PREFACE

Scientists have been working on the problem of current interruption in high-voltage networks ever since the introduction of high-voltage transmission lines. In recent decades, decided advances have been made in the development of circuit breakers which are reflected, for example, in an increase of breaker capacity per chamber or a decrease of the weight with respect to breaker capacity. The dynamic development of circuit breakers is also reflected in the diverse variety of quenching media now in use, such as oil, air, SF_6, or vacuum.

Recently, circuit-breaker development has been strongly influenced by research in the area of plasma physics. This research has provided methods which have made possible investigations of the underlying physical aspects of breaking phenomena. Development, which until now has proceeded on a purely empirical basis, thus receives important new impulses the effects of which are of an absolutely basic nature. One significant example, which is typical, is the question of determining the fundamental physical limits of the breaking process and its dependence on the most important parameters.

The more fundamentally such a question is treated, the more general are the conclusions and the less their dependence on specific design features. These general results are just those which have to be included in an overall systems approach to the investigation of current interruption in a high-voltage network. Numerous questions exist which can only be answered if the fundamental electrical phenomena occurring in the network and the physics involved in the breaking process are considered simultaneously.

These considerations contributed to the decision of devoting the 5th Brown Boveri Symposium to 'Current Interruption in High-Voltage Networks'. Leading scientists were invited to deliver detailed review papers intended to illustrate some of the most important aspects of this topic. The Sympsoium took place on September 29 and 30, 1977, at the Brown Boveri Research Center in Baden, Switzerland.

The Symposium began with an introductory paper by K. Ragaller and K. Reichert which gave a general overview of the problems involved in current interruption. The second part of the Symposium was devoted to the current-zero regime, that is, the thermal interruption mode, and the third part with the peak regime of recovery voltage, the dielectric interruption mode. Network specialists and plasma physicists dealt jointly with the relevant problems in each of these three sections.

PREFACE

The presentations were organized in such a way that the network papers (of M. B. Humphries, C. Dubanton and J. L. Diesendorf), together with the introductory paper, provide an overview of the current status in this area. Also, the authors attempted to provide a didactic structure, in order that these reports might be recommended as an introduction to this area.

A similar pattern of organization was followed in the papers concerned with physical aspects. In this case it must be mentioned that some of the underlying principles are extremely specialized and difficult. In many areas the research is still not complete. The point to be made here is that the roots reach quite far into various specialized physical disciplines and only the careful cultivation of interdisciplinary contacts will yield the general results desired. An example is the contribution of J. J. Lowke, who is investigating radiant-energy transport which directly influences the voltage limit of high-voltage circuit breakers. This work requires the current results of research in the area of atomic physics. Whereas, a prerequisite for general statements about the voltage limit is knowledge of radiant-energy transport, for the current limit the corresponding requirement is basic information about turbulent heat transport (G. Jones' paper).

As another example, any field engineer will attest to the fact that the switching process is incalculable, paradoxical, complex and sensitive to the minutest changes. Experience shows that careful scientific analysis can put things in order. Even such complex processes as are associated with the arc root can be understood by this methodology, as is well exemplified by J. Mentel's paper.

The papers by G. Frind, B. W. Swanson, W. Hermann and K. Ragaller show to what extent general conclusions can already be derived from fundamental phenomena for the thermal mode, whether in the form of limiting curves or theoretical descriptions of arc behavior in the network. The papers of T. H. Teich and W. Zaengl and of J. Kopainsky give an impression of the present state of research in the area of the dielectric mode. Here understanding has not progressed to the same extent as for the thermal mode.

Each individual paper was followed by very intensive and interesting discussion. An edited version of these discussions and comments, based upon written notes and tape recordings, have been included in this volume.

It is a pleasure to acknowledge the most gratifying cooperation of all authors with my editorial attempts to impose consistency of style and format on contributions from diverse scientific disciplines and nationalities. I am grateful for their responsive and considered support which made the timely completetion of this book possible. I also acknowledge the indispensable efforts of Mrs. U. Richter who typed the entire camera-ready manuscript. Thanks are also due to Dr. A. M. Escudier for her extensive help in the editorial work.

Baden, March 1978   K. Ragaller

CONTENTS

Foreword   v
Preface ix
List of Participants xiii
Abbreviations xix

INTRODUCTION AND SURVEY: PHYSICAL AND NETWORK PHENOMENA

    K. Ragaller and K. Reichert   1
    Discussion 26

Part 1: Current-Zero Regime / Thermal Interruption Mode

TRANSIENT RECOVERY VOLTAGE IN THE SHORT LINE FAULT REGIME

    M. B. Humphries 29
    Discussion 64

EXPERIMENTAL INVESTIGATION OF LIMITING CURVES FOR CURRENT INTERRUPTION

    G. Frind 67
    Discussion 93

THE INFLUENCE OF TURBULENCE ON CURRENT INTERRUPTION

    G. R. Jones 95
    Discussion 116

THE INFLUENCE OF ARC ROOTS ON CURRENT INTERRUPTION

    J. Mentel 119
    Discussion 134

THEORETICAL MODELS FOR THE ARC IN THE CURRENT-ZERO REGIME

    B. W. Swanson 137
    Discussion 179

INITIAL TRANSIENT RECOVERY VOLTAGE

    M. Dubanton 185
    Discussion 201

INTERACTION OF ARC AND NETWORK IN THE ITRV-REGIME

    W. Hermann and K. Ragaller 205
    Discussion 226

Part 2: Peak Regime of Recovery Voltage / Dielectric Interruption Mode

DETERMINATION OF THE PEAK TRANSIENT RECOVERY VOLTAGE

    J. Diesendorf, S. K. Lowe and L. Saunders 231
    Discussion 264

THE DIELECTRIC STRENGTH OF AN SF_6 GAP

    T. H. Teich and W. S. Zaengl 269
    Discussion 294

RADIATIVE ENERGY TRANSFER IN CIRCUIT BREAKER ARCS

    J. J. Lowke 299
    Discussion 327

INFLUENCE OF THE ARC ON BREAKDOWN PHENOMENA IN CIRCUIT BREAKERS

    J. Kopainsky 329
    Discussion 352

Author Index 355
Subject Index 357

LIST OF PARTICIPANTS

1. E. Arn
   Brown Boveri Research Center
   CH-5405 Baden, Switzerland

2. G. Bär
   Schluchseewerk AG.
   Rempartstr. 12-16
   D-7800 Freiburg, W. Germany

3. D. M. Benenson
   State University of New York
   at Buffalo
   Department of Electrical Eng.
   4232 Ridge Lea Road
   Amherst, New York 14226, USA

4. K. D. Bennemann
   VEW Vereinigte Elektrizitätswerke
   Westfalen AG.
   Postfach 941
   D-4600 Dortmund, W. Germany

5. A. N. Bird
   State Electricity Commission
   of Victoria
   Monash House
   15 William Street
   Box 2765 Y GPO
   Melbourne, Vic. 3001, Australia

6. J. E. Bird
   Ewbank and Partners Ltd.
   Prudential House, North Street
   Brighton, E. Sussex, BN1 5BD,
   England

7. W. Bötticher
   Institut für Plasmaphysik
   Technische Universität Hannover
   Callinstr. 38
   D-3000 Hannover, W. Germany

8. E. v. Bonin
   AEG-Telefunken
   Hochspannungsinstitut
   Postfach 100 129
   D-3500 Kassel, W. Germany

9. K. P. Brand
   Brown Boveri Research Center
   CH-5405 Baden, Switzerland

10. A. Brauch
    BASF-Ludwigshafen
    Leuschnerstr. 44
    D-6700 Ludwigshafen, W. Germany

11. T. E. Browne
    5 Woodside Road
    Pittsburgh, PA 15221, USA

12. J. Bruhin
    Ingenieurunternehmung AG der
    Schweizerischen Elektrizitäts-
    und Verkehrsgesellschaft
    Postfach
    CH-4010 Basel, Switzerland

13. G. Catenacci
    CESI
    Via Rubattino 54
    I-20134 Milano, Italy

14. M. D. Cowley
    University Engineering Lab.
    Trumpington Street
    Cambridge, CB2 1PZ, England

15. N. Cuk
    British Columbia Hydro
    B. C. Hydro Power Authority
    700 West Pender Street
    British Columbia, V6C 2S5, Canada

16. D. J. Cunningham
    The Electricity Commission of N.S.W.
    Box 5257, G.P.O.
    Sydney, N.S.W. 2001, Australia

17. G. C. Damstra
    Kema
    Utrechtseweg 310
    Arnheim, Netherlands

18. J. L. Diesendorf
    Transmission Line
    Electrical Design Dept.
    The Electricity Commission of N.S.W.
    Box 5257, G.P.O.
    Sydney, N.S.W. 2001, Australia

19. C. Dubanton
    Electricité de France
    Direction des Etudes et Recherches
    Service Matériel Electrique
    1, av. du Général de Gaulle
    BP 408
    F-92141 Clamart-CEDEX, France

20. A. Eidinger
    BBC Brown, Boveri & Company, Ltd.,
    Dept. AG.
    CH-5401 Baden, Switzerland

21. G. Frind
    General Electric Company
    Research and Development Center
    B. P. Box 8
    Schenectady, New York 12301, USA

22. R. Gampenrieder
    c/o Bayernwerk AG.
    Blütenburgstr. 6
    D-8000 Munich 2, W. Germany

23. I. Gasia
    Gerencia de Construccion
    Edificio Centro Electrico Nacional
    Avenida Sanz - El Marqués
    Caracas 107, Venezuela

24. W. F. Geist
    B. C. Hydro & Power Authority
    100 West Pender Str.
    Vancouver, B. C. V6C 2S5, Canada

25. R. Gröber
    FGH
    Postfach 810 169
    D-6800 Mannheim 81, W. Germany

26. P. J. R. Grool
    Planning & Research Department
    N. V. Electriciteitsbedrijf
    Zuid-Holland
    Alexanderstraat 16
    Den Haag, Netherlands

27. O. Hauge
    Electrical Department
    NVE - Statskraftverkens
    Middelthunsgt. 29
    Oslo 3, Norway

28. W. Hermann
    BBC Brown, Boveri & Company, Ltd.
    Dept. AGY
    CH-5401 Baden, Switzerland

29. G. Herrmann
    Grosskraftwerk Mannheim AG.
    Aufeldstr. 23
    D-6800 Mannheim, W. Germany

30. W. Hertz
    Gruppe Plasmaphysik
    Siemens Forschungszentrum
    Günther Scharowskystr. 2
    D-8520 Erlangen, W. Germany

31. E. Hoffmann
    Badenwerk
    Postfach 1680
    D-7500 Karlsruhe, W. Germany

32. P. Hummel
    BBC Brown, Boveri & Company, Ltd.
    Dept. KL
    CH-5401 Baden, Switzerland

33. M. B. Humphries
    Central Electricity Generating Board
    Planning Department
    Sudbury House
    15 Newgate Street
    London EC1, England

34. G. R. Jones
    The University of Liverpool
    Arc Discharge Research Project
    131 Mount Pleasant
    Brownlow Hill
    P. O. Box 147
    Liverpool, England

35. L. H. A. King
    Head Circuit Interruption Lab.
    GEC Switchgear Limited
    Stafford Works
    Lichfield Road
    Stafford, England

36. G. Köppl
    BBC Brown, Boveri & Company, Ltd.
    Dept. AGT
    CH-5401 Baden, Switzerland

37. G. Körner
    Brown Boveri & Cie. AG.
    Dept. SI/TE
    Postfach 351
    D-6800 Mannheim 1, W. Germany

38. U. Kogelschatz
    Brown Boveri Research Center
    CH-5405 Baden, Switzerland

39. G. Kogens
    BBC Brown, Boveri & Company, Ltd.
    Dept. KLS
    CH-5401 Baden, Switzerland

PARTICIPANTS

40. J. Kopainsky
    Brown Boveri Research Center
    CH-5405 Baden, Switzerland

41. P. Kulik
    Energie-Versorgung Schwaben AG.
    Goethestr. 12, Postfach 158
    D-7000 Stuttgart 1, W. Germany

42. J. J. de Leeuw
    Elektrotechnisch Studiebureau
    Gemeente Energiebedrijf Amsterdam
    Tesselschadestraat 1
    Amsterdam, Netherlands

43. C. Lindsay
    Ontario Hydro
    700 University Avenue
    Toronto, Ontario M5G 1X6, Canada

44. H. Lipken
    Rheinisch-Westfälisches
    Elektrizitätswerk AG., Dept. E-G
    Kruppstr. 5
    D-4300 Essen 1, W. Germany

45. J. J. Lowke
    School of Electrical Engineering
    University of Sydney
    Sydney, N.S.W. 2006, Australia

46. E. Lybeck
    Imatran Voima Osakeyhtioe
    P.O.B. 138
    00101 Helsinki 10, Finland

47. H. Maecker
    Lehrstuhl für Technische
    Elektrophysik
    Technische Universität München
    Arcisstr. 21
    D-8000 Munich 2, W. Germany

48. J. J. Maley
    Station Electrical Engineering Dept.
    Commonwealth Edison Company
    P.O.B. 767
    Chicago, Ill. 60690, USA

49. S. Manganaro
    CESI
    Via Rubattino 54
    I-20134 Milano, Italy

50. R. Markwalder
    Aare-Tessin AG.
    Bahnhofquai 12
    CH-4600 Olten, Switzerland

51. G. Mauthe
    BBC Brown, Boveri & Company, Ltd.
    Dept. AGS
    CH-5401 Baden, Switzerland

52. R. W. Meier
    Brown Boveri Research Center
    CH-5405 Baden, Switzerland

53. J. Mentel
    Ruhr-Universität Bochum
    Abt. für Elektrotechnik
    Universitätsstr. 150
    Gebäude IC 1/152
    D-4630 Bochum 1, W. Germany

54. J. J. Morf
    EPF-L
    Chair d'Installations Electr.
    16, chemin de Bellerive
    CH-1007 Lausanne, Switzerland

55. M. Naglik
    Rheinisch-Westfälisches
    Elektrizitätswerk AG.
    Kruppstr. 5
    D-4300 Essen 1, W. Germany

56. L. Niemeyer
    Brown Boveri Research Center
    CH-5405 Baden, Switzerland

57. H. O. Noeske
    General Electric Co.
    Corporate Res. & Development
    7500 Lindbergh Blvd.,
    Bldg. 23 A
    Philadelphia, PA. 19153, USA

58. T. Nyberg
    Brown Boveri Svenska AB
    Box 39062
    S-100 54 Stockholm 39,
    Sweden

59. P. G. Parrot
    Central Electricity Generating Board
    15, Newgate Street
    London, EC1A 7AU,
    England

60. Y. Pelenc
    Merlin-Gerin
    F-48041 Grenoble-CEDEX, France

61. P. Péraud
    Directeur Technique et Scientifique
    Cie. Electro-Mécanique
    12, rue Portalis
    F-75383 Paris-CEDEX 08, France

62. G. Pietsch
    Rheinisch-Westfälische
    Technische Hochschule
    Schinkelstr. 2
    D-5100 Aachen, W. Germany

PARTICIPANTS

63. N. Pinto
    Jefe Dptd. de Transmision
    Gerencia Sistema de Transmision
    Asociado - Uribante - Caparo
    Avenida Sanz - El Marqués
    Edf. Centro Electrico Nacional
    Caracas 107, Venezuela

64. H. P. Popp
    Lehrstuhl für Allgemeine und
    Theoretische Elektrotechnik der
    Ruhr-Universität
    Postfach 2148
    D-4630 Bochum, W. Germany

65. K. Ragaller
    Brown Boveri Research Center
    CH-5405 Baden, Switzerland

66. K. Reichert
    BBC Brown, Boveri & Company, Ltd.
    Dept. KCT
    CH-5401 Baden, Switzerland

67. P. A. Rein
    NESA A/S
    Parallelvj
    DK-2800 Lyngby, Denmark

68. W. Rieder
    Technische Universität Wien
    Gusshausstr. 25
    A-1040 Vienna, Austria

69. H. B. Riegelsberger
    Brown Boveri & Cie. AG.
    Dept. SI
    Postfach 351
    D-6800 Mannheim 1, W. Germany

70. W. G. J. Rondeel
    A/S NEBB
    Dept. Electr. Engineering
    Norwegian Institute of Techn.
    Trondheim, Norway

71. E. Ruoss
    BBC Brown, Boveri & Company, Ltd.
    Dept. A
    CH-5401 Baden, Switzerland

72. P. Rutz
    Motor-Columbus
    Consulting Engineers Inc.
    Parkstr. 27
    CH-5401 Baden, Switzerland

73. J. Salge
    Institut für Hochspannungstechnik
    der Technischen Universität
    Pockelsstr. 4, Postfach 3329
    D-3300 Braunschweig, W. Germany

74. J. di Salvo
    Aguy y Energia Eléctrica
    Av. Lavalle 1554
    3er Biso
    Buenos Aires, Argentinia

75. E. Schade
    Brown Boveri Research Center
    CH-5405 Baden, Switzerland

76. G. Schaffer
    BBC Brown, Boveri & Company, Ltd.
    Dept. KCT
    CH-5401 Baden, Switzerland

77. P. M. Schmidhuber
    Stadtwerke München
    Elektrizitätswerke
    Blumenstr. 28
    D-8000 Munich 2, W. Germany

78. A. Schmidt
    BBC Brown, Boveri & Company, Ltd.
    Dept. A
    CH-5401 Baden, Switzerland

79. H. G. Schütte
    Hannover-Braunschweigische
    Stromversorgungs AG HASTRA
    Betriebsdirektion Braunschweig
    Celler Str. 84
    D-3300 Braunschweig, W. Germany

80. L. Slama
    CERCEM Lab. du Bourget
    49, Rue du Commandant Rolland
    F-93350 Le Bourget, France

81. J. Sloot
    Hazemeyer BV
    Hazemeyer Research
    Postbus 23
    Hengelo, Netherlands

82. C. E. Sölver
    ASEA
    Fack
    S-77101 Ludvika 1, Sweden

83. K. Sollich
    Brown Boveri & Cie. AG.
    Postfach 351
    D-6800 Mannheim 1, W. Germany

84. A. P. Speiser
    Brown Boveri Research Center
    CH-5405 Baden, Switzerland

85. W. Spriegel
    Brown Boveri & Cie. AG.
    Dept. SI/SH
    Postfach 351
    D-6800 Mannheim 1, W. Germany

PARTICIPANTS

86. H. Suiter
    Rheinisch-Westfälisches
    Elektrizitätswerk AG., Dept. E-G
    Kruppstr. 5
    D-4300 Essen 1, W. Germany

87. J. A. Sullivan
    Merz and McLellan
    Amberley Killingworth
    Newcastle upon Tyne NE12 ORS,
    England

88. B. W. Swanson
    Heat Transfer & Fluid Dynamics Dept.
    Westinghouse Electric Corporation
    Research and Development Center
    1310 Beulah Road
    Pittsburgh, PA 15235, USA

89. H. P. Szente-Varga
    BBC Brown, Boveri & Company, Ltd.
    Dept. AGH
    CH-5401 Baden, Switzerland

90. T. H. Teich
    ETH-Z/UMIST
    ETL - H 33, ETH Zentrum
    CH-8092 Zürich, Switzerland

91. A. Thorsrud
    A/S Norsk Elektrisk & Brown Boveri
    P. O. Box 429
    Sentrum
    Oslo 1, Norway

92. C. Tschäppät
    Ingenieurunternehmung AG der
    Schweizerischen Elektrizitäts-
    und Verkehrsgesellschaft
    Postfach
    CH-4010 Basel, Switzerland

93. D. T. Tuma
    Electrical Engineering Dept.
    Carnegie-Mellon University
    Pittsburgh, PA 15213, USA

94. J. Urbanek
    Skeats High Power Laboratory
    General Electric Co.
    7500 Lindbergh Blvd.
    Philadelphia, PA 19153, USA

95. H. Vierfuss
    Brown Boveri & Cie. AG.
    Dept. SI/SH
    Postfach 351
    D-6800 Mannheim 1, W. Germany

96. R. P. Vogt
    BBC Brown, Boveri & Company, Ltd.
    Dept. AGV
    CH-5401 Baden, Switzerland

97. M. N. D. de Vries
    High Power Laboratories DZL
    N. V. KEMA
    Utrechtseweg 310
    Arnhem, Netherlands

98. F. Waltenberger
    c/o Bayernwerk AG.
    Blütenburgstr. 6
    D-8000 Munich 2, W. Germany

99. L. Westin
    Sydkraft AB
    Fack
    S-20070 Malmö, Sweden

100. O. A. Willson
     Ingenieurunternehmung AG. der
     Schweizerischen Elektrizitäts- und
     Verkehrsgesellschft
     Postfach
     CH-4010 Basel, Switzerland

101. K. Zückler
     Siemens AG, Schaltwerk
     Nonnendammallee 102-110
     D-1000 Berlin 13, W. Germany

ABBREVIATIONS

ac	alternating current
AEG	Allgemeine Elektrizitäts Gesellschaft
ANSI	American National Standards Institute
BBC	Brown Boveri Company
CEGB	Central Electricity Generation Board
CERL	Central Electricity Research Laboratories
CESI	Centro Elettrotecnico Sperimentale Italiano
cigré	Conference Internationale des Grands Reseaux Electriques à Haute Tension
ct	current transformer
cvt	capacitor voltage transformer
dc	direct current
ECNSW	Electricity Commission of New South Wales
EdF	Electricité de France
ehv	extra high voltage (> 300 kV)
ETH	Eidgenössische Technische Hochschule
GE	General Electric
GEC	General Electric Company
hv	high voltage
IEC	International Electrotechnical Commission
itrv	initial transient recovery voltage
KEMA	N. V. tot Keuring van Elektrotechnische Materialen
PD	partial discharge
psi	pound per square inch
p.u.	per unit
rms	root mean square
rrrv	rate of rise of recovery voltage
SECV	State Electricity Commission of Victoria
SIL	surge impedance load
slf	short line fault
TEA	transversely excited atmospheric pressure laser
tf	terminal fault
TNA	transient network analyzer
trv	transient recovery voltage
uhv	ultra high voltage
UK	United Kingdom
UV	ultra violet
vt	voltage transformer

INTRODUCTION AND SURVEY: PHYSICAL AND NETWORK PHENOMENA*

K. RAGALLER and K. REICHERT
Brown Boveri & Company Ltd., Baden, Switzerland

SUMMARY

The task of a circuit breaker is to interrupt short-circuit currents in a network. Two basic types of fault are discussed: the terminal fault and the short-line fault. Failure of the breaker may occur if the transient recovery voltage, which stresses the breaker, is too high. Two types of failure can be distinguished: dielectric failure, which is usually coupled with the terminal fault, and thermal failure which is associated with the short-line fault.

The most effective arrangement with which to interrupt the current is an electric arc burning in a gas at high pressure and with a strong axial pressure gradient. The high axial flow velocity of the arc plasma produced thereby results in a very thin arc channel at current zero. This conducting path is interrupted by flow turbulence. After current interruption a thin channel of hot gas continues to exist in the circuit breaker, thus limiting the allowable peak of the transient recovery voltage. For both the thermal and the dielectric interruption modes, limiting curves can be derived which together determine the range of application of a given circuit breaker.

1. INTRODUCTION

The title of this symposium 'Current Interruption in High-Voltage Networks' covers a very broad field of topics. The goal of the symposium cannot be to treat or even mention all the problems encompassed therein. Rather the emphasis will be on the area which is probably the most important from an economic and technical point of view, namely the interruption of short-circuit currents in high-voltage alternating-current networks. Not included in our discussions are such topics as high-voltage DC-networks and DC-circuit breakers, load breakers and isolators, as well as problems relating to current-limiting switches.

---

*) Presented at the symposium by K. Ragaller.

Our goal, in particular, is to compare the advances made in the understanding of the underlying physical processes with the latest investigations in the area of transient network phenomena during current interruption. Problems which require the aid of both disciplines for solution will be brought out and discussed.

For reasons of simplification and rationalization of operation, as well as to achieve a versatile application of the circuit breaker, it is desirable to describe the relation between circuit breaker and network with as few characteristic parameters as possible, such as rated voltage, short-circuit current and continuous current.

Not only the circuit breaker but also the network are characterized by these parameters and, if they agree or the values for the network are below those of the circuit breaker, then the application of the particular breaker becomes feasible for all possible breaking conditions. Increasing demands for circuit-breaker reliability and economy have forced the closer examination of this simple concept and the supplementation by investigations of the breaking limits. This necessitates a more precise understanding of both circuit breaker and network. Even if the above-mentioned parameters are identical, the transient network phenomena can be quite different.

This is also true for the circuit breaker. The consequences of such effects on the circuit-breaking capacity can be evaluated only if both circuit breaker and network are studied and, most importantly, if their interaction is investigated as well.

2. SURVEY OF NETWORK PHENOMENA

2.1. Introduction

One task of a circuit breaker is to switch off a faulted part of the network (Fig. 1), such as a transmission line.

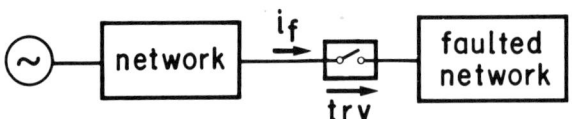

Fig. 1. Separation of a faulted part of the network by a power circuit breaker.

The circuit breaker is tripped by the protection system. The contacts open and an electric arc is formed. Under the appropriate arc conditions the circuit breaker interrupts the current at current zero. The network reacts to this current interruption by transient oscillations which give rise to a so-called transient recovery voltage (trv) across the circuit breaker. If the trv which stresses the circuit breaker after current zero is too high, the circuit breaker fails to interrupt the current.

For modern high-voltage circuit breakers two types of failure with different behavior have to be distinguished. If, immediately after current zero, the rate of rise of the trv, du/dt, frequently abbreviated as rrrv, is greater than a critical value, the decaying arc channel is reestablished by ohmic heating. This period, which is controlled by the energy balance, is called the thermal interruption mode. Figure 2 shows test oscillograms for this case which were taken for an SF_6-breaker at 72 kA. The rate of rise of the trv after current zero is about 2 kV/μs. Oscillogram (a) shows a successful interruption, whereas (b) represents a thermal failure. About 2 μs after current zero the voltage deviates in the second case from the trv, decreasing and approaching the arc voltage in a time interval lasting for several μs. The contributions in the first part of this book deal with this type of operation.

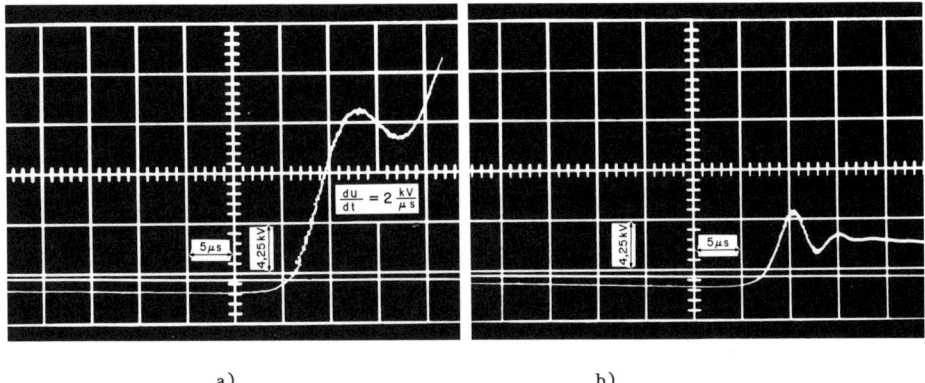

a) b)

Fig. 2. Test oscillograms for an SF_6-breaker showing 72 kA.
a) Successful interruption
b) Thermal failure

After a successful thermal interruption, the trv can reach such a high peak value U_c that the circuit-breaker gap fails through dielectric breakdown. This is called dielectric failure in the peak regime of recovery voltage. Figure 3 shows test oscillograms for this mode. The time and voltage scales are reduced in comparison with those of Fig. 2. The oscillogram on the left shows a successful interruption, whereas that on the right depicts the occurrence of a breakdown close to the peak. It is a feature typical of this failure mode that the voltage decay occurs so fast that it cannot be resolved on the oscillogram. This mode will be discussed in the second part of this volume.

The existence of these two basic failure modes, each having different properties, has important consequences for the range of application of a circuit breaker.

From the wide variety of possible fault locations and network conditions, only those types of fault are of interest which produce the highest stresses regarding the two crucial parameters du/dt and U_c together with the level of the short-circuit current I.

Practical experience shows that as far as the peak voltage is concerned, namely in the dielectric failure mode, the critical fault is usually the so-called terminal fault (tf), which occurs at the terminals of the circuit breaker.

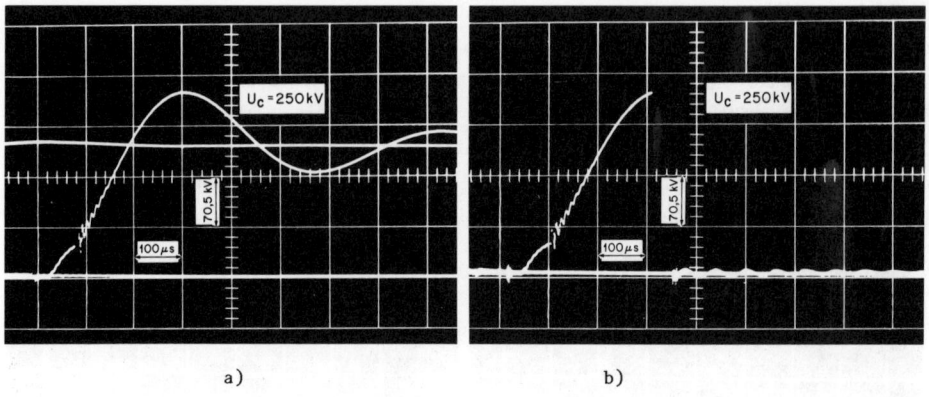

Fig. 3. Test oscillograms for an SF_6-breaker showing 50 kA.
a) Successful interruption
b) Dielectric failure close to the peak of the trv

For the thermal interruption mode, the critical fault is one which occurs on a line some distance from the circuit breaker. The most severe stresses occur in the case of relatively short lines some km in length. This fault is therefore called the short-line fault (slf).

In general, the electric arc that is produced in the breaker can influence the trv by its arc voltage and post-arc current. In order to characterize the network uniquely, we adopt the concept of an ideal circuit breaker without arc voltage or post-arc current. The trv initiated by such an ideal circuit breaker is called inherent trv. With modern circuit breakers, especially SF_6-breakers, a significant influence of the circuit breaker on the trv is observed only under special circumstances, such as the itrv-regime (see paper by M. Dubanton). The following sections are limited to a discussion of the inherent trv.

2.2. TRV for Two Elementary Networks

In this section the correlation between network parameters and the two important trv parameters U_c and du/dt is discussed. The best overview can be achieved by analyzing the behavior of two elementary model networks. In many circumstances these simple networks provide a good representation of a more complex network situation.

The simplest network for a discussion of the trv, shown in Fig. 4, consists of an imposed voltage source, an inductive impedance between the fault and the network, and a parallel capacitance with a damping resistor. The fault current i_f lags the source voltage U_g by 90°. If this network is used to model a one-phase fault in a three-phase system, the source voltage U_g is related to the interconnected nominal network voltage U_n by $U_g = U_n/\sqrt{3}$. In the general case a phase factor has to be included in this relation, as is discussed in the next section.

INTRODUCTION AND SURVEY

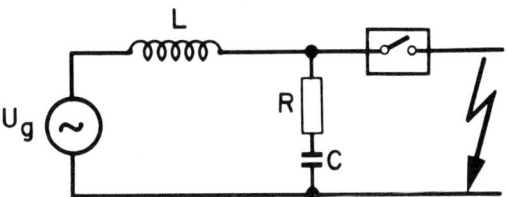

Fig. 4. Simplest model circuit for trv evaluation.

The trv after current interruption can easily be calculated, for example, by the method of superposition (current injection).

For $R \ll \sqrt{L/C}$ an approximate solution is found to be

$$u = \sqrt{2}\, U_g\, (\cos\omega t - \exp(-t/\tau) \cos\upsilon t) \tag{1}$$

with

$$i_f = \sqrt{2}\, I \sin\omega t, \quad I = U_g/\omega L,$$

$$\tau = 2L/R, \quad \text{and} \quad \upsilon = 1/\sqrt{LC} = \omega\, \sqrt{S_K/S_L}. \tag{2}$$

Figure 5 shows the trv in this case.

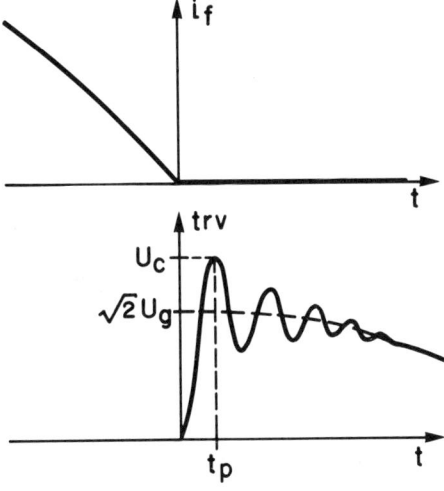

Fig. 5. Current and voltage oscillograms for current interruption in model circuit according to Fig. 4.

In equation (2), we have introduced the short-circuit power $S_S = U_g^2/\omega L$ and the charging power $S_C = \omega C U_g^2$.

The characteristic quantity for the dielectric failure mode, the peak voltage U_c of the trv, is given by equation (1) as:

$$U_c = k \sqrt{2}\, U_g, \qquad (3)$$

where the so-called peak (or amplitude) factor k takes values between 1 and 2 depending only on the magnitude of L/R.

The circuit diagram in Fig. 4 is a frequently used representation of a short circuit at the terminals of a circuit breaker (tf). It gives all the typical features of the trv for a variety of network conditions. An especially important feature given by equation (3) is that the peak voltage U_c is independent of the short-circuit current I. (In a real network a slight dependence of the peak factor on the short-circuit current can occur. See also Discussion of this paper.)

For modelling the trv in the first few microseconds the equivalent circuit of Fig. 4 is usually not a good approximation. As a general rule it can be stated that the trv in the tf has a time delay due to the capacitance of the substation. Therefore the rrrv is very small.

For the representation of a slf, the circuit diagram of Fig. 4 must be supplemented by a line as shown in Fig. 6.

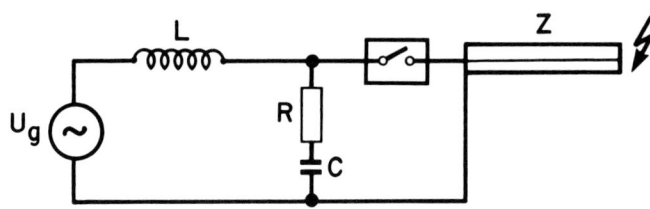

Fig. 6. Model circuit for trv evaluation in the case of a short-line fault.

In this case the trv consists of two parts: the source-side trv u_S and the line-side trv u_L. The latter is determined by wave-propagation phenomena on the line. For perfect reflection a triangular oscillation results.

$$u_L = \sqrt{2}\, IZ\omega \text{ for } 0 \leq t \leq 2\ell/c. \qquad (4)$$

Figure 7 shows the trv in this case. A typical feature is that the source-side voltage during the first linear ramp of the trv is very small. Therefore the rate of rise of the trv in the first reflection interval, $0 < t < 2\ell/c$ after current zero, is proportional to the short-circuit current I:

$$\frac{du}{dt} \sim \frac{du_L}{dt} = Z \frac{di_f}{dt} = \sqrt{2}\, \omega Z I = \sqrt{2}\, \omega U_g \frac{S_S}{SIL} \qquad (5)$$

INTRODUCTION AND SURVEY

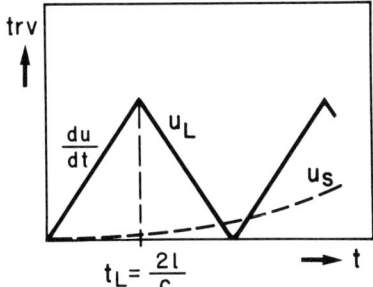

Fig. 7. Transient recovery voltage in the short-line fault.
u_S = source-side contribution
u_L = line-side contribution.

wherein c = propagation velocity (300,000 km/s for overhead transmission lines) and SIL = U_g^2/Z = surge impedance load.

The short-circuit current I in the slf is smaller than the one for the tf due to the additional impedance L_L of the line. The corresponding reduction factor

$$s = L/(L+L_L)$$

is called the slf factor. For cases of practical interest s has values between 0.6 and 0.9.

The peak voltage in this circuit is produced only by the source side because the line-side oscillation is damped out before the peak is reached (peak factor of line k_L = 1.5-1.7).

It can easily be shown that the source-side trv is reduced to:

$$u_S = \sqrt{2}\, U_g(\cos\omega t - L/(L+L_L)\exp(-t/\tau)\cos\upsilon t) \tag{6}$$

so that the peak of the trv is also reduced.

The following conclusions can be drawn from these simplified network diagrams:

As it was anticipated in Section 2.1 the stresses on the circuit breaker in the peak regime are higher in the tf than in the slf. The reason is that the current as well as the peak voltage are reduced by the slf factor s.

On the other hand, the rrrv in the slf is much higher than in the tf because in the thermally critical period the aforementioned time delay is still effective.

2.3. TRV for Real Networks

A real network differs from the examples discussed so far in its additional complexity. Figure 8 shows as an example a single-phase circuit diagram of a substation. A

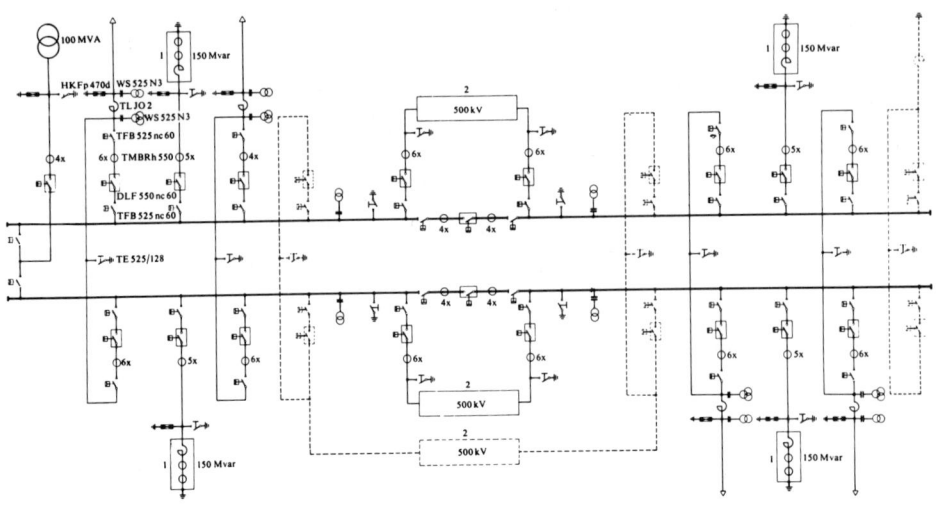

Fig. 8. Single-phase diagram of a 525 kV substation (El Chocon power-transmission system Argentina)[12].

network would consist of several such substations connected by transmission lines. In reality three phases have to be considered. The short circuit can involve one or more phases, and may or may not be grounded. Depending on the switching condition of the circuit breakers, the network configuration can differ considerably.

The determination of the trv requires the solution of the following problems:

- determination of the short-circuit current I, i.e. the current which has to be interrupted by the circuit breaker
- evaluation of an appropriate model for the trv calculation by means of transformation methods (sequence networks). This model must take account of the appropriate network topology and parameters, the switching and fault conditions (3-, 2- or 1-phase fault; first, second and third pole to clear, etc.)
- calculation of the trv by means of superposition or current-injection methods.

If faults close to generators have to be cleared, the determination of the circuit-breaker current at the fault-clearing instant should consider the decaying dc and ac components of the fault current (see Fig. 9).

The second step especially involves the art of trv-evaluation. Real power networks are complex systems, consisting of substations, lines, transformers, generators, measuring devices, etc. The electromagnetic field of the whole system contributes to the trv, and any modelling of this field by an equivalent circuit, consisting of such discrete elements as transmission lines, resistors, capacitors, inductances, etc., is an approximation to the problem. The complexity of the equivalent circuit

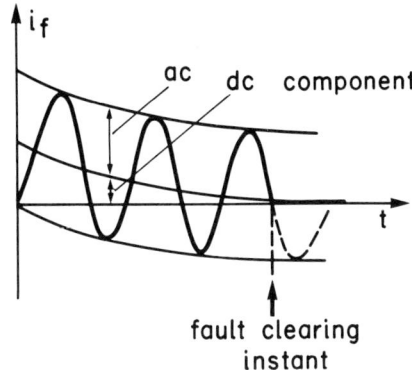

Fig. 9. Time dependence of the fault current.

depends on the network configuration, the accuracy required and the duration time T of the trv to be modelled.

The trv consists of oscillations between lumped elements and electromagnetic waves travelling with a velocity c from the circuit breaker into the network. Since these waves are reflected by discontinuities, the radius of the zone influencing the trv is approximtely cT/2. If the time period T of interest extends from current zero to the peak of the trv, all elements within a radius of 200-400 km from the circuit breaker have to be taken into account.

Thus all elements within this zone should be modelled as accurately as possible. Usually this consideration gives a zone radius which is much too large. In practical cases the network can be modelled only within smaller zones. In these cases elements such as lines leaving the zone have to be terminated by lumped elements determined by a reduction of the external network. A detailed discussion of these questions is given in the lecture by Diesendorf, Lowe and Saunders.

In spite of this complexity the circuit can in many cases be modelled by the simple circuit shown in Fig. 4, at least insofar as the peak voltage is concerned. The problem of calculating the trv is then reduced to the evaluation of the characteristic elements U_g, L, R and C. This can be done by means of the sequence transformation method. If U_n is the nominal interconnected network voltage, R_1, L_1, and C_1 the positive and negative sequence parameters, then one gets the following values for the elements of the model circuit given in Fig. 4 for the first pole to clear of a three-phase ungrounded fault:

$$U_g = \frac{U_n}{\sqrt{3}} \cdot 1.5, \quad R = 1.5\, R_1, \quad L = 1.5\, L_1, \quad C = C_1/1.5 \tag{7}$$

The factor 1.5 is called first-pole-to-clear factor or phase factor k_p.

Depending on the network configuration, the network parameters and the fault conditions k_p can take on values in the range $0.5 < k_p < 1.5$.

For the peak voltage in the general case we get the result

$$U_c = k_p \, k \, \frac{\sqrt{2}}{\sqrt{3}} U_n. \qquad (8)$$

Higher accuracy is needed to model the trv in the current-zero regime because even a voltage of some kV can influence the breaking behavior in a short time interval of about one to 10 µs after current zero. Consequently the radius of influence in the network in this case is considerably smaller than in the peak regime. This radius includes the substation where the breaker is located and all outgoing lines for a distance of several km. The representation of the substation itself has to be improved over that for the peak regime. The simple circuit of Fig. 6 serves again as reference. Problems arise from the need for a correct determination of Z. The source-side contribution can be different and the voltage ramp given by equation (5) can be delayed by capacitors in the substation.

The problem of trv calculation in this regime, especially for the short-line fault, is dealt with in the paper of Humphries.

In recent years the need has arisen for an improvement of the trv calculation in the first µs after current zero. This so-called initial transient recovery voltage regime is influenced by elements in the immediate neighbourhood of the circuit breaker (i.e. busbars, measuring transformers, insulators etc. of the substation). An accurate calculation of the voltage in the first µs thus requires representation of these elements. Problems related to this question will be covered by Dubanton.

Details of the problem of trv calculation are given in references[1-11].

3. SUMMARY OF THE PHYSICS OF CIRCUIT-BREAKING PHENOMENA

3.1. Basic Concepts

In much the same way as we discussed network phenomena in the previous section, we present here an overview of the physical processes occurring in high-voltage circuit breakers.

The basic concept is the same for all circuit breakers in use today, regardless of design and type, whether oil, air, or SF_6. An electric arc is generated which burns in a gas under high pressure, and with a strong pressure gradient along the arc axis. (A good survey over different designs of hv switchgear is given in refs.[13] and[14]).

The basic design, in which all conditions can be optimized is shown in Fig. 10 and can be described as follows:

There are two nozzles: the so-called double-nozzle design. High pressure p_o exists in the gas chamber surrounding the nozzle inlet, whilst the nozzle exits are con-

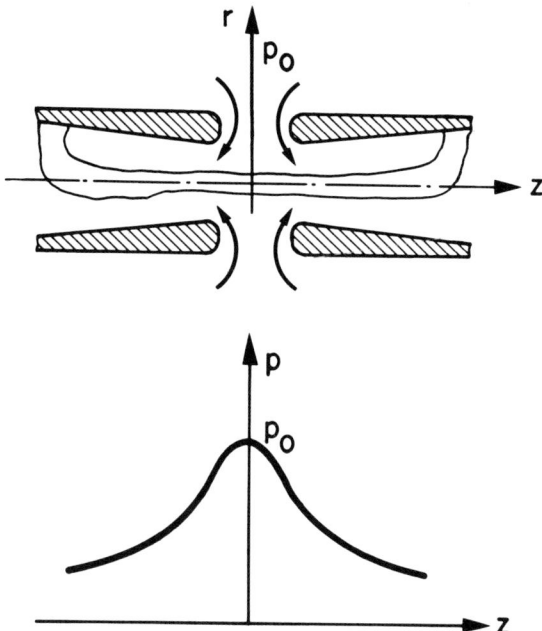

Fig. 10. Upper portion: Double nozzle design for a high-voltage gas-blast circuit breaker. Lower portion: Pressure distribution along the axis of this device.

nected to an area of low pressure. As a result of this pressure difference a flow is induced symmetrically into the nozzles on both sides. The lower portion of Fig. 10 represents the pressure distribution along the axis. A stagnation point, where the pressure = p_o, develops in the center region between the two nozzles. Originating from this stagnation point, a pressure drop occurs, resulting, in the case of supercritical pressure ratios, in a pressure in the nozzle throat equal to about half the stagnation pressure. Beyond the throat the pressure continues to drop to the exhaust value.

This configuration exists in a more or less pure form in air-blast and SF_6-circuit breakers. Oil circuit breakers usually have a somewhat different geometry because the arc itself must generate the high pressure gas by vaporizing the oil. However, in essence, here too the arc burns in a strong pressure gradient along the arc axis.

The significant mechanisms can now be discussed with reference to Fig. 10. After contact separation, which can be accomplished in various different ways, an arc is created, rapidly transported along the axis by the gas flow and the arc roots driven downstream into the nozzles as shown in Fig. 10.

Thus for the subsequent extinction process, there exists a well-defined and reproducible initial condition which in an optimum situation is largely independent of the contact-separation phase. One effect of the contact-separation phase which can influence the extinction process is that of metal vapor production during contact separation. This vapor can be stored in the quenching gas even up to the moment of extinction. Possible effects of the behavior of arc roots on the interruption process will be discussed in the paper by Mentel.

3.2. Important Arc Characteristics

Before describing the extinction process, we summarize some experimental results[15,16] for an arc in the configuration shown in Fig. 10. Two arc segments must be distinguished. In the area between the two nozzle throats there exists an approximately parabolic (see Fig. 10) axial pressure distribution, which results in a cylindrical arc ($dT/dz = 0$). The radial temperature profile for the steady-state arc is shown in the upper part of Fig. 11, whilst the lower portion shows a differential interferogram of

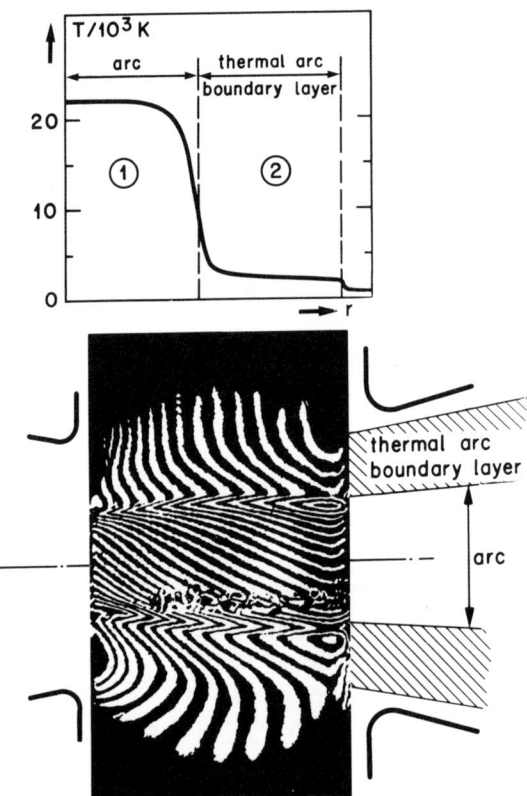

Fig. 11. Upper portion: Radial temperature profile of the arc section between the nozzles. Lower portion: Differential interferogram showing the arc and the thermal arc boundary layer at a current of 2 kA in nitrogen.

the arc between the nozzles. The arc boundary, with its steep temperature gradient, can be observed from the kink in the interference fringes.

An almost rectangular temperature profile with a temperature at the axis of about 20,000 K is created as a result of the very strong energy transport by radiation within the arc. Due to the low density of the gas at this temperature, the plasma is strongly accelerated by the axial pressure gradient: in the throat, the axial plasma velocity is 6,000 m/s.

A steady-state equilibrium is maintained in that the axially flowing plasma is fed by cold gas flowing radially into the arc. Absorption of the radiation emitted by the arc plays an important role in heating up this cold gas flow at the arc boundary. A balance between Joulean heating and the temperature increase of the gas flowing through the arc region (1) is a good approximation to the energy equilibrium and permits a simple, but accurate, calculation of the arc diameter as a function of current, pressure and the geometrical parameters.

A so-called thermal arc boundary layer is formed outside the steep temperature decay at the arc boundary. For the temperature profile in the upper portion of Fig. 11, this thermal arc boundary layer is the zone (2) adjacent to the arc with a temperature between the cold gas temperature (300 K) and about 2,000 K.

This zone can clearly be seen from the deformation in the fringes of the differential interferogram in the lower part of Fig. 11. This zone occurs because a fraction of the radiation emitted by the arc is absorbed by the adjacent cold gas layer. The spread and temperature of the thermal arc boundary layer is also determined by an energy balance. Energy is supplied by radiation which depends on the spectral composition and intensity of the emitted radiation as well as on the absorption characteristics of the gas surrounding the arc. Only convection through the flow field is significant for removing energy. The time constant involved in restoring the equilibrium of this energy balance is, compared to the arc, very large (order of magnitude ms). Therefore, the size of this zone depends mainly on the total current variation, and in particular on the high-current phase. The formation of the thermal arc boundary layer has little influence on the arc and its characteristics. Later, however, after passing current zero in the dielectric interruption mode, the temperature profile is controlled by the thermal arc boundary layer. Therefore, the radiation phenomena, which are significant in this region, will be treated separately in the paper of Lowke.

It should also be mentioned that other phenomena can also influence the magnitude and temperature of the thermal arc boundary layer. From the contact-separation phase, hot gas can build up in the vicinity of the arc. In the case of arc instability, a thermal arc boundary layer is also created due to a 'stirring' motion in the cold gas. These, and other phenomena increase the thickness of the thermal arc boundary layer. However, of interest from a physical and technological point of view is the fact that the minimum thickness is determined by radiation effects.

For the portion of the arc which is downstream of the nozzle throat, the energy balance is basically the same as outlined above. Since the pressure distribution is no longer parabolic, the arc is no longer cylindrical but expands into the nozzles.

A significant difference for the circuit-breaking process is that in this region the boundary layer between the arc and the cold gas is turbulent. The turbulence is created as a result of the difference in the axial flow velocities of the plasma and the cold gas. The instability of the resulting free shear layer can already be observed in the region between the nozzles: the formation of discrete vortices within the arc boundary layer, which is characteristic of instabilities in free shear layers, can be seen on one side of the differential interferogram of Fig. 11. The maximum intensity of this effect is not reached until the peak value of the axial velocity of the plasma has been attained, which is the case approximately in the nozzle throat.

3.3. Thermal Interruption Mode

As current zero is approached, the arc diameter can follow the current variation with a very small time constant. For the cylindrical arc segment between the nozzles we have

$$\tau = \ell/v_z, \qquad (9)$$

where ℓ is the length of the part of the arc from the stagnation point and v_z is the axial velocity. Given the above data, the order of magnitude for τ is 10 μs.

Figure 12 shows the arc diameter in the cylindrical arc region as a function of time (streak recording) in the period approaching current zero. The current decreases with a slope of 24 A/μs corresponding to a 50 Hz sinewave of 54 kA_{rms}.[17]

It is clearly seen how the arc diameter decreases with the current to less than 1 mm at current zero. However, the temperature in this arc segment is still very high (15,000 K) which means that this section of the arc is still a good electrical conductor and therefore the breaking capacity minimal.

Downstream of the throat, this thin arc channel is immersed in the turbulent zone and an intensive interaction occurs between the turbulent flow field and the arc channel. This is shown schematically in Fig. 13.

From the streak recording for the arc in this region, shown in Fig. 14, it can be seen that the arc is distorted by the turbulence and split up into individual globules. This phenomenon is particularly strong in a zone about 2 cm in length downstream of the nozzle throat. Further downstream the arc diameter increases again so that the intensity of this interaction is strongly decreased.

A measure of the significance of the various axial regions is the axial variation of the electrical field intensity for the period of current zero.

Figure 15 shows measured values of the electrical field strength as functions of the coordinate at different times. In the high-current phase the arc section upstream

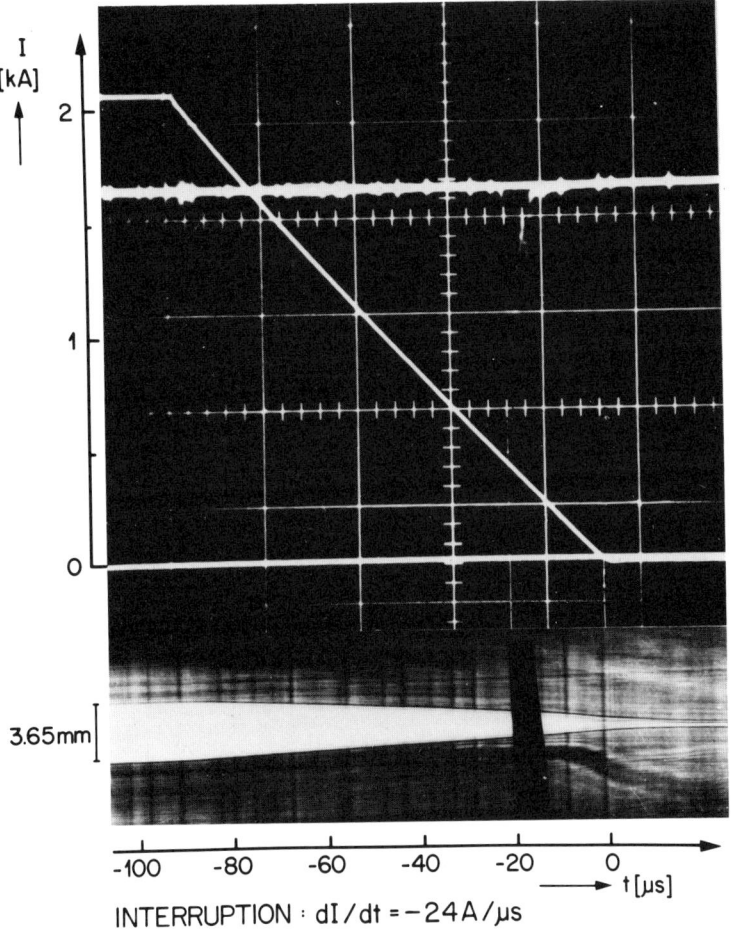

Fig. 12. Current oscillogram and streak record of the arc between the nozzles in the case of a current interruption.

of the nozzle throat has the highest field strength. Five µs before current zero there appears a maximum of the electrical field strength downstream of the nozzle throat at the position where the interaction between the thin arc column and the turbulent flow field has its highest intensity. This maximum is even more pronounced 2 µs after current zero. Various measurements which support the above picture have been made and the investigations are summarized separately in the paper of Jones.

The recovery voltage is the integral of the curve $E(z)$ in Fig. 15c. Whether this voltage leads to quenching or restriking of the current after current zero is deter-

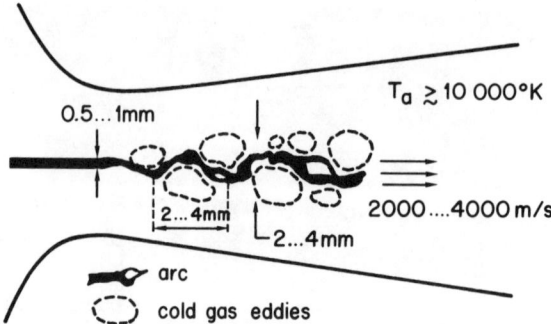

Fig. 13. Illustration of the effect of turbulence on the arc column close to current zero.

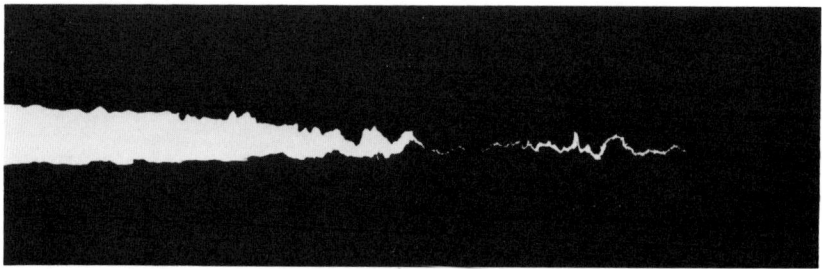

Fig. 14. Streak record of the arc downstream of the nozzle throat in the turbulent-flow section.

mined by the energy equilibrium existing in the effective regions of turbulence. The significant factors of this energy balance are represented in equation (10):

$$\frac{\partial}{\partial t}(A_1 \, \overline{(\rho h)}_1) = A_1 \sigma_1 E^2 - 2\pi r_1 W_1 \tag{10}$$

wherein

ρ = density
h = enthalpy
σ = electric conductivity
E = electric field strength
W = turbulent heat flow.

Subscript 1 refers to the arc zone and A_1 to its cross section.

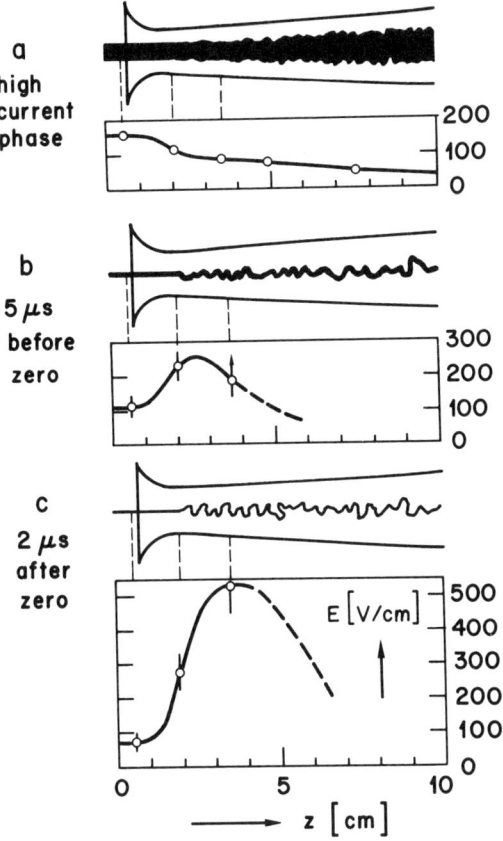

Fig. 15. Measured values of electrical field strength versus axial coordinate at different times. The arc is shown schematically.

The left-hand side of the equation reflects the changes in energy content with respect to time (thermal inertia). The first term on the right-hand side of the equation is the Joulean-heating factor and the second term is the turbulent heat transport. The expanded and complete equation is derived in Ref.[18].

The period of time for which the arc is controlled by this balance is defined as the thermal interruption mode.

The breaking capacity of a circuit breaker depends primarily on the type of gas and its pressure, as well as on the pressure distribution along the axis (geometry). These dependencies are generally represented in the form of limiting-curve diagrams in which the limiting current is plotted against the imposed linearly increasing voltage (slope du/dt) on a double logarithmic scale. Figure 16 shows such a curve for the case of SF_6 at a pressure of 14 bar.[18] In the area below the curve, current interruption takes place and above it so-called thermal failure.

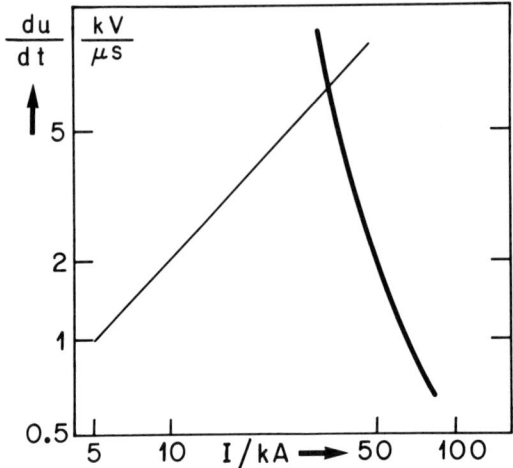

Fig. 16. Thermal mode limiting curve for SF_6 at 14 bar (thick line). Dependence of du/dt on current according to equation (4) (thin line). Point of intersection gives thermal current limit.

Since, as shown above, the time constant correlating the plasma region to the equilibrium value of a specific current is extremely small, the plasma region controlling thermal breaking is not influenced by the current variation of the high-current phase. Current intensity is only significant because the current slope is proportional to the current when approaching current zero.

For this reason thermal breaking limits can be investigated in experimental models where low current is used but with a current slope at current zero equal to that in the actual circuit breaker. However, as mentioned earlier, in actual circuit breakers metal vapor and other factors can influence the thermal arc boundary layer and hence the breaking capacity.

The time constant for restoring the state of equilibrium in the thermal arc boundary layer is significantly greater than for the plasma region, so that it is possible for effects from the high-current phase to exist, up to the point of current zero.

A thorough discussion of the limiting curve diagram is given in the paper of Frind.

Just as successful as the experimental investigations has been the development of a theoretical description of the phenomena involved in thermal breaking. By taking into account the physical phenomena outlined above, in particular turbulence, it has been possible to set up theoretical models which, for example, reproduce quite well the limiting curves as a function of pressure, current and type of gas.

Swanson's paper reviews the current state of development of the theoretical description of the thermal breaking process.

It is the theoretical description that permits the investigation of the interaction of circuit breaker and network by simultaneously solving the network and arc equations. In the case of SF_6-circuit breakers, significant interaction takes place only in the itrv-region. An example of such calculations will be presented in the paper by Hermann and Ragaller.

As shown in Section 2.2, after circuit breaking in the case of a short-line fault, the network supplies a recovery voltage at a rate proportional to the current. This relationship is shown in Fig. 16. The point of intersection of this line for the network and the limiting curve for the circuit breaker determines the critical current which can be broken by the breaker in a particular network. (The interaction between the circuit breaker and the short-line-fault network has been disregarded. This is a very good approximation, at least for SF_6-circuit breakers.)

3.4. Dielectric Interruption Mode

After successful thermal breaking, the contact gap is stressed because of a rise in the recovery voltage. The original arc channel still has an elevated temperature and, as a result, reduced density. Therefore, a limiting value for the allowable peak of the recovery voltage exists which is dependent on the switched current.

Between the area of thermal restriking and the purely dielectric interruption in the cold gas, a physically continuous transition without sharp boundaries may be possible. However, practical experience, especially with SF_6-circuit breakers, has shown that a significant time interval and also other important phenomenological differences exist between the actual thermal failures and the failures in the peak vicinity. It is therefore reasonable and necessary to distinguish between a thermal and a dielectric region.

The physical principle underlying dielectric interruption is no longer the energy balance, but the generation of electrons in the electrical field according to the Townsend equation:

$$\frac{\partial n_e}{\partial t} = - \frac{\partial n_e V_e}{\partial z} + (\alpha - \eta) n_e V_e \tag{11}$$

where α is the ionization coefficient and η the attachment coefficient. Both coefficients are functions of the electrical field and the gas density. If the amplification, either locally or in the entire region surpasses a given value, then the current increases rapidly and a spark is formed which is subsequently converted into an arc.[19,20]

The goal for the investigation of fundamental physical processes, as is also the case in studying the thermal aspects, must be the derivation of breakdown criteria for the entire contact gap starting with the basic equation (11).

Under the conditions present in a circuit breaker during peak recovery voltage it has not been possible as yet to establish definite criteria. However, progress is being made in this direction.

An important area of success in recent years has been the investigation of the breaking capacity of cold SF_6 in homogeneous and inhomogeneous fields. The large increase of α-η in the region α > η with the field produced relatively simple and extensive breakdown criteria. The corresponding results will be outlined in the paper by Teich and Zaengl.

In the application of this criterion to contact gaps, an understanding of the state of the hot residual channel is of utmost importance.

After the conductivity of the plasma channel has decayed, a process which is completed a few μs after zero, the thermal arc boundary layer determines the further decrease in temperature. There are two reasons why this cooling process is significantly slower than that of the plasma. One is that within the thermal arc boundary layer, the thermal conductivity has a value 2 orders of magnitude lower than in the plasma region. Thus, with the exception of a very thin region close to the axis (r < 1 mm), heat conduction plays no significant role and the isotherms are 'frozen' within the flow. Thus any reduction in the extent of the hot zone can only be accomplished by the flow. The second reason is that the flow speed corresponding to the greater gas density is significantly lower than in the plasma.

Specific investigations presented by Kopainsky show that a few 100 μs after current zero, at the time of peak recovery voltage, the temperature on the breaker axis is still about 2,000 K.

This slow cooling of the thermal arc boundary layer has another important significance. As mentioned previously, the expansion and temperature distribution in this region is primarily influenced by the radiative energy transport during the high-current phase. Due to the long decay time, the effects of the high-current phase are still of influence in the region of the dielectric interruption mode. For this reason simulation by means of experimental models with reduced current is not altogether feasible.

A streak recording for the reduction of the thermal arc boundary layer in a circuit breaker is shown in Fig. 17. The arc is shown on the left-hand side of the photograph. After extinction, the thermal arc boundary layer is made visible as a black zone by means of a special Schlieren technique. The relatively slow contraction of this zone due to convection is clearly evident: even after 500 ms a hot channel is still seen along the axis.[21]

This hot channel connects the two electrodes, as the arc did previously, and results in a reduction of the breakdown voltage compared with a purely cold as flow. Because of the elevated temperature, the density is decreased and, as already mentioned, the coefficients α and η in equation (11) are density dependent.

A further phenomenon becomes evident in addition to this density dependency. The hot residual gas still contains charge carriers, especially positive and negative

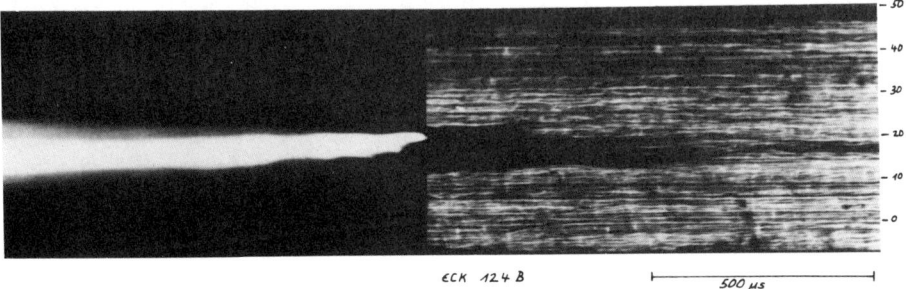

Fig. 17. Streak picture of decaying arc (left portion) and Schlieren streak record showing the decay of the thermal arc boundary layer after current zero for an SF_6-circuit breaker.

ions. Due to time-related relaxation effects, the density of these ions lies above the equilibrium density.[22]

In the electrical field, which produces the recovery voltage between the circuit-breaker contacts, the ions drift and produce a current. Because of the low charge carrier density, and the small mobility of the ions, the resulting current is very small but still sufficient to regulate the electrical field along the hot channel according to the conductance.

Since the coefficients α and η, and therefore the breakdown criterion, are field dependent, the flow of ions influences the breakdown voltage by distorting the electrical field. Just as in the case of the thermal mode, all these phenomena result in a limiting curve.

In Fig. 18 a streak recording of a dielectric failure is shown. After successful thermal extinction of the arc, no luminous pattern can be seen until after a few 100 μs around the peak recovery voltage when a breakdown occurs in the previous arc channel. The breakdown is initiated by a spark which is made visible as a result of a momentary overexposure.

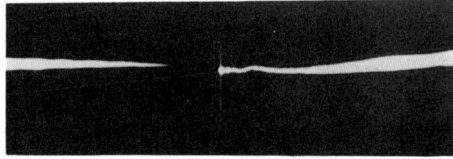

Fig. 18. Streak record of a dielectric failure in an SF_6-circuit breaker.

Compared with the thermal mode, the effects for the dielectric interruption mode are obviously more numerous and varied. In addition to the flow and pressure fields, the phenomena in the high-current phase and the configuration of the electric field, also play a relatively important role. Although these relationships have not yet been investigated in detail, the qualitative apsects of the limiting curves have been determined empirically.

Figure 19 shows such a dielectric limiting curve on a double logarithmic scale. The peak voltage, which can just be maintained, is plotted against the current. The typical curve is relatively flat at low current levels and drops off at higher currents. Below the limiting curve the breaker switches successfully, but fails above it. In addition to the circuit breaker curve, the network values can also be plotted in the limiting curve diagram, as was the case for the thermal mode. Since the peak voltage in a network does not depend on the short-circuit current, a horizontal line results as shown in Fig. 19. The point of intersection of the two curves yields a current limit, determined in this case by the dielectric mode.

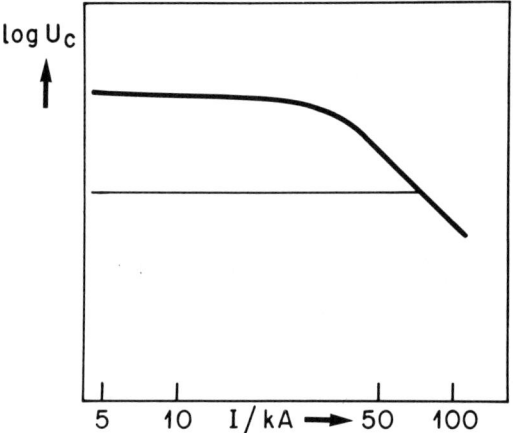

Fig. 19. Limiting curve for the dielectric interruption mode. The thin line indicates the network behavior. The point of intersection gives the dielectric current limit.

3.5. Diagram for Circuit-Breaker Application in the Network

Basically, the two limiting-curve diagrams described in previous sections define the area of application for a specific breaker chamber in a high-voltage network. In reality, of course, other influences are present which can only be mentioned here.

Test conditions are further complicated by additional requirements such as small inductive currents, or out-of-phase switching. Furthermore, caution is recommended in the application of physical limiting curves for actual breaker chambers. The configuration of contact separation as well as flow control can lead to relatively strong deviations of actual circuit-breaker chambers from the physical limiting curves which are of primary interest here.

However, with the aid of the limiting curves some general aspects may be discussed. In general, a circuit breaker must cope with the full peak voltage U_c in the terminal fault and the high du/dt values after current zero in the short-line fault. Therefore, in order to determine the area of application, the diagrams of Figs. 16 and 17 must be superimposed as shown in Fig. 20.

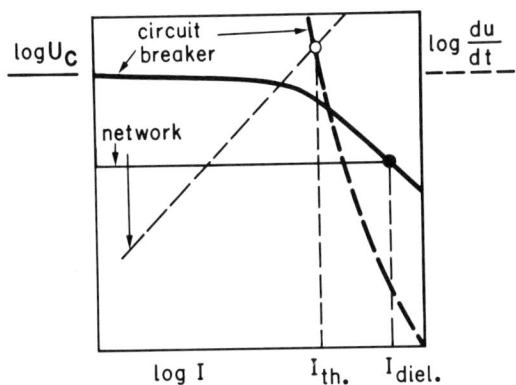

Fig. 20. Superposition of thermal and dielectric limiting-curve diagrams.

Several different possibilities may occur depending on the relative positions of the two breaker limiting curves and the two straight lines which characterize the network.

The two curves in Fig. 20 which characterize the circuit breaker are indicated by heavy lines, and the curves representing the network by thin lines. The du/dt scale is appropriate for the broken-line curves, and the U_c scale for the solid-line curves.

In the example, both broken-line curves, which describe thermal behavior, result in a limiting current I_{th}, which is lower than the corresponding limiting current for the dielectrical case I_{diel}. This leads to an area of application, which is limited in terms of voltage to the limiting current I_{th} by the dielectric limiting curve. The thermal mode, on the other hand, prescribes a current limit which is independent of voltage.

For another network, where the straight line for the peak voltage is higher, or the du/dt line lower, the dielectric limiting current can lie below the thermal value. The area of application is then determined completely by the dielectric limiting curve.

An interesting aspect of the interaction between circuit breaker and network can be illustrated quite well with the aid of the limiting-curve diagram. The area of application can be expanded by changing either the circuit breaker limiting curves (for example by increasing the pressure in the quenching chamber or the number of chambers) or the straight lines representing the network (for example, with parallel capacitors connected to the line).

The optimum solution can be found only by carefully taking into consideration both possibilities.

Using the diagram in Fig. 20 as a basis, a simplified diagram for application to a circuit breaker with an interconnected rated voltage U_n and a rated breaking current I_n can be designed.[23] This is shown in Fig. 21.

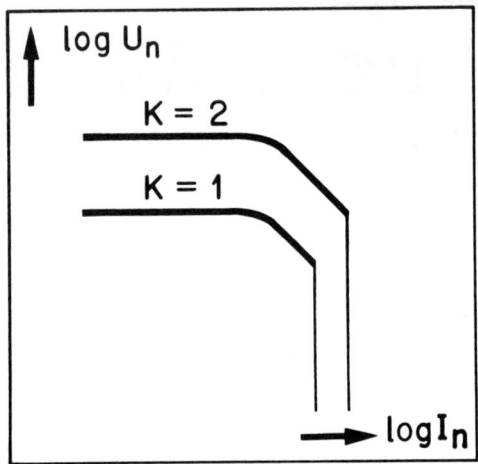

Fig. 21. General form of limiting-curve diagram in a double log plot with rated voltage on the ordinate and rated short-circuit current on the abscissa. Thick line: dielectric mode; thin line: thermal mode. K indicates the number of chambers in series.

Since the correlation between U_c and the rated voltage U_n is independent of the current (see equation (8)), the peak-voltage limiting curve according to Fig. 19 can be converted into a rated-voltage limiting curve by introducing a numerical factor (heavy solid line).

The thermal limiting current, which according to Fig. 20 is independent of the rated voltage (provided Z is independent of the voltage), results in a vertical straight line (thin line). The particular circuit breaker can be utilized in the entire area below the solid thick line and to the left of the thin limiting curve.

The region outside of the above-mentioned limiting curves can be covered by connection of several chambers in series. Under ideal conditions, both U_c in the dielectric mode, and du/dt in the thermal mode, can be increased by a factor corresponding to the number of chambers.

Figure 21 shows an additional limiting curve (K = 2) for a 2-chamber breaker. Because of the relatively steep slope of the thermal limiting curve in Fig. 16 the current limit is displaced less than proportionally when increasing the number of chambers.

This diagram contains in a very condensed form the physical and network aspects of current interruption. Many problems of the design and application of circuit breakers can be discussed with the aid of this diagram. We mention here only two examples:

- If a circuit breaker is operated close to one of the limiting curves, what can be done to extend the current range? Depending on which of the two limiting curves is involved, different measures in the circuit breaker or in the network will be effective. For example, the current limit in the short-line fault can be shifted to higher currents by a capacitor parallel to the line. However, this no longer works close to, or above, the intersection point of the two limiting curves because there one is limited by the dielectric mode.

- A second example refers to the problem of circuit-breaker testing. Over the years, the number of tests recommended by standards has tended to increase continuously. Only better knowledge of fundamental physics can be a counter-balance to that development with the aim of avoiding uneconomic test expenditures. For example, it makes little sense to prescribe a large number of short-line fault, and perhaps even itrv, tests for a breaker which is supposed to operate above the intersection point of the two limiting curves. Such a breaker will never be limited by the thermal interruption mode, unless it is used at a lower network voltage.

REFERENCES

1. R. Rüdenberg, 'Elektrische Schaltvorgänge', Springer-Verlag, Berlin-Heidelberg-New York (1974) 123
2. W. Wanger and J. K. Brown, Brown Boveri Mitteilungen 25 (1937) 283
3. G. Hosemann, W. Frey and D. Oeding, BBC-Nachrichten 43 (1961) 55
4. G. Köppl and P. Geng, Brown Boveri Review 53 (1966) 311
5. J. A. Adams, W. F. Skeats, R. C. Van Sickle and T. G. A. Sillers, Trans. IEEE PAS 61 (1942) 771
6. A. Braun, K. H. Hinterthür, H. Lipken, B. Stein and O. Völcker, ETZ A 97 (1976) 489
7. Y. Pelenc, Revue général d'électricité 86 (1977) 271
8. G. Catenacci, Electra 46 (1976) 39
9. A. Greenwood, 'Electrical transients in power systems', Wiley-Interscience, New York 1971
10. E. Slamecka and W. Waterschek, 'Schaltvorgänge in Hoch- und Niederspannungsnetzen', Siemens AG (1972)
11. E. Slamecka, 'Prüfung von Hochspannungs-Leistungsschaltern', Springer-Verlag, Berlin-Heidelberg-New York (1966)
12. I. Drganc, K. Müller, S. Shaikh and G. A. Gertsch, Brown Boveri Review 61 (1974) 64
13. C. H. Flurscheim, IEE Monograph Series 17, 'Power circuit breaker theory and design', Peter Peregrinus Ltd. 1975
14. C. J. O. Garrard, Proc. IEE 123 (1976) 1053
15. W. Hermann, U. Kogelschatz, K. Ragaller and E. Schade, J. Phys. D 7 (1974) 607
16. W. Hermann, U. Kogelschatz, L. Niemeyer, K. Ragaller and E. Schade, J. Phys. D 7 (1974) 1703
17. W. Hermann, U. Kogelschatz, L. Niemeyer, K. Ragaller and E. Schade, IEEE Trans. PAS 95 (1976) 1165

18. W.Hermann and K. Ragaller, IEEE Trans. PAS 96 (1977) 1546
19. A. Pedersen, IEEE Trans. PAS 89 (1970) 2043
20. W. Boeck, Bull. SEV 66 (1975) 1234
21. U. Kogelschatz, E. Schade and K. D. Schmidt, Brown Boveri Review 61 (1974) 488
22. B. Eliasson and E. Schade, Int. Conf. on Phenomena in Ionized Gases, Berlin (1977) 409
23. K. Ragaller, Brown Boveri Review 64 (1977) 264

DISCUSSION
(Chairman: W. Rieder, Technical University, Vienna)

D. T. Tuma (Carnegie-Mellon University)
I missed in your lecture some emphasis on the fact that research still needs to be done on the physical aspects of circuit breakers in order to improve their performance and predict their behavior.

K. Ragaller
I agree that many problems are still unresolved and many things still have to be done. What I wanted to emphasize is that the efforts in basic physical research have reached a point where a dialogue with network specialists is necessary and fruitful for both sides.

J. Urbanek (General Electric)
I have three remarks. First, there are many more breaking and switching requirements and, therefore, many more failure modes than dielectric and thermal, although these are, of course, the important ones.

Second, the simple model circuit you used for the trv evaluation in the terminal fault is valid only if you have the main contribution to the short-circuit current from local generation. But then the current is usually small. The highest currents occur when you have line contributions, and then you get an exponential-shape trv as described, for example, by ANSI and considered in IEC.

Third, your statement that the peak voltage does not depend on the short-circuit current is not exactly right. The peak voltage is higher for so-called percent current faults.

K. Ragaller
Your first remark refers to an often-discussed question: Can one simplify things so highly and discuss only two types of fault? As you said, there is a large number of requirements on circuit breakers besides these two: interruption of small inductive currents, capacitive currents, and many more. In principle, one could evaluate limiting curves in the U_n I-plane for all these conditions. However, for the development of circuit breakers the two limits discussed in my paper are by far the most important. And, in fact, they usually give the limits of the range of application. I agree with your second and third comments. The IEC standards specify a peak factor of 1.5, instead of 1.4, for lower short-circuit currents (10, 30 and 60 %).

J. J. Lowke (University of Sydney)
Your dielectric limiting curve is independent of the current at low currents and falls off at higher currents. Could you comment on the physical significance of these two sections? Is the breakdown voltage in the low-current region that of the cold gas, i.e. there is no hot gas remaining in this case?

K. Ragaller
The curve will be discussed in more detail in the paper of J. Kopainsky. Unfortunately there are very few measured curves available. One quantitative measurement of such a curve was made at Siemens (see discussion of Kopainsky's paper). From this measurement, it is found that the curve is practically horizontal at low currents whereas we have indications that the voltage could be increasing slightly in the direction of lower currents. This will, of course, depend on the time at which the peak occurs. However, the part of the curve which is only slightly dependent on the current is considerably below the cold gas value. The reason is that you still have a hot channel in the breaker at the time of the peak, even after interruption of relatively small currents.

G. Frind (General Electric)
The picture of the turbulent arc in the current-zero regime shows loops. I think that one would encounter problems in interpreting the driving force. My intuitive understanding is that the arc here would rather swim in the cold flow than shoot with the high velocity which it has at higher currents.

K. Ragaller
I agree that the extremely thin arc filament around current zero has no possibility of moving relative to the cold gas. It is transported along with the large-scale structure of the turbulence. It is distorted and attacked very strongly by the large- and fine-scale structure of the turbulent flowfield.

G. Frind
Do you think that turbulence persists over a significant amount of time? My colleague H. Nagamatsu (General Electric) feels that there might be a relaminarization after current zero.

K. Ragaller
Yes, it's a question of time. We have measurements where the turbulence does not decrease in intensity within a period of at least 10 µs after current zero.

D. M. Benenson (State University of New York at Buffalo)
Can you comment on the current dependency of the turbulent wiggles you showed in Fig. 11? Could you take similar pictures on the downstream side?

K. Ragaller

The picture in Fig. 11 was taken for the fairly high current of 2 kA. It is very difficult to detect turbulence in the region between the nozzles because the flow velocity is very small there. What I wanted to show is that the turbulence develops as a result of instability of the free-shear layer in the arc boundary. The important thing is that downstream of the nozzle throat this effect becomes one of extreme intensity because of the high velocity difference between the hot and cold gas. In the paper of G. Jones a Schlieren picture of the turbulent flow downstream is shown (Fig. 5 of G. Jones' paper).

W. Hertz (Siemens)

Do you include in your limiting-curve diagram any nozzle-clogging effects, or are these curves derived from model experiments with small arc diameters?

K. Ragaller

The so-called nozzle-clogging effect can have several consequences. One important effect is an increase of the pressure in a puffer-type breaker. This pressure effect can easily be included when deriving the limiting curves. There are other influences, such as metal-vapor contamination, which can cause a difference between the characteristics of a real breaker and the physical limiting curves. It is clear that there can be several phenomena in a real breaker which reduce the limiting curves compared with those for an ideal situation.

H. Noeske (General Electric)

The double-nozzle device you showed is very often used in breakers, but not always. Depending upon the design, the effects mentioned by W. Hertz become even more important.

K. Ragaller

You're right. The circuit-breaker engineer has several degress of freedom in designing a circuit breaker. The point I want to make is that there exist basic physical limits which lead to very general statements about limiting curves and their dependence upon current and voltage. A real circuit breaker deviates from these basic limits to the extent that the designer was unsuccessful in avoiding such negative effects as metal vapor.

TRANSIENT RECOVERY VOLTAGE IN THE SHORT LINE FAULT REGIME

M. B. HUMPHRIES
Central Electricity Generating Board, London, England

SUMMARY

The basic phenomenon of the short line fault transient is outlined in relation to the line and source side oscillations. The concept of the typical sawtooth line oscillation is analysed in detail with respect to the line high frequency parameters. The equivalent line surge impedances are then considered in depth for both sequence and modal analysis and the effects of the line physical characteristics, such as conductor clashing, are examined.

The effect of line terminal capacitance is shown to reduce the severity of the transient, although the benefit is partially offset by the presence of line traps. Other related aspects, such as arc resistance and current asymmetry, are briefly considered. Additional terminal capacitance in the form of a short cable termination or shunt capacitor may be used to control the short line fault severity. Finally the relative severities of the line oscillation for single and three phase earthed faults are examined in relation to the ratio of the source 50 Hz zero and positive sequence impedances.

1. INTRODUCTION

It is generally recognised that some of the most severe transient recovery voltage conditions imposed on circuit breakers are caused by faults occurring on transmission overhead lines relatively close to the clearing circuit breaker. These faults may have currents close to the circuit breaker rating and hence have an appreciable source side transient, and in addition have a line side transient given by the short length of line between the circuit breaker and the fault. This latter component is the well known sawtooth-shaped transient.

This paper sets out the basic phenomena of the short line fault (slf) transient and provides an understanding of the contribution made by other plant connected between the circuit breaker and overhead line. As the control of the slf is relevant to the application and required rating of circuit breakers, consideration is given to the ways in which this may be achieved.

2. SIMPLE CONCEPT OF LINE SIDE OSCILLATION

When a circuit breaker interrupts fault current due to a fault on its associated overhead line, the instantaneous voltage to earth of the breaker line terminal (U_o) will return to zero by a series of travelling waves reflected back and forth along the line between the circuit breaker and the fault. This produces a transient voltage, u_L, on the line side in the form of a damped sawtooth oscillation as shown in Fig. 1. When combined with the source side transient, u_s, the resultant trv, as illustrated in Fig. 2, appears across the circuit breaker terminals.

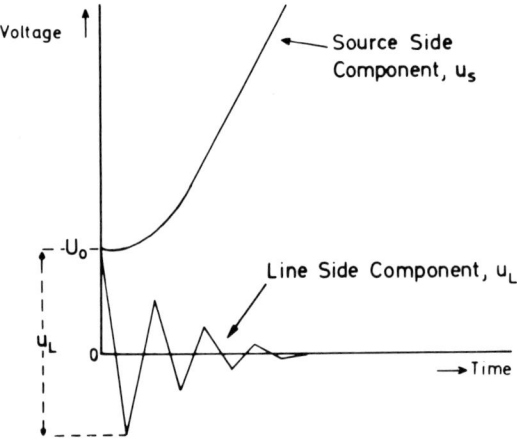

Fig. 1. Components of trv

The voltage to earth of the line terminal, U_o, at the instant of arc extinction depends on the power frequency impedance of the line to the fault and the fault current, and is given by:

$$U_o = L_p \, di_L/dt$$

where L_p = power frequency inductance of the line to the fault
di_L/dt = rate of change of fault current at the instant of arc extinction.

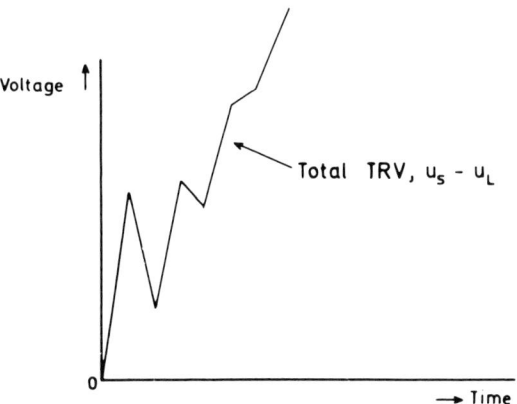

Fig. 2. Total trv

Since the fault current is largely reactive, U_o is also related to the crest value of the power frequency voltage to earth (U_m) by the approximate expression:

$$U_o = U_m(1-s)$$

where s is the slf factor defined in the paper by K. Ragaller and K. Reichert:

$$s = I_L/I_n$$

where I_L = short line fault current
I_n = rated fault current.

Following current interruption, the transient voltage excursion on the line side will rise at a rate of $Z \cdot di_L/dt$ until the reflected wave returns at time t_L. Neglecting the effects of any terminal capacitance or inductance, the peak of this excursion, U_L, can then be defined as

$$U_L = Z \cdot di_L/dt \cdot t_L$$

where Z = effective surge impedance of the line
t_L = twice the transit time of the line from the breaker to the fault.

t_L can be obtained from the velocity of propagation of the line, which to a 1st approximation can be assumed to be the speed of light, i.e. $3 \cdot 10^8$ metres/sec, and the distance of the fault from the breaker. It can also be obtained from the expression

$$t_L = 2\sqrt{L_h \cdot C}$$

where L_h = high frequency inductance of line
C = capacitance of line.

The surge impedance Z can be obtained from

$$Z = \sqrt{L_h/C}$$

However, a more detailed examination of the surge impedance and propagation speed will be made later.

Thus

$$U_L = \sqrt{L_h/C} \cdot di_L/dt \cdot 2\sqrt{L_h \cdot C}.$$

Now

$$di_L/dt = \frac{d}{dt}(\sqrt{2} \cdot I \sin\omega t) = \sqrt{2} \cdot I \cdot \omega$$

at current zero where I = rms fault current
ω = angular frequency of current waveform ($2\pi f$)

$$U_L = 2\sqrt{2}\, L_h\, I\, \omega$$

The excursion or rated peak factor (k_L) for the transient voltage is given by the ratio of the peak transient voltage to the circuit breaker voltage to earth at current zero

$$k_L = U_L/U_o = \frac{2 \cdot L_h\, di_L/dt}{L_p\, di_L/dt} = 2 \cdot L_h/L_p$$

Thus the excursion factor or degree of overshoot of the line side voltage following current interruption is directly related to the difference between the high and normal frequency inductances of the line. However, as an overhead line inductance varies significantly with the frequency of oscillation it is very necessary to use the inductance pertaining to the natural frequency of the faulted line. Table 1 below illustrates the different inductances for one circuit of a double circuit 400 kV line when the other circuit is energised normally.

As the fault location moves away from the circuit breaker, the fault current decreases and hence the rate of rise of transient voltage decreases. However the line surge impedance increases slightly as the natural frequency of the faulted section of line decreases, as implied from Table 1. At the same time, the time to peak of the line side oscillation, t_L, increases.

TABLE 1
High Frequency Parameters of One Circuit of Double Circuit 400 kV Line -
Normal Bundle Spacing

Frequency Hz	Inductance mH/km		Surge $\sqrt{L/C}$ Impedance			Velocity of Propagation $1/\sqrt{LC}$-m/µs	
	L_{h1}	L_{ho}	Z_1 ohms	Z_o ohms	$\dfrac{2Z_1+Z_o}{3}$ ohms	+ ve seq	0 seq
50	.883	2.515	259	561	360	293	223
100	.883	2.418	259	550	356	293	227
500	.882	2.178	258	522	346	293	240
1,000	.878	2.070	258	509	342	294	246
5,000	.871	1.848	257	481	332	295	260
10,000	.869	1.770	257	470	328	295	266
50,000	.864	1.642	256	453	322	296	276
∞	.857	1.515	255	435	315	297	287

C_1 = 0.0131 µF/km L_{h1} = Positive Sequence Inductance
C_0 = 0.0080 µF/km L_{ho} = Zero Sequence Inductance
Z_1 = Positive Sequence Surge Impedance
Z_0 = Zero Sequence Surge Impedance

The overall effect of varying the length of the faulted short line is illustrated in Fig. 3 which shows the line side oscillation for different line lengths. No account has been taken of terminal capacitance or other modifying factors apart from a certain degree of damping after the first peak. Although not very apparent, the rate of rise of the line oscillation is higher for the shorter line lengths due to the higher interrupted fault currents, although the lower surge impedances reduce this effect. These slf transients have been combined with a source trv envelope in Fig. 4 to show the overall effect of different short-line faults. In this illustration the same basic source trv is used for all the slf conditions and modified according to IEC 56-2.

From Fig. 4 it can readily be seen that the higher rate of rise of trv occurs with the shorter line lengths to the fault location although the initial voltage peaks are lower.

3. SIMPLE CONCEPT OF THE SOURCE SIDE WAVEFORM

IEC Publication 56-2 shows that the source side waveform for the slf should be equivalent to that of a single phase-earth terminal fault with modifications to account

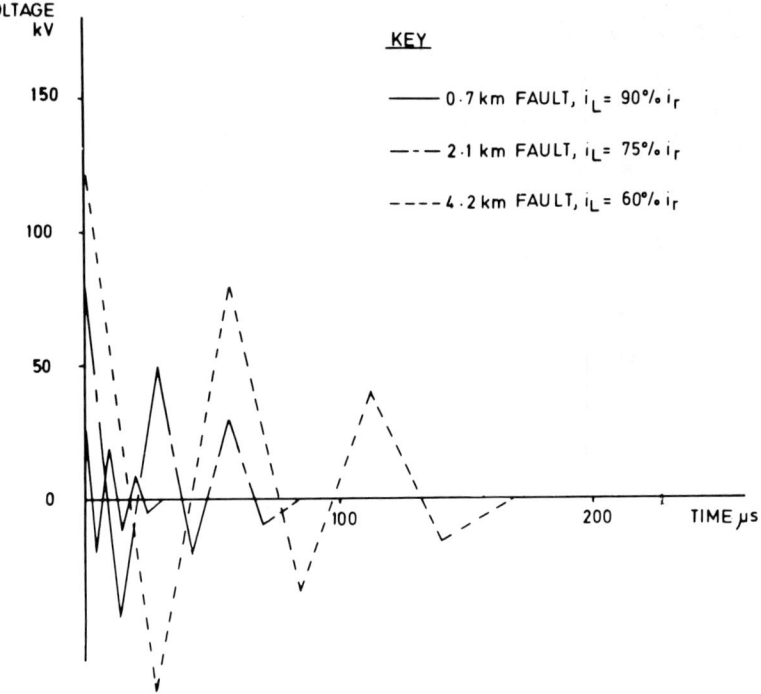

Fig. 3. Line side oscillations for different fault locations on a 400 kV line

for the initial voltage to earth, U_0, prior to current interruption. As the source side voltage peaks are proportional to current for the same source network and occur at the same time as for terminal faults, the absolute value of the 1st peak voltage remains constant at U_1 and hence the source side rate of rise

$$du_s/dt = (U_1-U_0)/t_1$$

where U_1 = 1st peak voltage for terminal fault
t_1 = time to 1st peak for terminal fault.

For the short line fault, the source side crest voltage U_{c1} is then given by

$$U_{c1} = U_0 + s\, U_c$$

where U_c = crest voltage for terminal fault
U_{c1} = crest voltage for slf

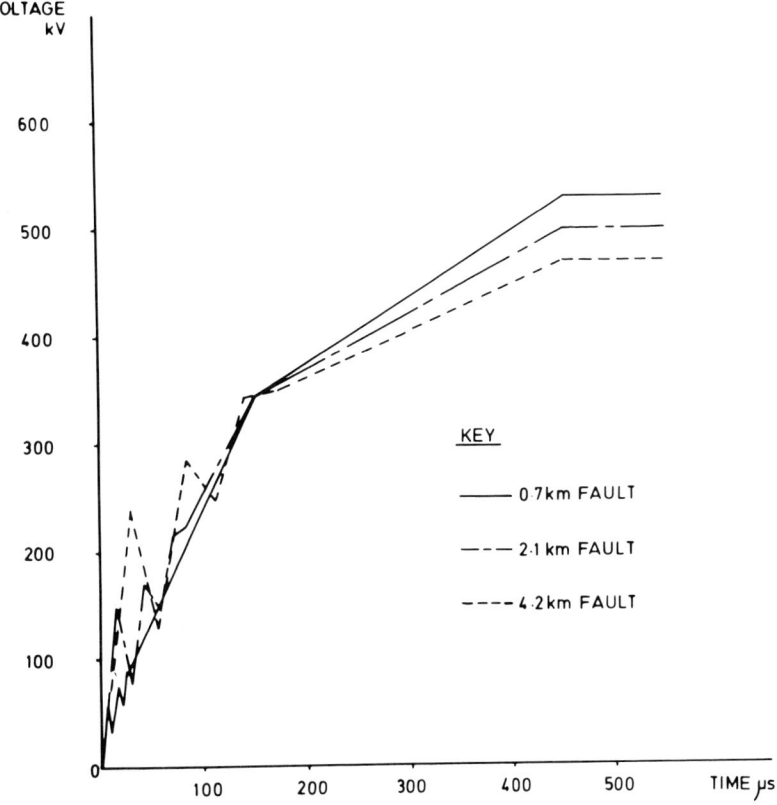

Fig. 4. Line and source side waveforms combined to give trv outlines for different fault locations on a 400 kV line.

If
$$U_c = kU_m$$

where k = amplitude factor
U_m = peak 50 Hz voltage

then
$$U_{c1} = U_m(1-s) + s\,kU_m = U_m[1+(k-1)s]$$

This is equivalent to formula (5) in the introductory paper. Fig. 4 shows the overall trv across the circuit breaker contacts for different slf's on a 420 kV circuit.

4. LINE PARAMETERS

4.1. Surge Impedance

Reference was made earlier to the effective surge impedance of the line and the influence it has on the rate of rise of the line oscillation. When a circuit breaker interrupts fault current the associated line surge impedance, Z, depends on the state of the adjacent phase conductors, the pole opening sequence of the breaker and the location of the conductor on the tower of the phase being opened.

In the sequence component method of representation, the line surge impedance, Z_{L1}, for the first pole to open during a 3 phase earthed fault, as illustrated in Fig. 5(a), is given by

$$Z_{L1} = 3Z_0 Z_1 / (Z_1 + 2Z_0),$$

where Z_1 = positive sequence surge impedance of line
Z_0 = zero sequence surge impedance of line.

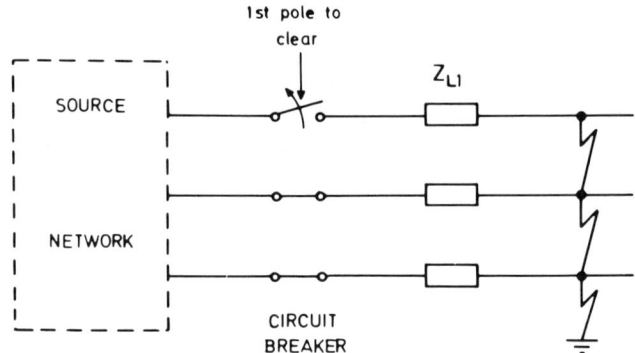

Fig. 5(a). Representation of 1st pole to clear a 3-phase earthed fault.

A similar expression can be derived for the line surge impedance, Z_{L3}, for the third or last pole to open during a 3 phase earthed fault, as illustrated in Fig. 5(b) and is given by

$$Z_{L3} = (2Z_1 + Z_0)/3.$$

For the same fault current, the relative severities of these two conditions is illustrated by the ratio of

$$Z_{L3}/Z_{L1} = 1/9 \, [2/x + 5 + 2x]$$

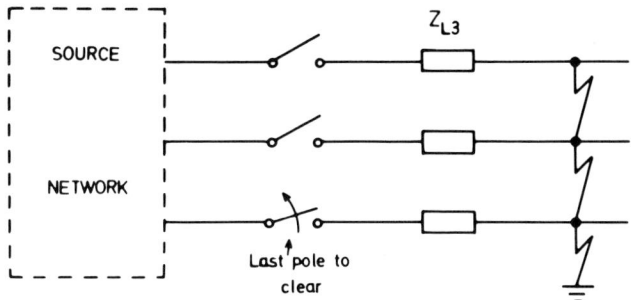

Fig. 5(b). Representation of last pole to clear a 3 phase-earthed fault.

where $x = Z_0/Z_1$. Thus the relative severity of the two pole opening conditions depends on the ratio of Z_0/Z_1 and can be plotted graphically for the practical values Z_0/Z_1. The resulting diagram is shown in Fig. 6. As the practical ratio of the surge impedances varies from 1.3 to 2.5, the last pole to clear a 3 phase earthed fault is seen to be 1.02 to 1.1 times the severity of the first pole to clear.

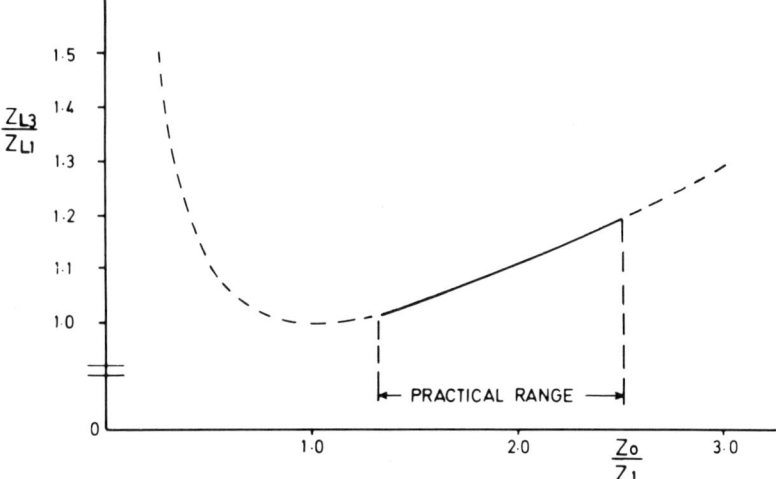

Fig. 6. Variation of Z_{L3}/Z_{L1} with Z_0/Z_1

The positive and zero sequence surge impedances Z_1 and Z_0 are derived directly from the positive and zero sequence inductance and capacitance of the line at the particular frequency concerned, i.e.

$$Z_1 = \sqrt{L_1/C_1} \qquad Z_0 = \sqrt{L_0/C_0}$$

where L_1, L_0, C_1 and C_0 are the positive and zero sequence inductances and capacitances.

Considering, for the moment, a single circuit line, the sequence inductance, L_1 and L_0 can be derived from the expressions

$$L_1 = [L_{11} + L_{22} + L_{33} - (L_{12} + L_{13} + L_{23})]/3$$

$$L_0 = [L_{11} + L_{22} + L_{33} + 2(L_{12} + L_{13} + L_{23})]/3$$

where L_{11}, L_{22}, L_{33} = self-inductances of the three phases
and L_{12}, L_{13}, L_{23} = mutual inductances between the three phases.

These self and mutual inductances at the required frequencies can be calculated from knowledge of the conductor construction, bundle spacing and spatial disposition on the tower. The sequence capacitances do not vary with frequency and are calculated from the constructional detail given above. Hence Z_1 and Z_0 can be calculated from the line and tower details.

A single phase earthed fault being cleared by the last pole to open represents substantially the same network conditions as given by Z_{L3} and hence the same surge impedance is used. As the single-phase earthed fault is the most common type of fault which occurs on ehv systems, it is usual practice to use the last pole to clear surge impedance Z_{L3} for short line fault calculations.

An alternative approach is to use modal analysis and thereby produce a 'phase surge impedance matrix' of an overhead line at a particular frequency. In theory a multiconductor transmission line will have as many modes of surge propagation as it has phase and earth conductors. By considering only the conductor for which the pole is opening, the line can be represented by an equivalent single-phase surge impedance as derived earlier using the sequence parameters. This representation of surge impedance is more accurate however, since unlike sequence parameters, it does not assume the lines are completely transposed. The other errors, i.e. due to neglecting the effects of all modes of propagation simultaneously in the conductor system, introduced in the solution of trv problems using this assumption are not significant and comparisons with field test results indicate that waveforms calculated in this way are realistic. These surge impedances can be calculated for each conductor position on the tower and for each pole opening sequence of the circuit breaker concerned. Thus the first part of Table 2 below shows these various surge impedances for a 4 x 400 mm^2 double circuit 400 kV overhead line.

The surge impedances in Table 2 are for the same line as used to calculate the sequence parameters given in Table 1 and so a comparison can be made between these different parameters at 50 kHz. From Table 1 the first and last pole to clear surge impedances Z_{L1} and Z_{L3} are calculated to be 299 and 322 ohms respectively.

The equivalent surge impedances from the modal analysis for the same line conditions lie between the 1st/4th pole impedances and the 3rd/6th pole impedances respectively. Thus it can be seen that the single value of surge impedance derived from the sequence parameters represents an average value for the line. For some particular studies the average value may be insufficient if the conductor location on a tower is considered to be significant.

TABLE 2
Surge Impedances at 50 kHz for a Double Circuit
4 x 400 mm^2 kV Overhead Line

Pole Opening Sequence	Conductors Normal			Conductors Clashed		
	Conductor Position			Conductor Position		
	Top ohms	Middle ohms	Bottom ohms	Top ohms	Middle ohms	Bottom ohms
1st Pole to clear	305	293	285	418	407	397
2nd Pole to Clear						
Top Cleared		305	287		417	398
Middle Cleared	317		298	429		408
Bottom Cleared	306	306		419	417	
3rd Pole to Clear	321	321	302	432	430	410
4th Pole to Clear	318	297	289	430	412	400
5th Pole to Clear						
Top Cleared		314	292		426	403
Middle Cleared	336		305	444		413
Bottom Cleared	320	313		432	425	
6th Pole to Clear	343	335	311	449	441	418

In the CEGB, these parameters are calculated using a computer program based upon a multi-conductor modal approach while frequency dependent earth return characteristics are determined by Carson's method. System tests and measurements carried out on the UK ehv system have been used to validate the calculation methods so that the line surge impedances used in the associated transient studies are considered accurate to $\pm 5\%$.

4.2. Effect of Conductor Clashing

For a circuit with a single conductor per phase, the value of surge impedance to use in short line fault calculations is that given by the normal spatial arrangement of the conductors in relation to the towers and ground plane. Account should be taken of the variation of surge impedance with the frequency of oscillation if the higher degree of accuracy obtained is justified. However, for short line faults up to 15 km from the circuit breaker, the normal surge impedance for a single phase to earth fault will only vary from 315 - 328 ohms according to Table 1, and hence for most purposes a single value should be appropriate.

A more significant change can occur to the surge impedance of lines with bundled conductors if the fault current is high enough and persists long enough to cause the conductors of the phase bundle to touch. Tests at CESI, KEMA and CERL (UK) have shown that fault currents greater than about 25 kA will cause bundled conductors to clash in less than 100 ms. This applies to the typical conductor configurations in common use and hence it is necessary to use the corresponding surge impedance for the line side oscillation during short line faults.

System tests have validated calculations which show that at 50 kHz the surge impedance for clashed conductors of a quad bundle of a double circuit 400 kV line is 32 - 39 % higher than the normal value, depending on the pole opening sequence adopted. Thus the values of the earth fault surge impedance $(2Z_1 + Z_0)/3$ given in Table 1 should be multiplied by about 1.35 at the higher frequencies. While this higher surge impedance should be used for the line side oscillation, it is incorrect to adopt it for the source side as it is highly improbable that the lines in the source network would have a sufficiently high fault current to cause the conductor bundle to clash.

One other effect of the clashed conductors is to raise the 50 Hz impedance of this line by about 20 % and hence the short line fault current will reduce as the conductors clash. Thus for the 90, 75 and 60 % short line faults mentioned in Section 3 the actual currents would be 88.2 %, 71.4 % and 55.6 % for the same fault locations. Alternatively, this higher impedance can be considered to lead to a shorter length of line for the 90, 75 and 60 % slf's (this refers to the factor s) and hence higher frequencies and lower amplitudes for the line side oscillations.

TABLE 3
50 Hz Parameters for a Double Circuit
4 x 400 mm^2 400 kV Overhead Line

	Conductors Normal	Conductors Clashed
+ ve Sequence - ohms/km Impedance	0.019 + j 0.277	0.019 + j 0.389
0 Sequence - ohms/km Impedance	0.105 + j 0.790	0.105 + 0.902
+ ve Sequence - µF/km Capacitance	0.0132	0.0093
0 Sequence - µF/km Capacitance	0.0080	0.0063

Tables 2 and 3 compare the surge impedance and 50 Hz parameters for normal and clashed conductors of a double circuit 4 x 400 mm^2 400 kV overhead line. In Table 2, the surge impedances are given for the different conductor positions on the tower and the order of pole opening assumed.

4.3. Effect of Line Height

In general the surge impedance of lines increases with the conductor height above ground and while the variation may be insignificant for normal line routes using the same type of tower and conductor configuration, it can be appreciable for river crossings.

TABLE 4
Effect of Small Variations in Line Height on Surge
Impedance of Top Conductor, Last Pole to Clear

Height to Top Conductor m	31.0 (Standard)	35.3	41.0
Surge Impedance 1 kHz - ohms	356	360	366
Surge Impedance 100 kHz - ohms	342	347	351

Using the computing facilities mentioned earlier, calculations have been made to illustrate these effects. Table 4 above shows the difference for small height variations of the same 4 x 400 mm^2 400 kV circuit used before with normal subconductor spacing and it can be seen that the impedance varies by less than 3 % for more than 30 % difference in conductor height above ground. The effect of very high towers for a river crossing is illustrated in Table 5 below for a 2 x 400 mm^2 ACSR 275 kV line where the bottom conductor height was varied from 12.2 m to 83 m. In this latter case the surge impedance increases by about 10 % for the top conductor and 20 % for the bottom. These differences are caused by the changing influence of the ground plane between the two conditions. For this high river crossing the bottom conductor has the highest surge impedance of the three which is the reverse of that for a normal tower.

TABLE 5
Effect of Extra High Towers on 100 kHz Surge Impedance
(Subconductors Touching) for a 2 x 400 mm^2 275 kV Overhead Line

Line Crossing	Bottom Conductor Height Above Ground/Water Metres	Surge Impedance (Last Pole to Open)		
		Top	Middle	Bottom
River Crossing	83	492	510	517
Standard Line	12.2	451	449	430

4.4. Effect of Earth Resistivity

For the same standard 4 x 400 mm^2 ACSR 400 kV line, the surge impedance varies according to Table 6 below with different values of earth resistivity. The values show that surge impedance varies by 5.7 % at 1 kHz and 3.2 % at 100 kHz for earth resistivity changes from 10 to 1000 ohm-metres.

The variation in surge impedance is mainly due to changes in the high frequency inductance caused by the different earth penetration effects.

TABLE 6
Effect of Earth Resistivity on Top Conductor Last Pole to Clear
Surge Impedance for a 4 x 400 mm^2 ACSR 400 kV Line

Earth Resistivity ohm m	10	20	30	100	300	1000
Surge Impedance - 1 kHz-ohms	353	356	358	363	368	373
Surge Impedance - 100 kHz-ohms	341	342	342	344	347	352

4.5. Effect of Tower Footing Resistance

Only the sending end tower footing resistance need be considered for the following reason. Near this tower, the current fed into the phase conductor returns through the earth and earth wires in a ratio determined by their surge impedance and the earthing resistance of the terminal tower foot. At subsequent towers along the line, the ground and aerial mode travelling waves arrive almost simultaneously and hence the current through their footing resistances will be very small. Thus for the short lengths of line considered here, the division of the travelling wave between the earth and earth wire will be determined largely by the sending end tower footing resistance.

The variation in surge impedance with sending end tower footing resistance is illustrated in Table 7 below, assuming that other tower footing resistances are zero and the earth resistivity is constant at 20 ohm metres.

TABLE 7
Effect of Sending-End Tower Footing Resistance on Top Conductor,
Last Pole to Clear Surge Impedances for a 4 x 400 mm^2 ACSR 400 kV Line

Tower Footing Resistance - ohms	0	10	20	30	50	70
Surge Impedance - 1 kHz-ohms	356	363	369	375	388	398
Surge Impedance - 100 kHz-ohms	342	350	356	362	373	383

For the range of sending end tower footing resistance from 0 - 70 ohms the surge impedances are seen to increase by more than 10 %. However in the large majority of cases, the tower footing resistances will lie in the range 0 - 10 ohms as the terminal tower will be connected to the low resistance substation earthing mats.

4.6. Which Surge Impedance to Use?

As the surge impedance of the short line to the fault has a direct bearing on the severity of the line side transient, it may be prudent to take account of all the effects outlined above in determining the resultant surge impedance for particular applications if the physical characteristics, such as earth resistivity, are known. For more general use, it is generally satisfactory to use some form of average or normal value. However, with the high levels of possible fault current, it is now accepted practice at the higher voltage levels to assume clashed bundle conductors with the resulting higher surge impedance.

The IEC have adopted this course and specify a standard surge impedance for short line faults of 450 ohms. For the line assumed in Table 2, it can be seen that this just covers all the possible surge impedances with clashed conductors at 50 kHz, but makes no allowances for variations in line height, earth resistivity or tower footing resistance which are assumed to be 31 metres to the top conductor, 20 ohm-metres and 0 ohms respectively. However this value represents a good compromise and adequately covers most applications.

5. SPEED OF PROPAGATION

It was mentioned earlier that there are as many modes of transient propagation along a line as there are parallel phase conductors and earth conductors. Thus for a double circuit line with a single earth wire there are seven different modes for the surge to travel along and for a rigorous treatment each must be considered separately. As a first approximation the voltage at every point in the earth conductor is assumed zero, and hence the number of modal propagations may in effect be reduced to six. Thus in the 3 phase travelling wave digital computer programs six different velocities will be used, five of which will be close to the speed of light, i.e. 2.93 - 2.99 x 10^8 metres per second at 10 kHz and represent the propagation between earth and the phase conductors. At higher frequencies, the velocities are closer to the speed of light, i.e. 3×10^8 metres per second.

For single phase modal representation using the parameters of Table 2 it is usual to assume the speed of light for short line fault calculations as only minor errors are introduced by so doing.

Using the sequence calculation methods individual speeds of propagation are used for the positive and zero sequence networks according to values given in Table 1 which are derived from the expression $1/\sqrt{LC}$.

6. EFFECT OF PLANT AT THE LINE TERMINATION

6.1. Lumped Capacitance

It is well understood that the source side trv will experience a time delay prior to its initial rate of rise due to the lumped capacitance of the substation. The same phenomenon applies to the line side oscillation due to the lumped capacitance associated with Current Transformer Bushings, Capacitor Voltage Transformers, Line Trap Bushings, Circuit Breaker Bushings, etc.

The simple expression given earlier for the peak of the line side transient at the time the reflected wave returns from the fault is:

$$U_L = Z \cdot di_L/dt \cdot t_L$$

and as a function of time (t) this becomes

$$U_L(t) = Z\sqrt{2}\, \omega\, I_L\, t.$$

With the addition of line side lumped capacitance, C, a simple expression for the transient up to the time at which the first reflected wave arrives back at the circuit breaker is:

$$U_{L0}(t) = Z\sqrt{2}\, \omega\, I_L\, (t - ZC + ZC\, \exp(-t/ZC))$$

As indicated before, the first reflected wave from the fault arrives back at the circuit breaker after twice the travel time of the line and hence beyond this time the expression for the voltage due to the reflection of this wave from the circuit breaker terminal is:

$$U_{L1}(t) = Z\sqrt{2}\, \omega\, I_L\, [4ZC - 2t - 2(t+2ZC)\, \exp(-t/ZC)]$$

Similarly the voltage due to the second reflected wave which arrives at the breaker after four times the line travel time, is given by

$$U_{L2}(t) = Z\sqrt{2}\,\omega\, I_L\, [2t-8ZC+(8ZC+6t+2t^2/ZC)\exp(-t/ZC)]$$

Successive reflections can be handled in a similar manner.

Thus up to the time of the third reflection the total line side transient is given by

$$U_L(t) = U_{L0}(t) + U_{L1}(t) + U_{L2}(t)$$

These successive reflections and the resultant voltage are illustrated in Fig. 7. The total line side component is thus composed of the initial transmitted voltage and reflected voltages added at multiples of twice line travel times.

A more rigorous analysis of this line oscillation can be carried out using the full travelling wave equations in digital computer programs.

Although Fig. 7 shows substantial delays and subsequent rounding of the oscillation, this has only been achieved through the use of a lumped capacitance of the order of 10,000 pF for a fault current of about 40 kA and line travel time of 5 microseconds. With the plant components in common use on ehv systems, this amount of capacitance is unlikely to be realised in practice. Using the normal capacitances associated with line side plant in the UK as indicated below for 400 kV plant,

Oil filled Current Transformer (CT)	- 1570 pF
Gas (SF_6) filled Current Transformer (CT)	- 300 pF
Capacitor Voltage Transformer (CVT)	- 2000 pF
Circuit Breaker (Total)	- 300 pF

line side oscillations have been computed for different combinations of plant and the first peaks are shown in Fig. 8. The most common terminal capacitance for 400 kV circuits in the UK is that given by a gas filled CT and a CVT with a combined capacitance of 2300 pF, excluding the effect of the line side circuit breaker terminal and line side post insulators. For this condition, curve 3 in Fig. 8 shows that the line side oscillation time delay is approximately 1 µs and the overall rate of rise has fallen by about 6 %, using the IEC method.

From the expanded time base, computer drawn, curves shown in Fig. 9(a) and (b) for lumped capacitances of 300 pF and 3570 pF (Cases 2 and 4 of Fig. 8) the IEC time delays, t_d, for the line oscillations are seen to be 0.24 µs and 1.26 µs respectively. The reduction in the overall rates of rise can also be determined and for Cases 2 and 4 amount to 2.2 % and 12 % respectively. Thus the simple theoretical sawtooth waveform discussed earlier is modified by the lumped capacitances of the terminal plant normally connected on the line side of the clearing circuit breaker. The oscillation is seen to be delayed and made less severe depending on the magnitude of the overall terminal capacitance.

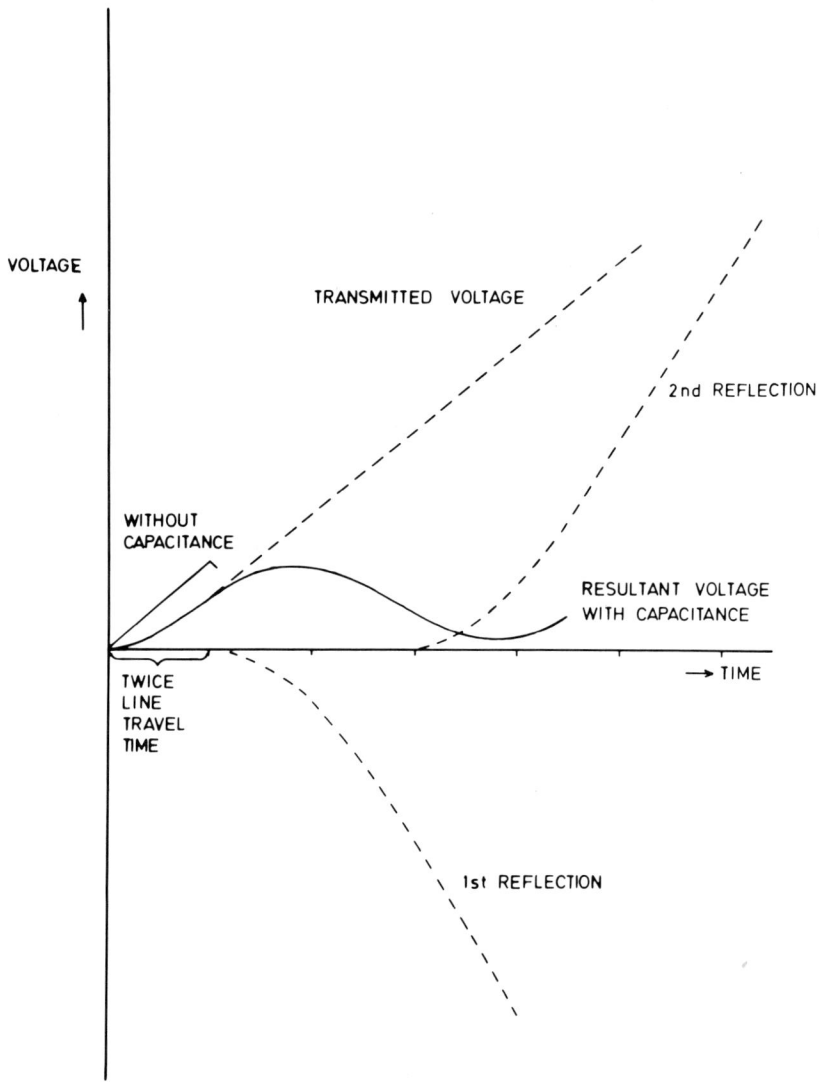

Fig. 7. Line side oscillation with added capacitance

SHORT LINE FAULT

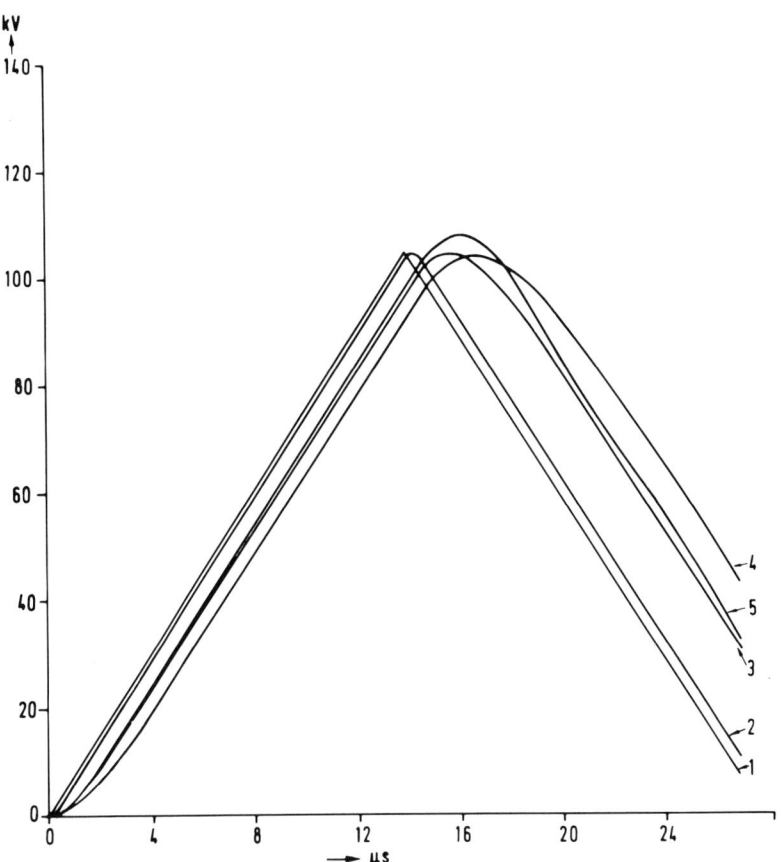

KEY:
1. NO TERMINAL CAPACITANCE.
2. 300 pF TERMINAL CAPACITANCE - GAS FILLED CT.
3. 2300 pF TERMINAL CAPACITANCE - GAS FILLED CT + CVT.
4. 3570 pF TERMINAL CAPACITANCE - OIL FILLED CT + CVT.
5. 2300 pF TERMINAL CAPACITANCE + LINE TRAP.

Fig. 8. Line side oscillations for 2.1 km, 37.5 kA fault on 400 kV line

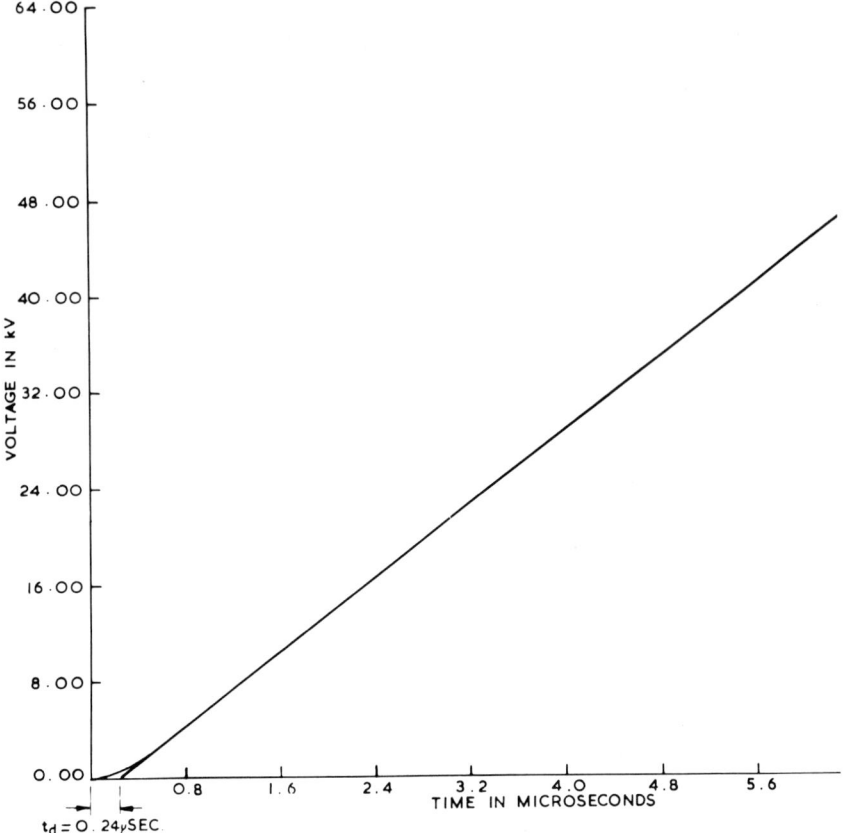

Fig. 9(a). Initial line side oscillation with gas filled CT (300 pF)

6.2. Line Traps

Although on ehv overhead line networks the lumped capacitances mentioned earlier are usually connected to all three phases, it is common practice to connect line traps to only two of the phases and hence their effect is not associated with all three phases. The line traps are used to inject high frequency signals onto the line for power line carrier protection and signalling purposes and generally operate in the 100 kHz - 550 kHz waveband. In many cases the line traps are mounted on bushings, which themselves constitute lumped capacitances, while in the UK they are often installed on the top of capacitor voltage transformers.

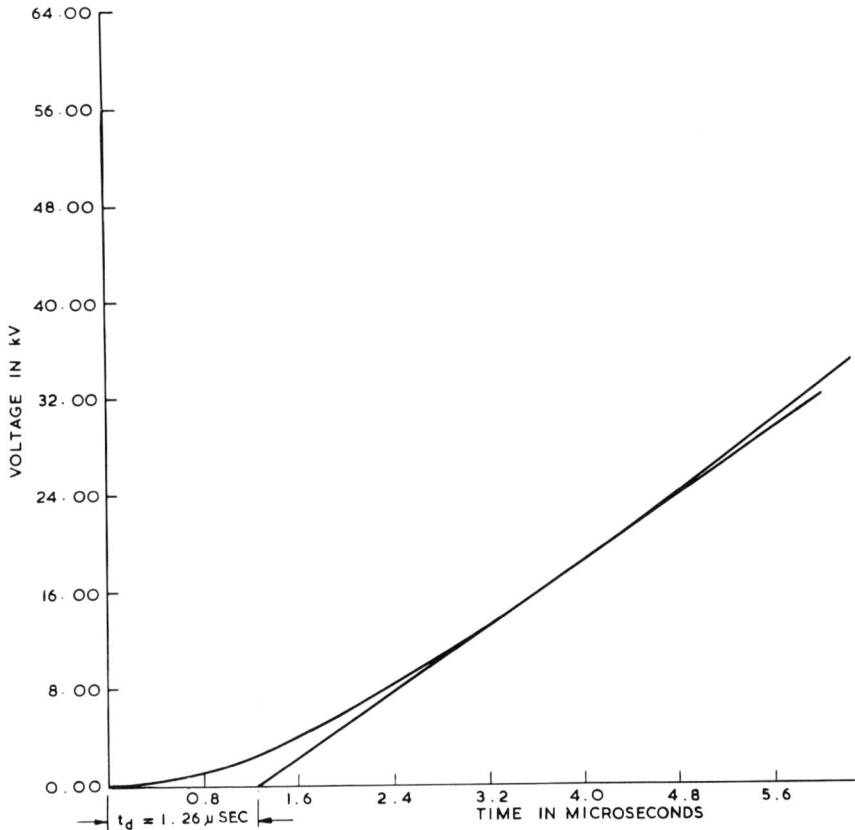

Fig. 9(b). Initial line side oscillation with oil filled CT and CVT (3570 pF)

Line traps react predominantly as an inductance although their equivalent circuit is complex as shown in Fig. 10(a). This illustrates a typical UK Line Trap, in this case tuned to 230 kHz, and although American Line Traps are somewhat different to the European ones, they show similar characteristics to ramp current injections for the same operating frequency. Fig. 10(b) illustrates the circuit connections to the line trap. The response of this line trap to a ramp current during a short line fault is illustrated in Figs. 8 and 11. Fig. 8 shows that compared to the case with no line trap, the line side oscillation reaches a higher peak voltage with a slightly increased rate of rise, while Fig. 11 indicates that the time delay, t_d, has reduced from 0.96 μs to 0.80 μs.

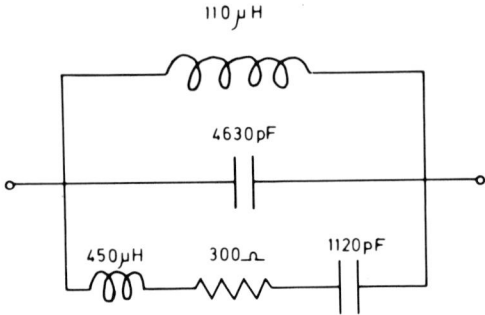

Fig. 10(a). Equivalent circuit of a 400 kV line trap.

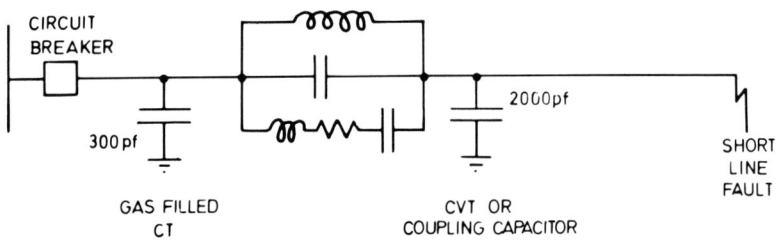

Fig. 10(b). Location of line trap.

The reduction in time delay, t_d, arises due to the effect of the series inductance in the line trap. Fig. 11 shows a slight ripple in the line side waveform due to an oscillation within the line trap, but this rapidly dies out. The increased peak voltage is also due to the line trap inductance at the time of the first reflection at the source and the increased rate of rise is due to the combined effect of reduced time delay and higher peak voltage even though the latter is slightly later.

The main effect of a line trap therefore is to reduce the line side time delay and marginally increase the rate of rise and peak voltage.

6.3. Effect of Substation Connections

In most designs of substations it is usual to have a short length of busbar connecting the overhead line termination to the circuit breaker. This connection is often up to 20 metres in length and may contain an off-load isolator. The surge impedance of

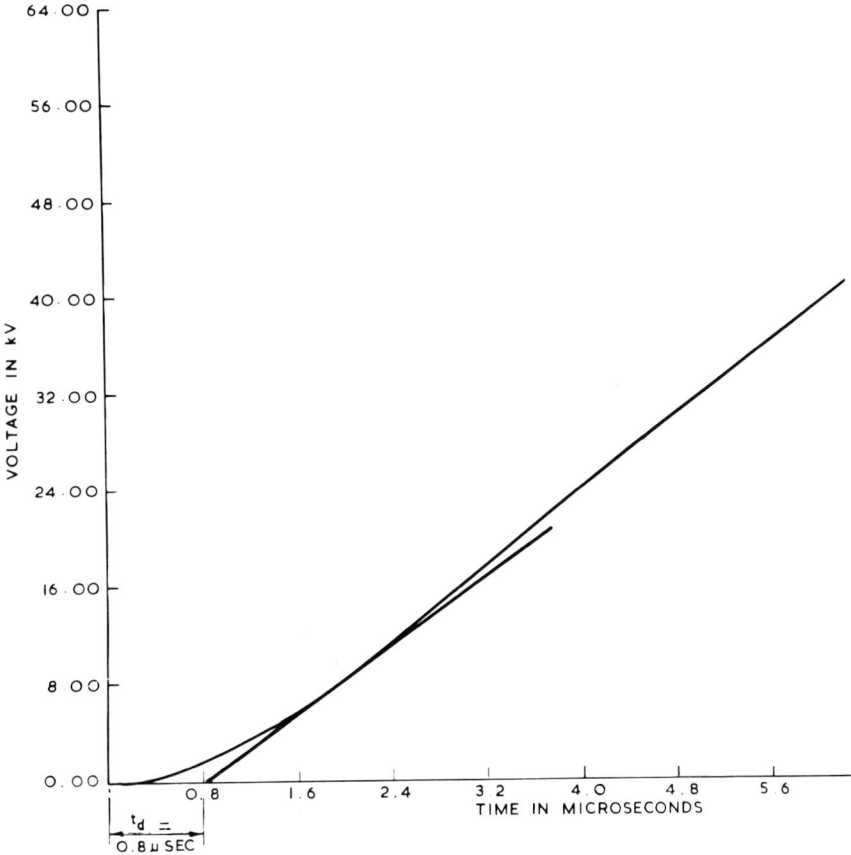

Fig. 11. Initial line side oscillation with line trap, gas filled CT and CVT (2300 pF).

the busbar connection is generally taken to be 260 ohms (see the chapter on itrv) and in theory should be taken into account for the short line fault. As it has a lower surge impedance than the line, it should give a lower rate of rise for the duration of its travel time. However, the latter is so short, i.e. 0.07 µs, that this reduced severity can be ignored for all practical purposes for the line oscillation.

Naturally the effect of the busbars should be taken into account for the source side oscillation as given in the chapter on itrv. However, the slf current should be used in this case rather than the full terminal fault current to determine the itrv.

7. SYSTEM ASPECTS OF THE SHORT-LINE FAULT

The use of standardised circuit breakers in a system makes methods of controlling trv conditions important to avoid the need for special purpose designs for particular applications and to achieve greater flexibility in system design.

In theory there are several ways to control the short line fault condition, some of which are not practical or economic. The severity of the fault on a circuit breaker depends on the level of fault current, the line surge impedance, current asymmetry, line termination plant, the fault location and the modifying characteristics of the circuit breaker arc. Some of these aspects are now discussed in relation to the electrical system.

7.1. Fault Currents

The magnitude of the available fault current for a circuit breaker terminal fault is an integral feature of the overall system design philosophy. Although a network may be designed to a certain kA rating it does not follow that this level of current can occur for terminal faults throughout the network. The U.K. 400 kV system, for instance, has been designed to a 3-phase short-circuit rating of 50 kA but right from its inception up to a projected capability of 4 times its present load, no more than 10 % of circuit breakers controlling overhead line feeders have potential terminal fault currents in the range 40 - 50 kA. The distribution of these fault currents for different system development stages is shown in Fig. 12. Hence the system configuration itself provides some degree of current control for the short-line fault. However, with solidly earthed systems the phase-earth fault current can be higher than the equivalent three phase value due to the effect of the low zero sequence 50 Hz impedances of transformers. In the U.K., this results in earth fault currents up to 10 % greater than the 3 phase currents on the 400 kV system and up to 25 % greater on the 275 kV and 132 kV systems. Apart from network sectioning, fault currents can be controlled using series impedance such as reactors, but these may introduce more onerous source trv conditions. Earth fault currents can be reduced by increasing zero sequence impedances of transformers through opening tertiary windings, installing neutral reactors or running transformers unearthed. However each of these methods introduces other technical problems outside the scope of this paper. Thus deliberate fault current control needs to be approached with great care.

Single phase overhead line faults can be categorised into solidly earthed faults due to line earths being left on the line when it is re-energised and arcing faults caused by flashover between a conductor and tower or earth. The former have effectively zero impedance to earth while the latter have a 50 Hz resistance governed by the length of the fault arc and magnitude of fault current. An expression derived by A. C. Van Warrington for the arc resistance is

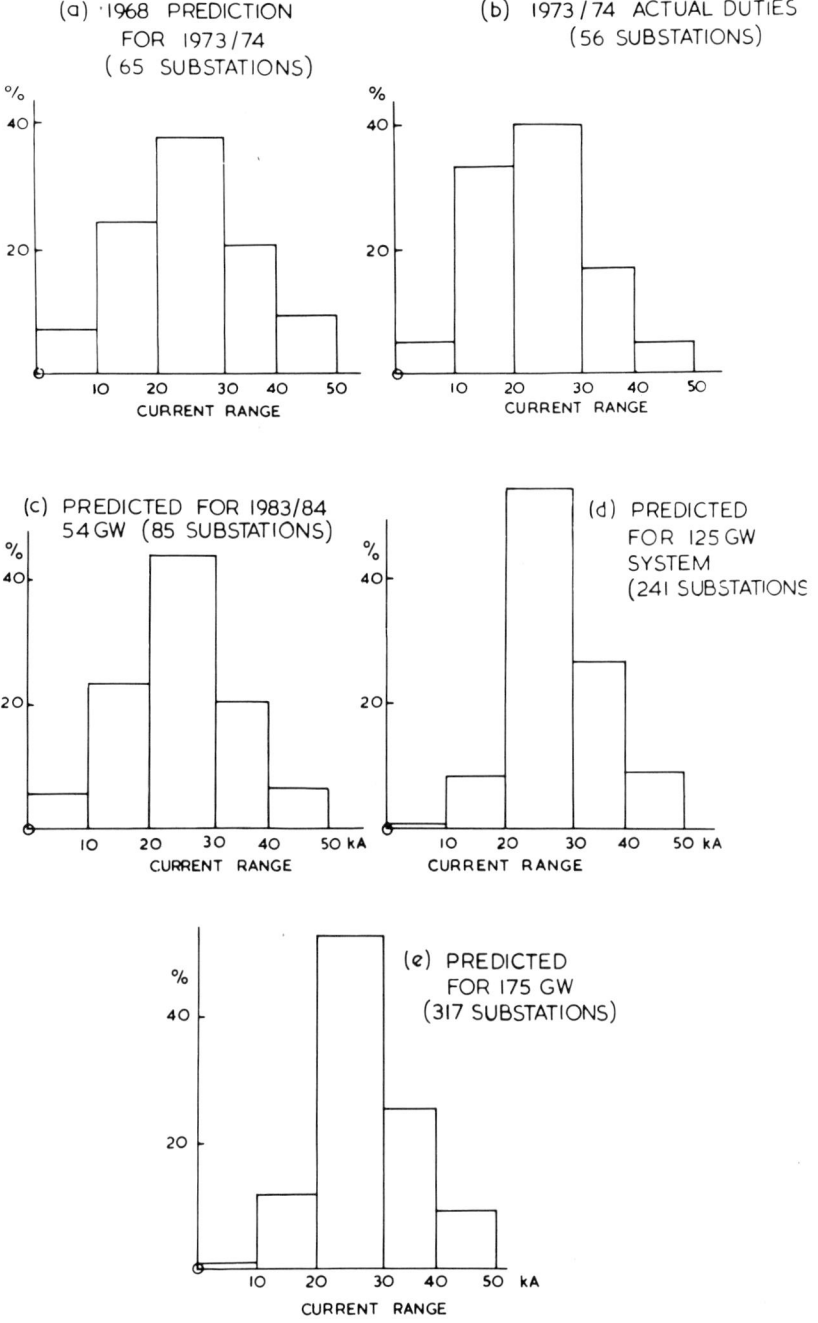

Fig. 12. Distribution of switchgear duties (excluding switchgear controlling transformers to 132 kV and lower voltage networks)

$$R_{arc} = 8750 \ \ell/I^{1.4} \text{ ohms}$$

where ℓ = length of arc in feet
         I = current in Amps.
In metric terms, this is equivalent to

$$R_{arc} = 27810 \times \ell/I^{1.4}$$

where ℓ is in metres.

For a 37.5 kA fault current, the arc resistance would be 0.03166 ohms on the CEGB system where the line arcing horns are set to 2.794 metres (110 inches). This resistance has a negligible effect on the level of current and the surge propagation along the line and hence for the short line fault with high values of current, an arcing fault can be considered in the same way as a solid fault.

7.2. Line Surge Impedance

By definition, the short line fault has a relatively high fault current and hence the appropriate surge impedance should be used. In systems with terminal fault currents of 40 kA or more, the values of clashed bundle conductors should be used, while for lower values of fault current, the lower surge impedance for unclashed conductors may be satisfactory depending on the expected fault current. Where the clashed conductor surge impedance is used then naturally the higher value of 50 Hz reactance should be used for the short line to determine the correct fault location and hence line transit times. If the specific line and terrain characteristics are known, then it may be necessary in particular instances to take account of high earth resistivity or extra high towers in determining the appropriate surge impedance.

7.3. Current Asymmetry

The point on the 50 Hz voltage wave at which a fault occurs together with the Reactance/Resistance (X/R) ratio of the fault current largely determine the degree of D.C. offset and its decrement. Faults which are caused by insulation breakdown under normal operating conditions usually occur at or near the peak value of the system voltage and hence the fault current will have little asymmetry in it as it lags the voltage by 75 - 90° due to the system 50 Hz impedance (>> R). The current is therefore near a zero crossing at the time of the fault. Flashovers caused by lightning strikes occur at any point on the voltage wave and hence the resulting fault currents may have a degree of asymmetry. This also applies to faults where pressurised head circuit breakers close onto earthed lines as the line energisation can happen at any point on the supply voltage wave.

The effect of D.C. asymmetry in the fault current is to ease the slf duty if current interruption happens before the asymmetry has disappeared. Fig. 13 illustrates

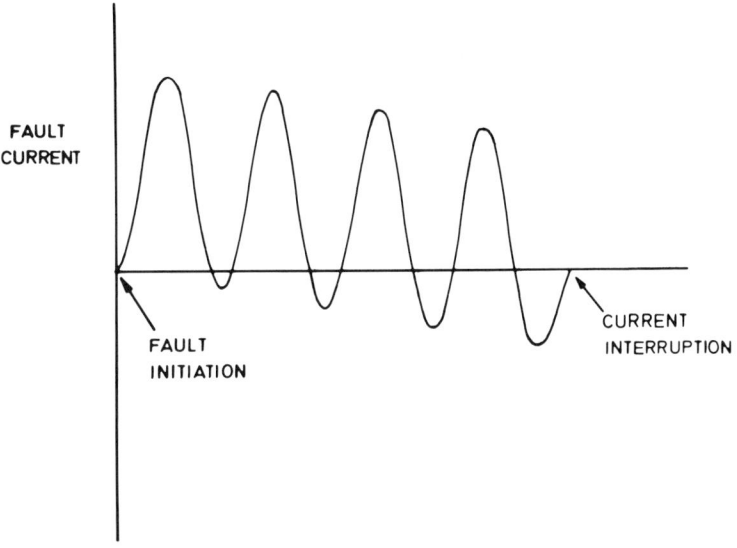

Fig. 13. Asymmetrical current interruption

a current zero interruption of an asymmetrical current and although the sinusoidal current has the same peak-to-peak amplitude as the equivalent non-asymmetrical current, the rate of change of current, di/dt at current zero is reduced. Since the line side transient is directly proportional to di/dt, the transient is less onerous to the circuit breaker. However the energy in the breaker arc is somewhat higher than normal. With system X/R ratios of up to 20 for short line fault currents, the severity of the line transient will not be significantly reduced as the time constant of the D.C. decrement would be less than 64 ms and the time to arc interruption in general would be greater than this.

Thus, although easier, the slf duty for asymmetrical currents is virtually no different from the symmetrical case.

7.4. Protection Operation and the Short Line Fault

It is common practice to install one or two sets of distance protection on overhead line circuits and due to its inherent characteristics for faults close to one end of a circuit, it is unlikely that the circuit breaker with the potential slf duty would be subjected to the full duty in a meshed system. This circuit breaker would open first, but its source-side trv will be eased as the 50 Hz recovery voltage will be less than the normal voltage as the system voltage will remain partially depressed until the fault is finally cleared by the remote circuit breaker. Under these conditions the line side transient will not be affected. The easement of the source trv will depend on the degree of external coupling between the local source busbar and the remote busbar still feeding the fault and other system conditions. With long tie-lines, no easement can be expected. Where unit protection is used such as power line carrier either circuit breaker may open first and hence no easement may occur.

7.5. Fault Location

As most O/H line faults are transient in nature and leave little evidence to pinpoint the exact fault location, it is difficult to determine from fault statistics those which can be classed as short line faults. However, an examination of distance protection indications for the 400 kV system in the U.K. over the period 1969 - 1974 shows that about 72 % probably occurred within the central 60 % section of the lines. Thus in practice fewer faults seem to occur close to the line circuit breakers than would be expected. This may be due to the fact that switching substations are not normally sited in fully exposed locations and hence the first few kilometres of lines are shielded from climatic faults to some extent.

8. CONTROL OF SLF SEVERITY

The effect of terminal plant capacitance on the line side transient was outlined in Section 6.1 and showed that the sawtooth waveform became rounded at the peaks and was subject to a time delay, t_d. For the practical case with lumped capacitance of 2300 pF the time delay was shown to be about 0.8 µs and the rate of rise reduced by about 6 %. This suggests therefore a means of reducing the severity of the line side transient by adding additional capacitance on the line side of the circuit breaker.

This additional capacitance may be in the form of purpose made shunt capacitors or a short length of cable to terminate the overhead line. To illustrate this, a short section of 400 kV cable has been added to the circuit of Case 3 in Fig. 8 (2300 pF terminal plant capacitance). The susceptance of the cable is assumed to be 20 % per km on 100 MVA base giving a cable capacitance of 0.4 µF per km. Thus a short length of 25 metres would have a capacitance of 10.000 pF and, due to its short length, can be considered as a lumped element for calculation purposes. Fig. 14 shows its effect on the initial part of the line oscillation. The time delay is increased from 0.8 µs to 2.6 µs, and the rate of rise reduced from 7.0 kV/µs to 5.1 kV/µs, a reduction of 27 %. Thus a significant reduction in severity is produced by this relatively short cable section. Naturally longer lengths will reduce the severity even more. In the calculations the very short transit time and low surge impedance of the cable have been neglected as they will have no noticeable effect.

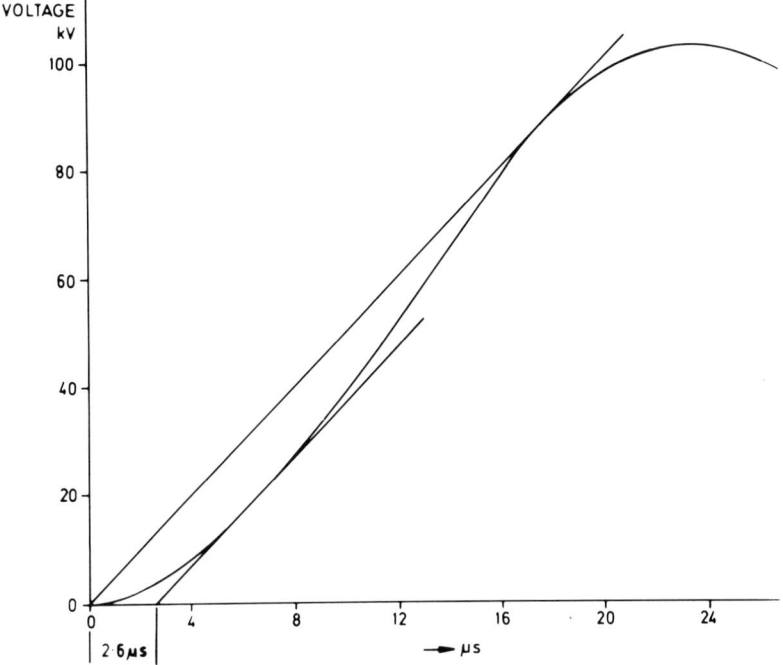

Fig. 14. Initial line side oscillation with additon of 25 metres 400 kV cable (10.000 pF)

The same reduction in servity can be achieved using the equivalent shunt capacitance of 10.000 pF. However, this solution may not be favoured due to problems in maintaining the integrity of the shunt capacitor units.

At voltages below 400 kV the length of cable required to produce the same reduction of severity remains substantially the same as the cable capacitance per unit length is similar.

The line terminal capacitance needed to reduce the severity of the line side oscillation is shown graphically in Fig. 15 for a single phase 400 kV fault about 2.1 km away with a fault current of 37.5 kA.

9. TOTAL TRV FOR THE SHORT LINE FAULT

As briefly indicated earlier, the total trv across the circuit breaker contacts is made up of the source and line side oscillations. In practical networks, the source side transient can have a variety of shapes depending on the system configuration while

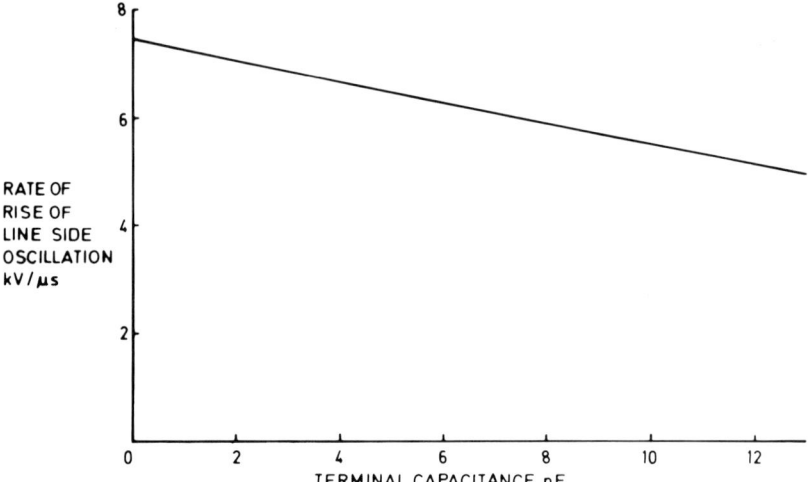

Fig. 15. The Reduction of line side rate of rise with terminal capacitance for a 400 kV, 2.1 km slf.

the line-side transient is modified as outlined previously. The line side transient has a very high effective rate of rise compared to the source side and hence tends to be the dominant feature immediately following arc interruption. However, the line-side transient is damped out before the crest of the source side trv is reached and hence it is the latter which dominates the later part of the total waveform. This is illustrated in Fig. 16 using a computer drawn source side transient with a manually calculated line side transient for a fault current of 37.5 kA and a fault location 2.1 km from the circuit breaker. As outlined earlier, the line side transient will have rounded peaks and a delay time, neither of which can be shown in the diagram.

10. RELATIVE SEVERITY OF 3 PHASE AND 1 PHASE SHORT LINE FAULTS

Hitherto, the discussion about the slf has centered around the 1 phase fault. However it would be useful now to compare the relative severity of the slf transient caused by the first pole to clear a 3 phase earthed fault.

For the 3 phase balanced fault, the current is determined by the positive sequence impedance, X_1, to the fault compared to the earth fault impedance

$$(2X_1 + X_0)/3$$

for a single phase fault (X_0 is the zero sequence 50 Hz impedance). Thus the 3 phase fault current is independent of the zero sequence impedance for the 1st pole

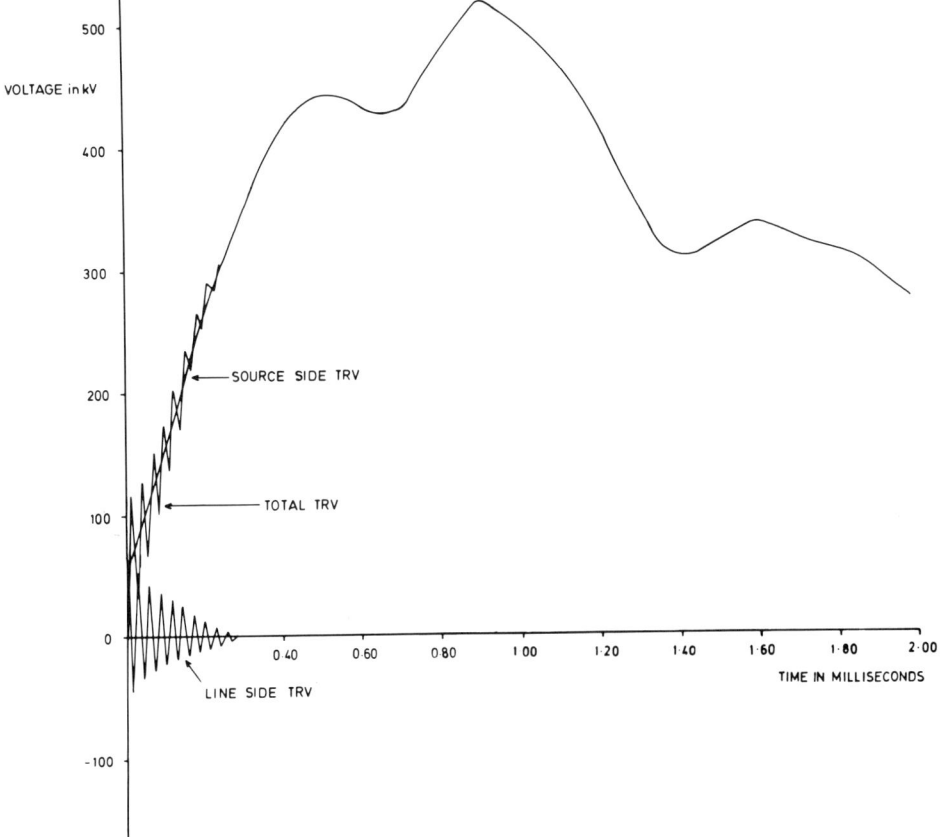

Fig. 16. Total trv for a 37.5 kA 400 kV fault, 2.1 km from the circuit breaker.

to clear. From the 50 Hz parameters given in Table 3, it is clear that the 3 phase fault current is reduced less by the line impedance than the 1 phase fault current, as X_0 is 2 - 3 times X_1. The overall effect of this is illustrated in Fig. 17 which shows the variation of fault current with distance along a 400 kV line for 3 phase and 1 phase faults for the same 50 Hz positive sequence source of 0.28 % on 100 MVA (about 50 kA) with clashed conductors. The single phase currents are shown for different ratios of source X_0/X_1.

The ratio X_0/X_1 for the source network is seen to have a significant effect on the single phase fault current and for networks with ratios greater than 1, the single phase slf current will always be lower than the 3 phase 1st pole-to-clear current. The last pole-to-clear current for a 3 phase fault is similar to the 1 phase earth fault current.

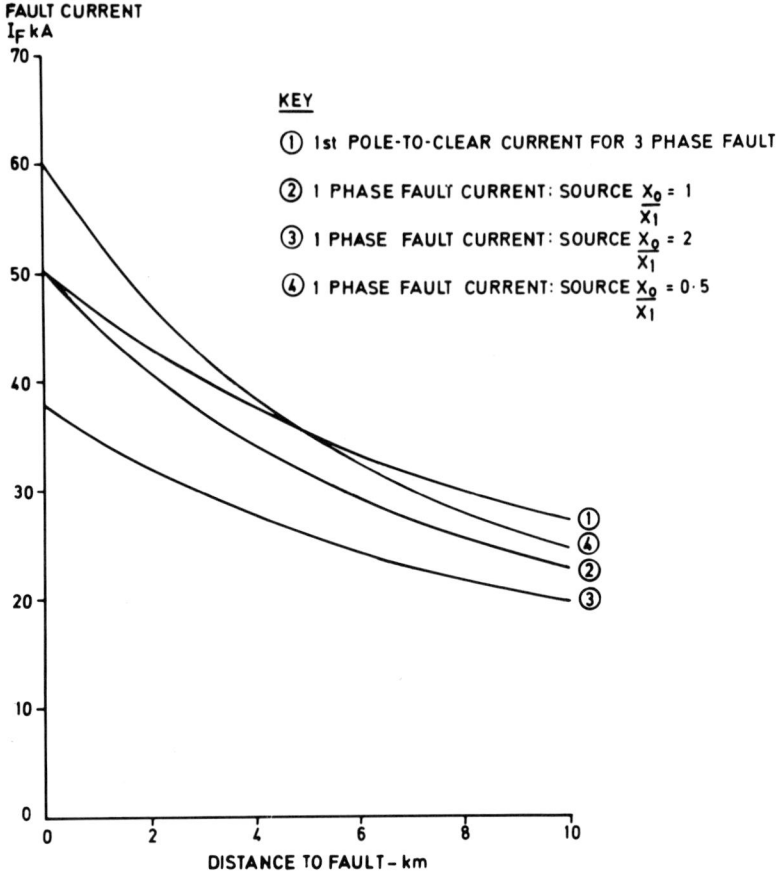

Fig. 17. Variation of fault current with distance to fault for 400 kV line with clashed conductors.

This variation of fault current will naturally affect the rate of rise of the trv and is illustrated in Fig. 18 for the line oscillation. Surge impedances of 430 ohms and 449 ohms respectively have been extracted from Table 2 and used for the 1st pole-to-clear and last pole-to-clear conditions. As these surge impedances are different, the curves giving the rates of rise are different. Fig. 19 illustrates the variation of 1st pole voltage, U_L, with distance to the fault assuming the velocities of propagation are 295 and 286 metres/μs respectively for the 3 phase and 1 phase faults.

These graphs show that in a system where the source $X_0 > X_1$, then the single phase to earth fault line oscillation is less severe than the 1st pole-to-clear 3 phase

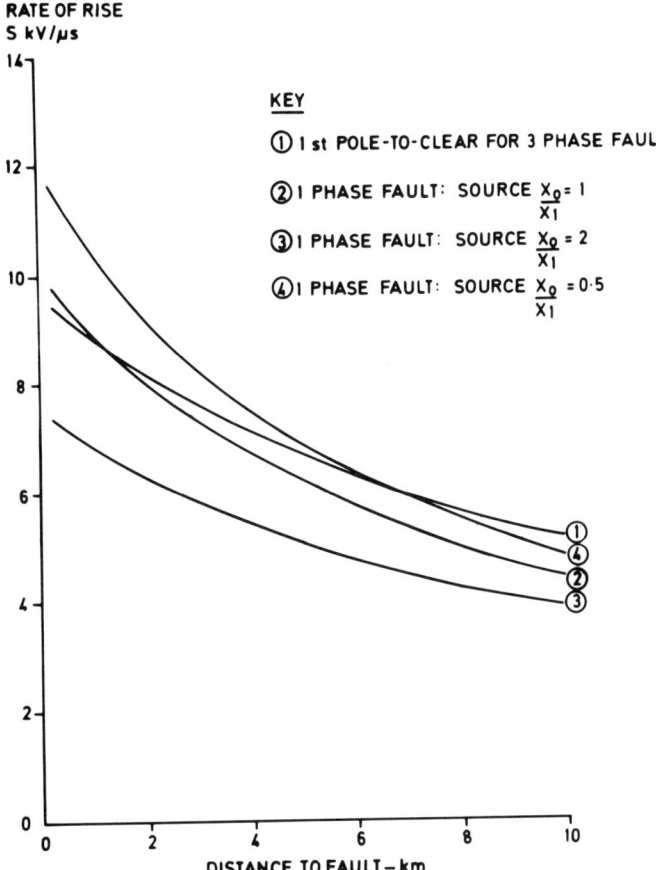

Fig. 18. Variation of line side rate of rise with distance to fault for 400 kV line with clashed conductors.

fault oscillation (see curves (3) and (1)), while if the source $X_0 < X_1$ then the line transient relative severities will vary depending on the precise fault location. Thus, in the U.K. system where the source X_0 is often less than X_1, it is satisfactory to base the slf duty on a single phase to earth fault. For systems where the source $X_0 > X_1$ then it may be preferable to base the line transient on the 1st pole-to-clear condition for a 3 phase earthed fault.

Other fault conditions, namely 2 phase earthed (or 2nd pole-to-clear a 3 phase earthed fault) can be studied and can be shown to produce line oscillations of similar magnitudes to those in Figs. 17 - 19 depending on the source X_0/X_1 ratio.

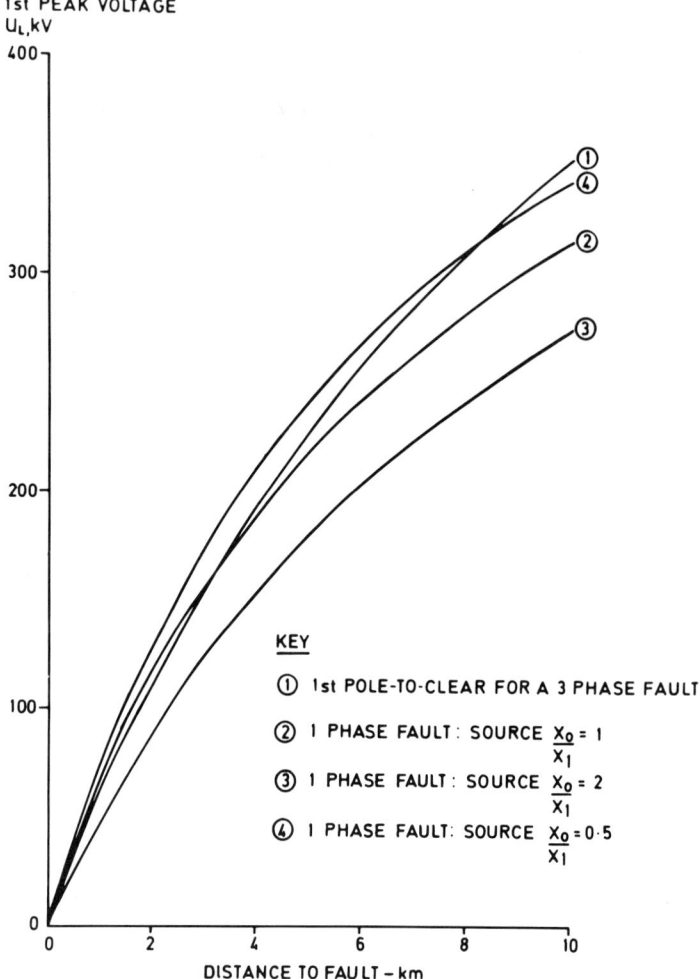

Fig. 19. Variation of 1st peak voltage with distance to fault for 400 kV line with clashed conductors.

11. CONCLUSIONS

The slf duty for the clearing circuit breaker is composed of a source side transient equivalent to a reduced terminal fault transient (according to the fault current), together with a rounded sawtooth oscillation on the line side. The latter is characterised by a very high rate of rise, determined by the fault current and line surge impedance, for a period of time equivalent to twice the travel time to the fault. The speed of propagation approaches the velocity of light, while the line surge impedance depends on the magnitude of fault current, conductor configuration, line earthing and pole opening sequence postulated.

The line side waveform is rounded and delayed by terminal capacitance adjacent to the circuit breaker. For a practical terminal capacitance equivalent to a Capacitor V.T. and gas-filled C.T., the time delay is approximately 0.8 μs and the overall rate of rise reduced by about 6 % for a 400 kV overhead line fault 2.1 km away from the clearing circuit breaker. The addition of a 25 metre cable termination to the line (10.000 pF) increases the time delay to 2.6 μs and reduces the rate of rise by 27 %. Thus the slf line side transient can be reduced in severity by the addition of lumped capacitance adjacent to the line terminal. The effect of line traps is to reduce the time delay and increase the peak voltage due to their inherent inductance, thus partially offsetting the effect of terminal capacitance.

Although power systems may be designed for fault levels of say 50 kA, it is unlikely that a large number of its circuit breakers would have potential slf duties in the range 40 - 50 kA. DC asymmetry in the fault current is seen to have virtually no detrimental effect on the slf transient. A majority of faults on 400 kV systems are single phase to earth in nature but the slf duty associated with them depends to some extent on the source network 50 Hz zero/positive sequence impedance ratio. The duty imposed by a 3 phase to earth fault, however, is independent of the zero sequence 50 Hz impedance and will usually be more onerous than the single phase fault duty where the source X_0/X_1 ratio is greater than unity.

ACKNOWLEDGEMENT

The author gratefully acknowledges the help and understanding given by his colleagues in the CEGB in the preparation of this manuscript.

REFERENCES

1. E. Bolton, M. J. Battisson, J. P. Bickford, M. G. Dwek, R. L. Jackson, M. Scott, Proc. IEE 117 (1970) 771
2. R. G. Colclaser, L. E. Berkebile, D. E. Buettner, IEEE Trans. PAS 90 (1971) 660
3. 'Surge Impedance of Overhead Lines with Bundle Conductors during Short-Line Faults'. Paper presented by CIGRE Working Group 13-01, Electra No. 17
4. R. G. Colclaser, J. E. Beehler, T. F. Garrity, IEEE Trans. PAS 94 (1975) 1943
5. H. Ohno, K. Nakanishi, Electrical Engineering in Japan 88 (1968) 79
6. L. Ferschl, H. Koppling, H. H. Schramm, E. Slamecka, J. D. Welly. 'Theoretical and Experimental Investigations of Compressed Gas Circuit Breakers under Short-Line Fault Conditions', CIGRE Paper No. 13-07, 1974
7. M. G. Dwek, E. Bolton, D. Birthwhistle, P. Bownes, G. W. Routledge. 'Overhead Line Parameters for Circuit Breaker Applications', CEGB Internal Report PL-ST/22/72

8. R. L. Jackson, J. G. Steel, 'The Attraction of Bundle Overhead Line conductors under Fault Conditions', CERL Report No. RD/L/R 1656, 1970
9. R. H. Galloway, W. B. Shorrocks, L. M. Wedepohl, Proc. IEE 111 (1964) 2051

DISCUSSION
(Chairman: W. Rieder, Technical University, Vienna)

J. Urbanek (General Electric)

I have two questions regarding the method of calculation. First, did you look at both grounded and ungrounded faults? And second, did you take the surge impedances beyond the fault into account?

M. B. Humphries

In the 3-phase earthed condition one does not need to consider the line behind the fault. If you look at the unearthed condition, on the other hand, you should take it into account. I haven't mentioned anything about the unearthed fault condition because it is not generally considered for short-line faults.

K. Ragaller (Brown Boveri)

I would like to comment on the problem of reducing the severity by using a capacitance parallel to the line. In our general limiting curve diagram (Fig. 1) this results in a shift in the vertical current limit to higher values. The important thing is that the gain depends on the type of circuit breaker. For example, for SF_6 the increase is much greater than for air. More generally, it depends on the thermal limiting curves. It is a question where the circuit and circuit breaker have to be considered simultaneously.

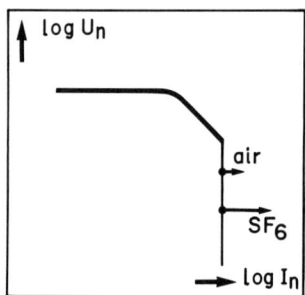

Fig. 1. Influence of a parallel capacitor on the general limiting curves of a circuit breaker.

H. Noeske (General Electric)

You showed the effect of a capacitance on the rate of rise of the recovery voltage and you mentioned a time delay which is caused by the capacitance. The breaker reacts quite differently to these two effects. In the case of a larger time delay, for example, it can withstand a higher du/dt.

M. B. Humphries

I have used the IEC-method to determine the rate of rise for the full wave trv. In this method the time delay causes a reduction of the rate of rise. If you neglect the time delay, the slope of the line-side transient is similar to what it would be without the additional capacitance.

P. G. Parrott (CEGB)

Surge impedances can be evaluated with an accuracy of roughly $\pm\ 5\ \%$. Can the physicists comment on the influence of this?

K. Ragaller

The influence of a Z-variation on the breaking limit can be discussed with the aid of Fig. 16 in the paper of Ragaller and Reichert. Varying Z changes the slope of the thin line representing the network. Because of the steepness of the circuit-breaker limiting curve, the influence on the current limit is very small. An accuracy of 5 % in Z is certainly sufficient.

EXPERIMENTAL INVESTIGATION OF LIMITING CURVES FOR
CURRENT INTERRUPTION OF GAS BLAST BREAKERS[*]

G. FRIND
General Electric Corporate Research and Development
Schenectady, N.Y., USA

SUMMARY

This contribution reviews recent measurements of initial or thermal recovery speed of gas blast interrupters and addresses thus a problem which is critically important for transmission line circuit breakers (short line fault, itrv).

Most measurements were made on gas blast interrupter models of reduced size and with frequencies of the power current significantly higher than 60 Hz. Theoretical arguments and some experimental evidence are presented in support of such a model testing procedure.

Rate of rise of recovery voltage was investigated in its dependence on several important design and circuit parameters: the rate of fall of current dI/dt before current zero, which is representative of the current level, the kind of gas used, the gas pressure and to some extent the electrode nozzle geometry. The variable found to be affecting rrrv most sensitively is dI/dt, especially for SF_6 at moderate currents below 30 kA. Also different gases, such as SF_6, CF_4, N_2, etc. show great differences in recovery speed. The dependence of rrrv on pressure is linear or somewhat steeper. These results agree in general well with predictions of theory. There are, to be sure, still some discrepancies between the experimenters and also between theory and experiment. It is suggested that future experiments with increased attention to both the detailed structure of the flow field and to the wave shapes of current and voltage very close to current zero (microstructure) will assist in resolving these uncertainties.

---

[*] Work supported in part by Electric Power Research Institute under Contract RP 246-1.

1. INTRODUCTION

This contribution reviews recent experimental work aimed at a better understanding of the upper current limit of alternating current gas blast interrupters. Although circuit breakers have to perform a great variety of duties, some of them quite difficult, high current interruption is one of the key challenges for the breaker designer. The scope of this paper will be restricted to a discussion of investigations concerned with the initial or thermal phase of the recovery process. Results on the consecutive dielectric recovery phase will be reviewed in part 2 of this volume.[1]

The thermal recovery phase of an interruption process can be characterized by the fact that failure occurs by a rather steady reheating process. When cooling in the current zero process is insufficient, with substantial electrical conductance remaining, reheating of the plasma is possible. If, for instance, arc resistance 1 μs after current zero is of the order of 1000 Ω and the system recovery voltage is 5 kV (rate of rise of recovery voltage 5 kV/μs) then the power input into the arc is at that moment 2.5×10^4 watts, which is in many cases sufficient to reheat the plasma and cause the interrupter to fail. The example illustrates the typical time scale for the thermal recovery period, which is of the order of a few μs immediately following current zero. It also points out the second factor important for thermal recovery problems, namely the occurrence of a very fast recovery transient of the order of 1-10 kV/μs. As discussed in the previous contribution,[2] such steep voltage rises occur typically with the fault located a short distance from the breaker on a transmission line (short line fault). Other breaker duties which impose high rates of rise of recovery voltage are reviewed in the paper by M. Dubanton.[3]

For the purpose of this review, it is useful to discuss interrupter response with respect to both circuit and interrupter design parameters. The only circuit parameter discussed in detail will be the current level, or more specifically the rate of fall of current before current zero. In addition, the importance of the fine structure of the current wave shape immediately before current zero, which is caused by the impedance parallel to the interrupter (resistor, capacitor, stray capacitance), is pointed out.

Design parameters reviewed are the gas pressure, the gas composition, and to some extent the geometry of the nozzle-electrode arrangement.

Most of the recovery studies of interest have been made on interrupter models of reduced size rather than on full-size devices. The technique of modeling gas blast circuit interrupters is, however, in its infancy. We will, therefore, discuss some background information and test results relevant to gas blast interrupter modeling in the following section.

LIMITING CURVES

2. EXPERIMENTAL MODELING OF GAS BLAST INTERRUPTERS

2.1. General

Today the use of models is a way of life for both theorists and experimenters. The theorist is often confronted with a situation, as in the case of turbulent plasma heat transfer, where the physics of the phenomenon is not completely understood or where the mathematics is too complex. He then tries to simplify the problem by accounting for its most important features in an approximate way.

The experimentalist has other, but equally valid, reasons for using a model rather than a full-size device. Focusing on the problem at hand, gas blast interruption studies can be made on full-size breakers. One can investigate different currents, different pressures, different nozzle geometries, and even different gases on a full-size device. However, with the many independent parameters involved, and taking into account the statistical character of the information obtained, one arrives at an astronomical cost for the project. Cost is a powerful motivator and therefore forces serious consideration of the use of interrupter models and of power sources with significantly reduced size.

The smaller interrupter has a number of additional advantages, besides a significantly lower cost. These include a faster turnaround time between experiments which require apparatus modification or repair. Another attractive feature is the advantage of using the sheltered atmosphere of a standard physics laboratory for some of the more sophisticated and often fragile experiments, rather than the relative rough and tumble of a high power test station.

The use of interrupter models of reduced size is, however, not an unmixed blessing. Similar to theory, one moves away from the real object to an approximation and one should therefore expect that the test results also are only approximate. The quality of the approximation will depend on the foresight and care with which the model is chosen. We will, in the following, discuss some of our thoughts on that matter.

2.2. The Weil Circuit and Its Modifications

It is common knowledge that the rate of fall of current dI/dt before current zero is a very sensitive parameter for gas blast interrupters. One of the important conditions for proper testing with the current injection type synthetic circuit (Weil circuit) is therefore the use of the correct value of dI/dt for the injected current. As a matter of fact, the practical use of the Weil circuit taught that in many cases the amplitude of the 60 Hz (or 50 Hz) current did not have a dramatic effect on the interruption event. During the early phases of a development program, development engineers used a significantly reduced 60 Hz amplitude (see Fig. 1) in order to prolong the useful life of their test rig. Test severity was varied by varying the ampli-

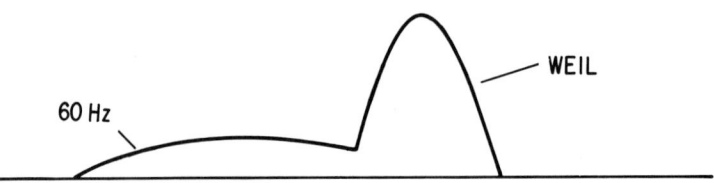

Fig. 1. Schematic of modified Weil circuit

tude and with this the dI/dt value at current zero of the Weil current. The 60 Hz current serves in such tests to allow time for contact opening, and it is therefore sometimes called 'holding current' or 'keep alive current'.

A convincing test series showing the usefulness and correctness of such a test method for their particular case was recently published by Nishikawa, et al.[4] These authors tested the thermal recovery speed of a double flow SF_6 interrupter. In their tests the 60 Hz current was varied between 25 % and 125 % of the 36 kA_{rms} rating of the interrupter and the Weil current was kept constant. Figure 2 presents their

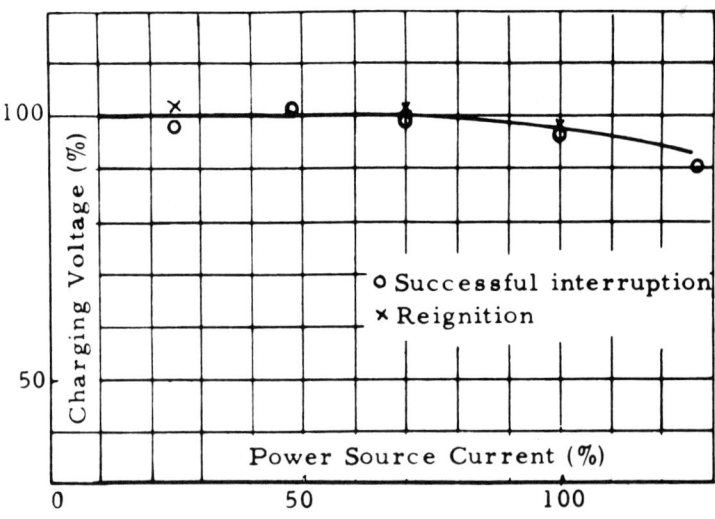

Fig. 2. Effect of power source current on interrupting capability (experiment); ordinate defines charge voltage on Weil capacitors.[4]

results, which show that the 60 Hz current has practically no effect on arc recovery. Although this test result cannot be considered a proof for the general validity of gas blast breaker tests with the modified Weil circuit, it certainly underlines the lesser importance of the 60 Hz current for thermal interruption tests as compared with that of the Weil current for such tests.

In the experiments described above, the 60 Hz holding current is used to provide sufficient time to open the contacts. It is therefore only a small further step to omit it completely, to bring the contacts into the desired 'open' position and to start the current by burning a very thin wire bridge (1 mil tungsten) or by electrical breakdown of the gap or part of it. Both methods, the modified Weil circuit and the high frequency current without holding current, have been frequently used in model tests with encouraging results,[5,6] although it must be admitted that the theoretical basis for the methods is still weak. We will, therefore, discuss in the following sections some theoretical arguments and some test evidence which are in support of the usefulness of such procedures.

2.3. A Typical Model Interrupter

A typical model interrupter used in a number of our own experiments[5] is shown in Fig. 3. It is of the single-flow, dual-pressure type and uses a Laval nozzle with 1/2-inch-diameter throat and an expansion half angle of 15°. Upstream pressures up to 1200 psi can be used. In most tests the downstream pressure is one quarter of the upstream pressure.

Power is derived from two independent capacitor banks; one is 20 kV - 840 µF, the other 20 kV - 140 µF. The larger bank is often used to provide the 'holding'

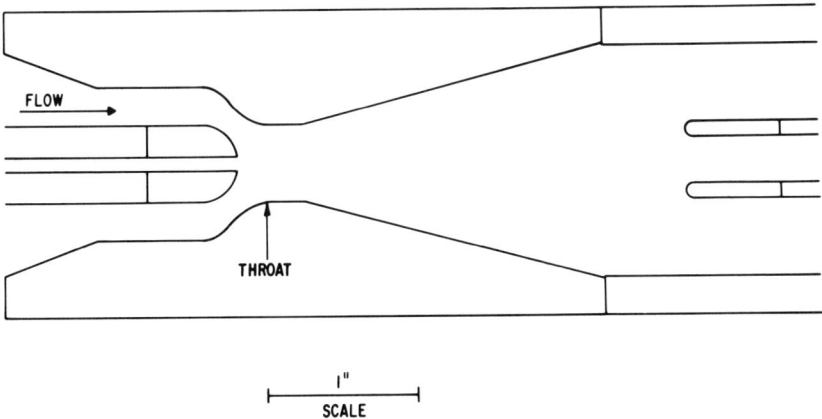

Fig. 3. Typical interrupter nozzle used in model experiments.[5]

current for a modified Weil test, as discussed before. In such a test, the smaller bank provides the Weil current. Both banks can be divided into smaller units to arrange for different power frequencies or voltage levels. Figure 4 shows a typical circuit arrangement.

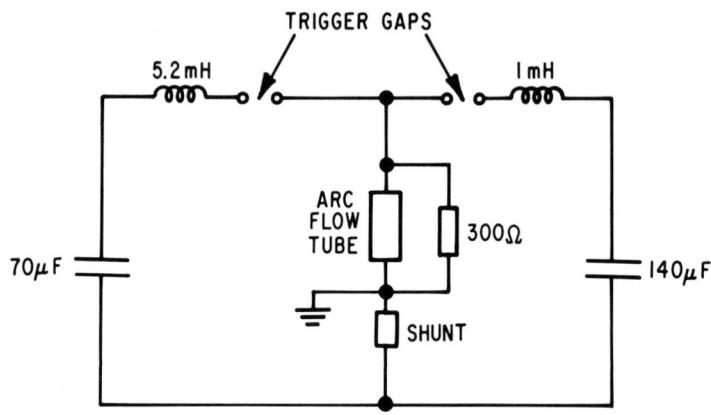

Fig. 4. Power circuit for modified Weil test.[12]

2.4. Investigation of the Modeling Technique

The experiment of Nishikawa, et al.[4] and the experience of other groups with Weil circuit testing showed that there are cases in which the peak value of the main current wave does not affect thermal recovery behavior significantly. At least two high current effects are known, however, which can affect recovery speed significantly. They are nozzle blocking and plasma contamination with electrode vapor.

Most of the recent thermal recovery measurements on gas blast interruptors have been made with current amplitudes small enough to stay under the blocking limit, which is approximately 7 kA_{peak}/cm^2.[7] Also, interruption theory, as of today,[8-11] does not take blocking effects into account. We will therefore discuss here what can be considered as the asymptotic case without blocking and leave the case with inclusion of the blocking effect for a later discussion.

2.4.1. Effect of Electrode Vapor

The effect of electrode vapor on thermal interruption speed was recently investigated experimentally.[12] The experiments were made with a single flow air blast model of 1/2-inch throat diameter which is shown in Fig. 3; pressure was 550 psi; dI/dt was kept constant to 14 A/µs. Electrode vapor was introduced through the erosion products of the upstream electrode, which had graphite, copper, or copper tungsten tips. The strategy for varying the amount of electrode contamination was to use sinusoidal current waves of very different intensities. The heaviest current was 12 kA$_{peak}$ for the power frequency 200 Hz, the lightest current was 600 A$_{peak}$ for the frequency 4000 Hz, Fig. 5. It was expected that graphite contamination, with

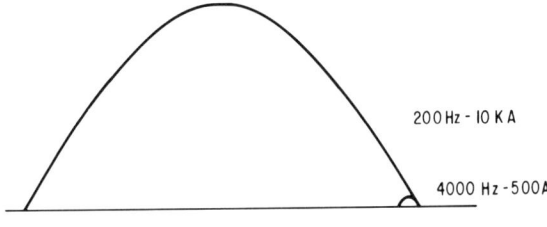

Fig. 5
Comparison of 200 Hz and 4000 Hz current wave with identical dI/dt.

COMPARISON OF 200Hz AND 4000Hz CURRENT WAVE

its ionization potential close to that of oxygen, which is plentiful in air, would not affect thermal recovery speed significantly, but that copper vapor with its much lower ionization potential would. Copper tungsten was used for two reasons; (1) it is a practical electrode material and can therefore serve as a bench mark; (2) it has a lower erosion rate than copper and could therefore assist in understanding the metal vapor influence. The result of the experiments is shown in Fig. 6, in which the thermal recovery speed for each of the three materials is plotted over power current frequency and the equivalent peak current.

As expected, and still surprisingly, the rate of rise of recovery voltage (rrrv) curve for graphite is flat over the whole frequency range, showing that carbon vapor has no effect on rrrv and that also, with carbon between 600 and 12.000 A we do not encounter any other degrading effect in our experiment. Also, the copper curve is flat, with a recovery speed about one half that of graphite, indicating the degrading effect of copper vapor ionization under conditions of our experiment. The Cu-W curve shows an intermediate behavior, demonstrating the effect of a lower metal concentration than found with copper electrodes. We interpret the fall off at high currents for Cu-W electrodes as caused by blocking.

Returning to our discussion of modeling effects, the experiment described shows that even very large differences in both peak current and arc duration do not produce a dramatic effect on thermal recovery speed. The tests also suggest a method of evaluating and eliminating plasma contamination effects (graphite).

2.4.2. Effect of Flow Relaxation

The experiments desribed in the last section provide a measure of confidence in the use of high power current frequencies for thermal interruption testing. The critical

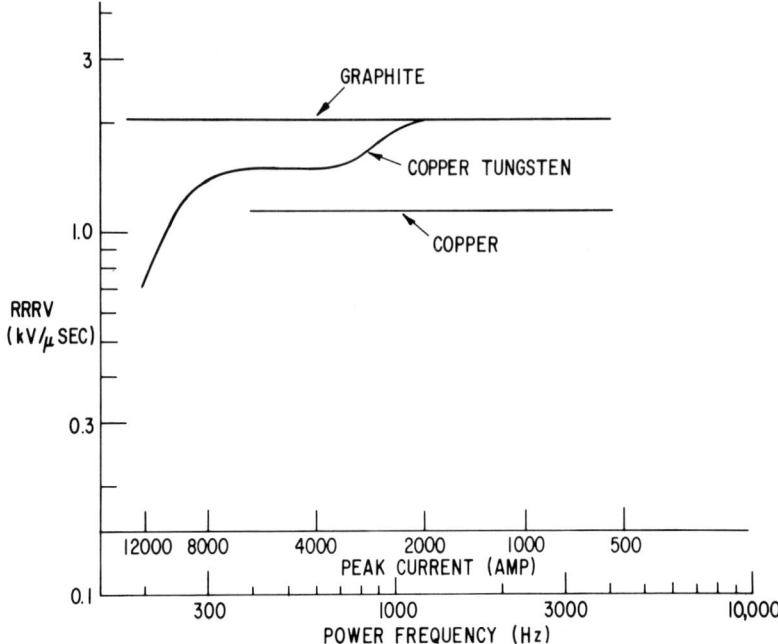

Fig. 6. Effect of electrode contamination on rate of rise of recovery voltage (rrrv).

importance of flow velocity for arc cooling and arc constriction, together with the relative slowness of gas motion in nozzle flow, suggest however that test frequency cannot be increased indefinitely. Slepian showed many years ago quantitatively[13] that the speed of arc constriction in a nozzle is limited by the finite flow velocity and that the plasma constriction time constant is of the order of 50 μs. This result suggests that in a model test the linear approach to current zero should not be tampered with for at least the last 50 or better 100 μs (2 flow time constants) before current zero.

We made an experiment to check this theoretical result. In order to sharpen discrimination, the interruption tests were not made with sinusoidal but with dc currents, which were ramped to zero, Fig. 7. The dc current will produce an arc in flow equilibrium, even for a time point t_1 for Fig. 7 which is closer to current zero than the flow time constant. Arcs which come down the linear ramp towards t_1 from high values of current however should show a flow relaxation effect. The arc diameter will therefore be larger and recovery slower.

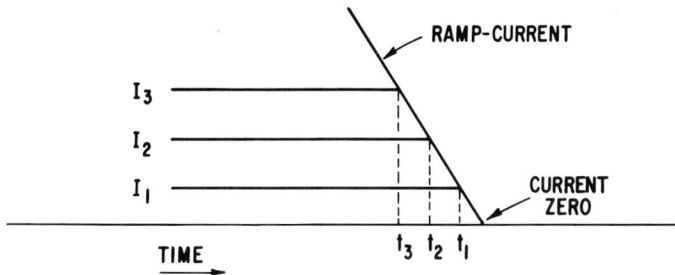

Fig. 7. Schematic of ramp experiment.

In our experiment, the ramp time was varied from 6 to 100 μs. The result is shown in Fig. 8. Current ramps as long as 30 μs are needed to reach asymptotically constant recovery speeds. For shorter ramps recovery speed is significantly higher, confirming Slepian's concept.

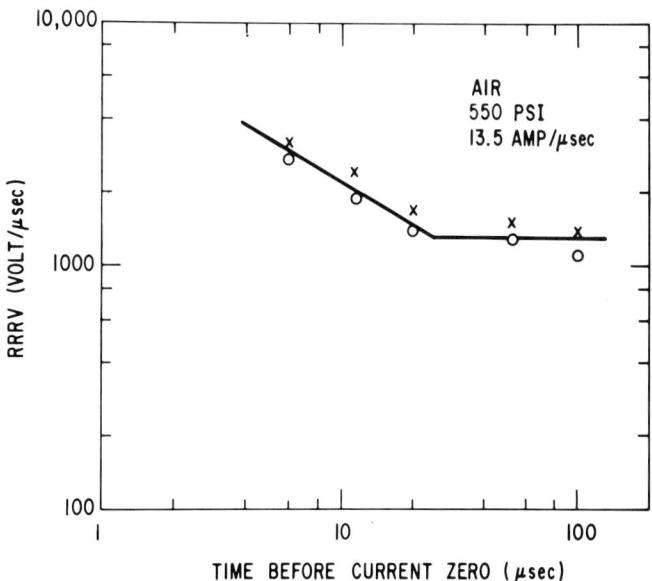

Fig. 8. Speed of recovery depends on duration of ramp current.

From the result on the relaxation time, the highest useful sinusoidal power current frequency can be calculated. We allow for the deviation of the sine function from linearity 10 % and obtain for a relaxation time of 30 µs a highest allowed frequency of approximately 4000 Hz.

Looking closely at the measured points in Fig. 8, one might question the constancy of rrrv for longer ramp times. Unfortunately, the experiment was made with copper-tungsten rather than with graphite electrodes and for larger currents will, therefore, as indicated earlier in Fig. 6, show a slight decrease in rrrv by plasma contamination with electrode vapor.

2.4.3. Effect of Nozzle Diameter

The experiments with electrode vapor contamination and with flow relaxation were concerned primarily with the allowed range of power frequencies for model tests with gas blast interrupters. Of equal concern is the question of how strongly and in which way nozzle diameter will affect thermal recovery speed. We have therefore made a test series, in which nozzle throat diameter was varied from 1/4 inch to 3/4 inch. Nozzle shape was kept similar. The expansion half angle was kept the same for all nozzles, 15°, and the diameter of the electrodes was made equal to the throat diameter; gas was air.

For all three nozzles, the optimum electrode distance was determined and the value of recovery speed, measured for that distance, was plotted over the throat diameter D in Fig. 9. The smaller nozzles show a moderately faster recovery speed than the larger ones, which can be described by the relationship:

$$\text{rrrv} \sim 1/D^{0.4} \quad \text{air; range } 1/4 \text{ inch} \leq D \leq 3/4 \text{ inch}$$

Our measurements on the effect of the throat diameter on recovery speed are not yet extensive and must therefore be considered as preliminary. They also have to be checked for larger nozzles up to full size and for different gases, especially for SF_6. They define however a first principal, though moderate, deviation of model tests from full-size tests.

3. MEASUREMENTS OF LIMITING CURVES

Two kinds of different interruption tests with limiting curves will be discussed. In one group, circuit parameters will be changed, and the extent to which thermal recovery speed is a function of the circuit parameter will be determined. In a second group of measurements, interrupter design parameters will be varied and the dependence of thermal recovery speed on these changes will be determined.

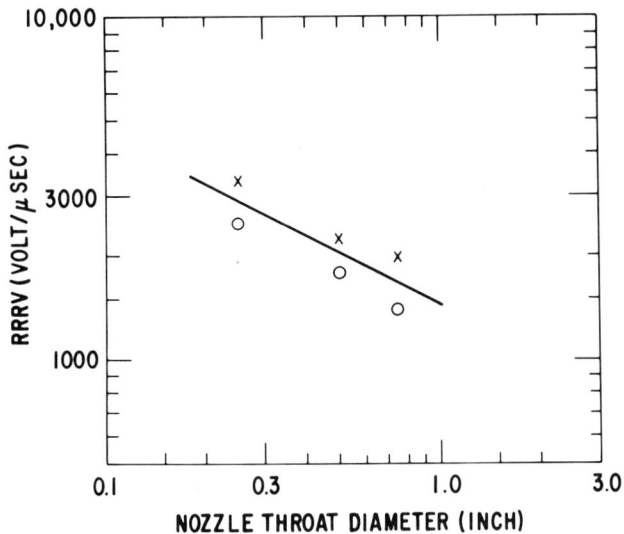

Fig. 9. Dependence of rrrv on nozzle throat diameter.

3.1. Circuit Parameters

Circuit parameters of known importance for thermal interruption speed include, as discussed above, the current level, or more specifically, the rate of fall of current before current zero. In addition, the fine structure of the current and voltage waves in the immediate neighborhood of current zero, is relevant. 'Fine structure' is defined here as the effects caused by capacitive and resistive impedances parallel to the interrupter. They will in general deform the current wave from its 'ideal' linear approach to current zero into a trailing one and in addition delay the rise of the voltage wave a short time, as shown in Fig. 10. The fine structure in both waves gives the interrupter a certain respite by diminishing the power input as compared to ideal conditions without parallel capacitors and resistors. As shown in two recent papers,[6,14] the effect of both capacitors and resistors on thermal interruption speed can be calculated. However, the thermal interruption measurements discussed in the following section did not define the 'fine structure' parameter in any detail, but were limited to an exploration of the more dramatic first order effects caused by changes in the current (or dI/dt) level. The only circuit parameter discussed explicitly in our contribution is therefore dI/dt.

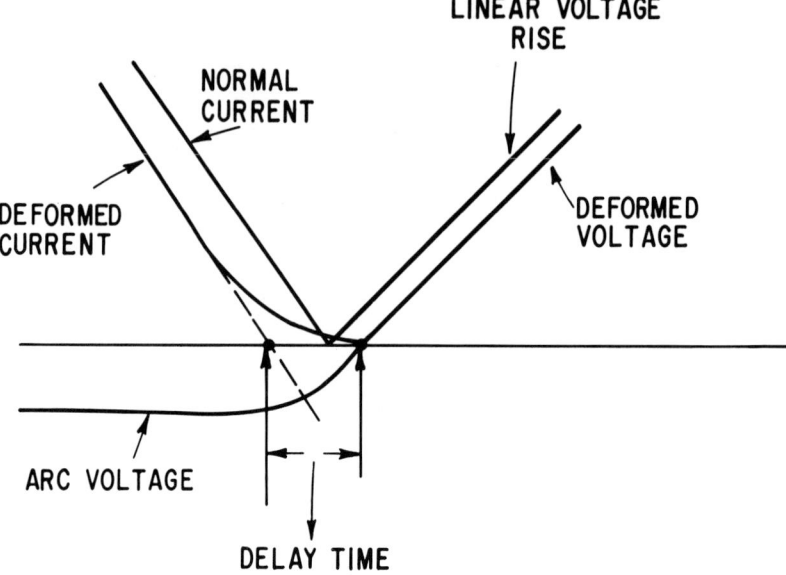

Fig. 10. Schematic illustrating current deformation before current zero by impedance.

3.1.1. Effect of dI/dt

The increasing difficulty, which gas blast breakers experience with growing current levels, is well known to the designer community. The catchword is shortline fault, which many airblast breakers can only master with rate controlling resistors parallel to the interrupter unit.

Significant progress in understanding this problem was only recently made by Swanson, Roidt, and Browne,[8] who derived a quantitative relationship for rrrv:

$$rrrv_{critical} \simeq 1/(dI/dt)^m \qquad (1)$$

with m = 1 or 3/2 depending on plasma cooling processes, which the authors could not specify at that time. At any rate, a well defined and strong dependence of the critical rate of recovery speed on the rate of current fall at current zero was predicted, meriting an experimental check. We decided therefore to make an experiment in order to discriminate between the two choices for m left open by theory.

The measurements were made in air and in SF_6 with the 1/2 inch single flow interrupter described briefly above.[5] The measured dependence of the rate of rise of recovery voltage (rrrv) on the rate of current fall dI/dt before current zero, as shown in Fig. 11, is even stronger than predicted by theory. The slope is $m = 2.0$

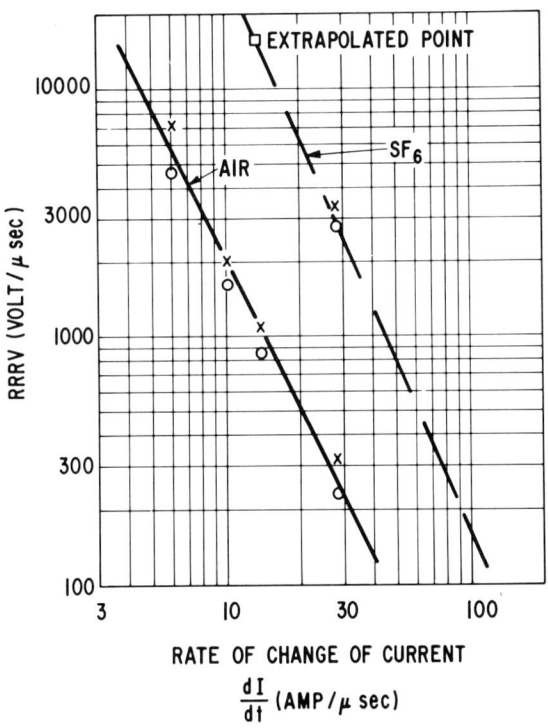

Fig. 11. Dependence of the recovery speed of air and SF_6 on dI/dt.

for air and $m = 2.3$ for SF_6. Since our experiment was performed, a number of additional measurements of the sensitivity of thermal recovery speed on current level (or the equivalent dI/dt value) have been published.[4,6,15-20] These measurements were made on different circuits and with different frequencies of the power current, with nozzles of different size and shape and also with different ratios of upstream to downstream pressure. In spite of these differences, we assembled the results as well as we could extract them from double logarithmic curves, in Fig. 12 for SF_6 and in Fig. 13 for air. Also added for comparison are the recent theoretical curves published by Hermann and Ragaller.[9] All results presented in the SF_6 figure exhibit a very strong dependence on dI/dt, which appears to be steepest for low currents; see the Noeske, BBC, and Marchwood curves. Also the absolute values of rrrv de-

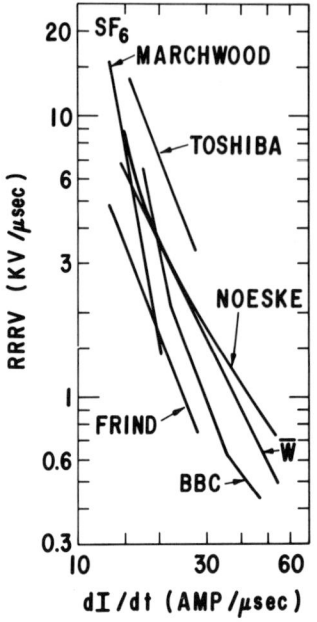
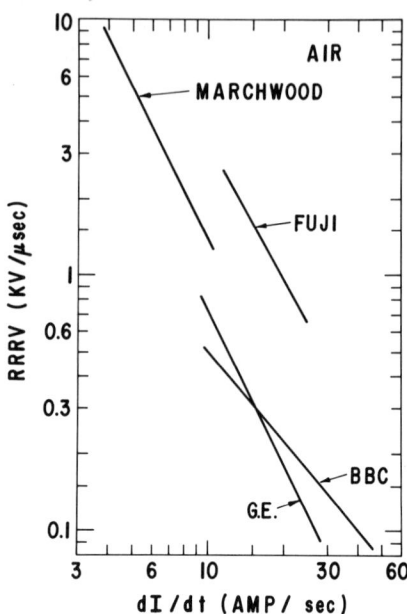

Fig. 12
Summary of measurements and calculations of recovery speed in SF_6 in dependence on dI/dt. Pressure was 15 bar. Marchwood[18]; Toshiba[4]; Noeske[12]; W[6]; BBC[9]; Frind[5].

Fig. 13
Summary of measurements and calculations of recovery speed in air in dependence on dI/dt. Pressure was 15 bar. Marchwood[18]; Fuji[19]; BBC[9]; GE[5].

termined by different authors are not very far apart. The highest curve, that marked Toshiba, was measured on a double flow interrupter, which is well known to be faster than a single flow device.

For air, Fig. 13, the experimental curves also show a strong dI/dt dependence. The theoretical curve marked BBC, in contrast, has a significantly flatter slope. None of the curves turns more steeply at small currents, as was observed for SF_6.

The agreement of the various curves with respect to absolute values of rrrv appears to be in air less good than found for SF_6. It should be noted, though, that also in air the highest curve, that of Fuji, was taken on a faster double flow device.

3.2. Design Parameters

Of the great variety of possible design parameters we will consider three here: gas pressure, the kind of gas used, and, to some extent, flow field structure as affected by electrode-nozzle geometry and by the upstream to downstream pressure ratio.

3.2.1. Effect of Pressure

In the past, the advantage of working with high gas pressures has not escaped the notice of interrupter designers. Circuit breakers with ever increasing pressure levels attest to this knowledge.[7,21,22] It was, however, only the above cited recent theoretical paper by Swanson, Roidt, and Browne,[8] which suggested a quantitative relationship between the variables du/dt and pressure:

$$du/dt \sim p^n \qquad (2)$$

with n = 1 or 5/4 still undetermined, pending more knowledge on the physics of the current zero cooling process. A consecutive measurement of thermal recovery speed for air and SF_6[5] in the pressure range 100 - 550 psi (6.8 - 37.5 bar) agreed well with that prediction, Fig. 14. Clearly, however, SF_6 depended more on pressure,

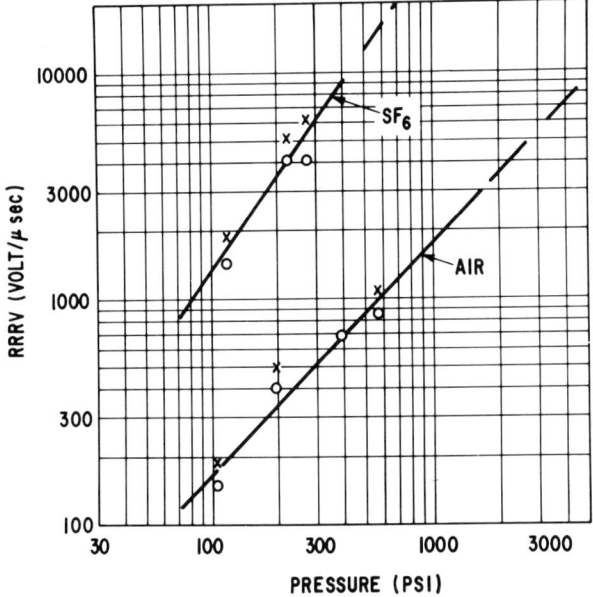

Fig. 14. Recovery speed of air and SF_6 in its dependence on pressure;[5] dI/dt was 13.5 A/µs.

with the exponent n = 1.4 to the air value of n = 1.0. The strong pressure dependence of the thermal recovery speed of SF_6 was confirmed and extended to the very high pressure of 1500 psi (102 bar), by Garzon,[16] Fig. 15.

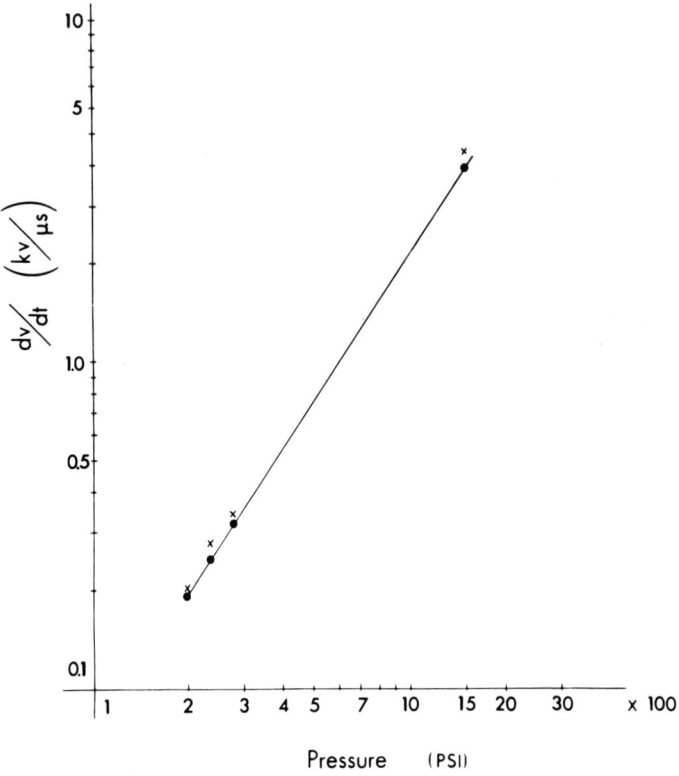

Fig. 15. Dependence of recovery speed in SF_6 upon pressure[16].

In Figs. 16 and 17 we have collected results of a number of recent measurements and calculations on the pressure dependency of thermal recovery speed in air (or N_2) and SF_6. All of the curves, perhaps excepting the SF_6 result of Noeske obtained with a very small nozzle (1/4 inch), exhibit a slope close to that predicted by Swanson, et al.[8] The theoretical result of Hermann et al.[9] appears to discriminate even correctly between the somewhat steeper pressure dependency of SF_6 and that of air.

3.2.2. Effect of different gases

In addition to the well investigated gases, air (or N_2) and SF_6, in the last few years a major number of other gases and gas mixtures have been tested for their thermal recovery capability.[12,23-26] The most extensive comparison of recovery speeds has perhaps been made by Noeske. Figure 18 shows a diagram summarizing interruption speeds measured by Noeske for a major number of gases and gas mixtures.[12] The curves represent the measured dI/dt dependence of the gases in a major range of dI/dt values, 13 to 56 A/μs, representative of 60 Hz current levels of

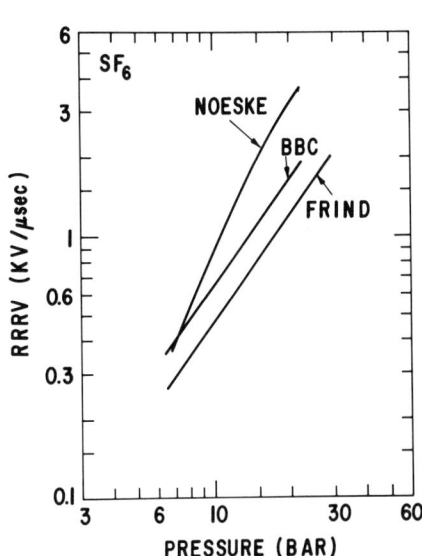
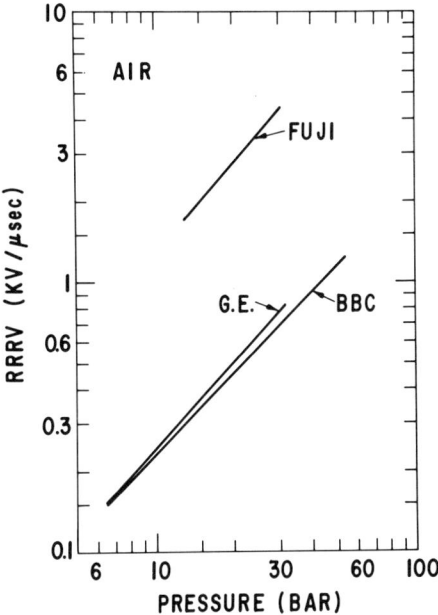

Fig. 16
Summary of measurements and calculations of recovery speed in SF_6 in dependence upon pressure. dI/dt was 27 A/μs. Noeske;[12] BBC;[9] Frind.[9]

Fig. 17
Summary of measurements and calculations of recovery speed in air in dependence upon pressure. dI/dt was 13.5 A/μs. Fuji;[19] BBC;[9] GE.[5]

approximately 25 - 100 kA_{rms}. We note that the slope of most of the gases is similar to that of SF_6. In addition, the figure presents interruption speeds of a major group of slower gases, including nitrogen and argon, for the dI/dt value 14 A/μs. In their performance relative to SF_6 some of these gases, notably nitrogen, are slower than expected from other available data.[5,18,28]

In Fig. 19, we show Noeske's result for the pressure dependency of the thermal recovery speed of a major number of gases. This figure also shows a narrow grouping of some of the fast gases, such as SF_6, CF_4, C_2F_6 and mixtures of these. The slope for many of the curves, excepting that for SF_6, is close to 2; The SF_6 slope is even steeper. We find, therefore, that the 1/4 inch nozzle produces steeper pressure dependencies and also greater differences between fast and slow gases than expected from our experience with the 1/2 inch nozzle.

A calculation of the relative recovery speed of the gases tested by Noeske was recently made by Kinsinger,[27] who in his theory focused essentially on the very different amounts of disscoiation energy stored in different gases in the temperature

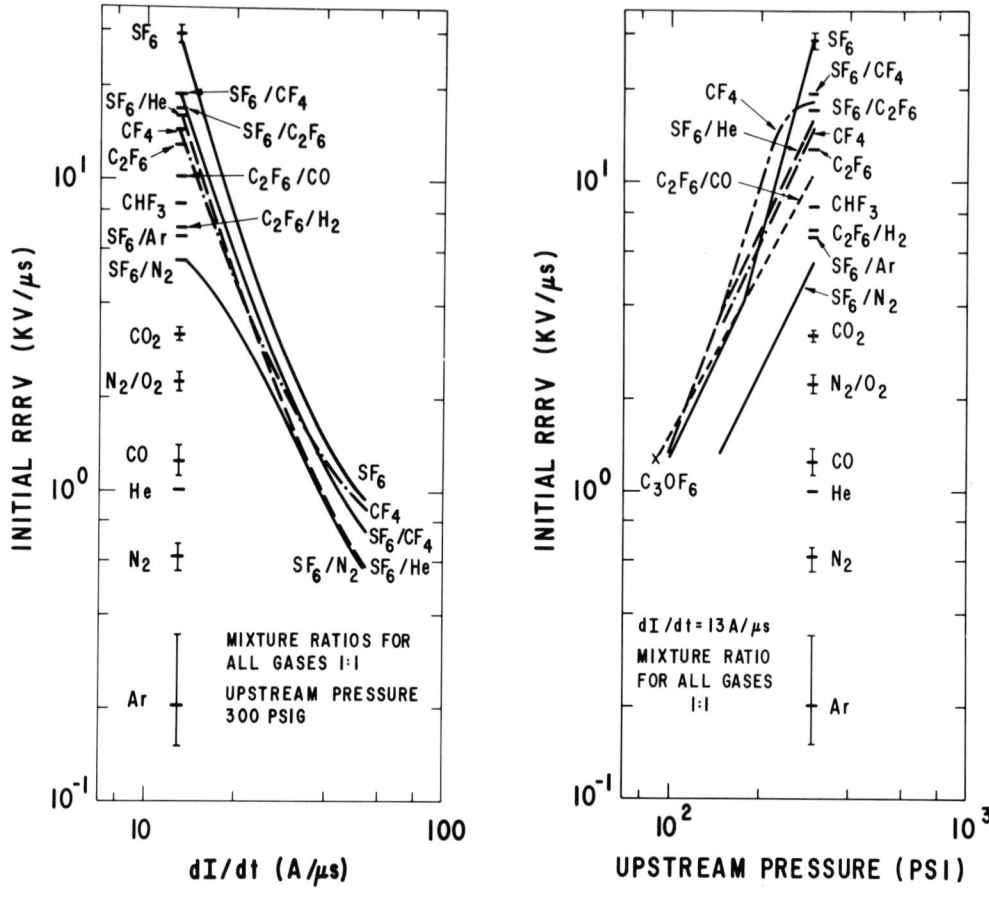

Fig. 18
Dependence of recovery speed on dI/dt for various gases; Noeske.[12]

Fig. 19
Dependence of recovery speed on pressure for various gases; Noeske.[12]

range relevant for thermal interruption. Weakly bound gases, such as SF_6, recombine at temperatures lower than the thermal recovery range. Thermal recovery deals therefore only with the relatively small amount of kinetic energy stored in these plasmas. Strongly bound gases, such as nitrogen, recombine at temperatures at which the plasma still has a significant electrical conductivity. A great amount of recombination energy must therefore be dissipated in addition to the kinetic energy. This delays significantly thermal recovery in strongly bound gases.

Table 1 shows a comparison of Kinsinger's results with Noeske's measurements, normalized to an SF_6 value of 100. The agreement is good - for most gases within a factor of 2. The exceptions are nitrogen and carbon monoxide, for which the experimental values are low. Kinsinger's relative theoretical data for nitrogen and air are close however to values measured with larger nozzles.[5,18,28]

TABLE 1
Relative Recovery Speeds from Experiment and Theory

Gas	Experiment (Relative rrrv)	Theory (Relative rrrv)
SF_6	100	100
SF_6/CF_4 (1/1)	70	51
CF_4	66	89
SF_6/C_2F_6 (1/1)	63	50
SF_6/He (1/1)	59	98
C_2F_6	46	86
CHF_3	39	52
C_2F_6/H_2 (1/1)	31	52
C_2F_6/CO (1/1)	36	51
SF_6/Ar (1/1)	23	
SF_6/N_2 (1/1)	20	41
CO_2	11	19
N_2/O_2 (1/1)	8	14
CO	4	11
He	4	
N_2	2	12
Air		10
Ar	1	

3.2.3. Effect of Flow Field Structure

The effect of different nozzle geometries on interruption capability has been investigated since the early days of gas blast breakers. In 1932, Kopeliowitsch had already found an optimum value of the distance from the upstream electrode to the nozzle throat,[29] an effect which has since been investigated in increasing detail by many other workers.[30,31] In Figs. 20 and 21 we show as examples some results for air and SF_6 obtained recently in our own group.[12] Both gases exhibit a maximum of rrrv with respect to the distance of the upstream electrode to the throat of the nozzle. In air, the maximum value of rrrv is about double that of the plateau found for rrrv at longer distances. In SF_6, the relative increase is even higher than that, particularly for the pressure 300 psi.

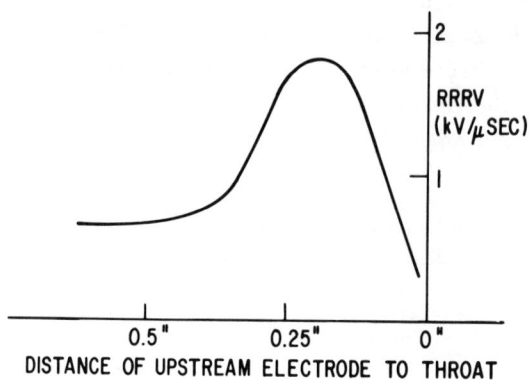

Fig. 20
Dependence of recovery speed on the distance of the upstream electrode from the nozzle throat; for air, 1/2 inch nozzle.[12]

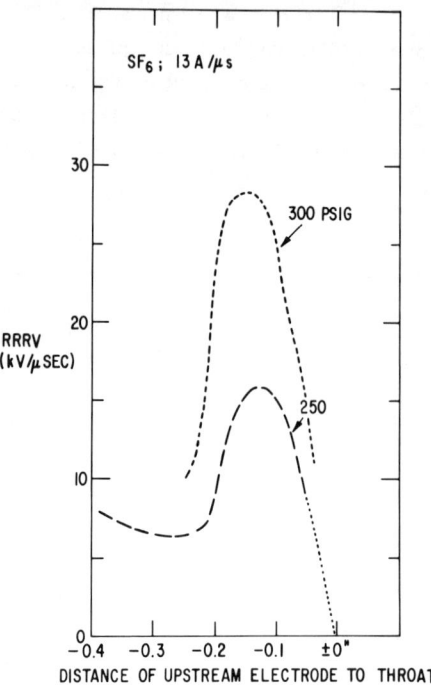

Fig. 21
Dependence of recovery speed on the distance of the upstream electrode from the nozzle throat; for SF_6, 1/4 inch nozzle.[12]

Flow field structure is strongly affected not only by nozzle geometry but also by the ratio between the upstream to the downstream pressure. First, in order to generate sonic flow conditions in the nozzle throat, the upstream pressure should be at least approximately double the downstream value. However, this pressure condition determines essentially only the upstream flow field before the throat but leaves conditions in the downstream nozzle part still largely open. For a given nozzle geometry, the downstream flow field depends strongly on the upstream to downstream pressure ratio. This fact is known from nozzle theory and, for the purpose of the interrupter application, was pointed out recently by Nagamatsu.[12,32] It is difficult to fill nozzles with large expansion angles properly with a shockfree supersonic flow field. As used in many of our own experiments, a nozzle with an expansion half* angle of 15° needs a pressure ratio of approximately 50. Even the very slim nozzle

---

*) Strictly speaking, the relevant parameter is the area ratio, not the expansion angle.

used by Hermann et al.[33] in their excellent interruption studies needs a pressure ratio of the order of 10. In our experiments and in many practical circuit breakers, when smaller pressure ratios are employed, the flow field separates from the nozzle walls at some point downstream of the nozzle throat and a shock wave system is set up. This condition leads to a supersonic but significantly more complicated flow field structure.[12,32]

A group of thermal recovery measurements investigating the effect of various pressure ratios on interruption speed was made recently by Campbell at al.[34] Their result, depicted in Fig. 22, shows that, for a given upstream pressure, higher pressure ratios result in significantly faster recovery. The authors also point out the change in the slope of the curves in Fig. 22. Higher pressure ratios lead to significantly steeper slopes.

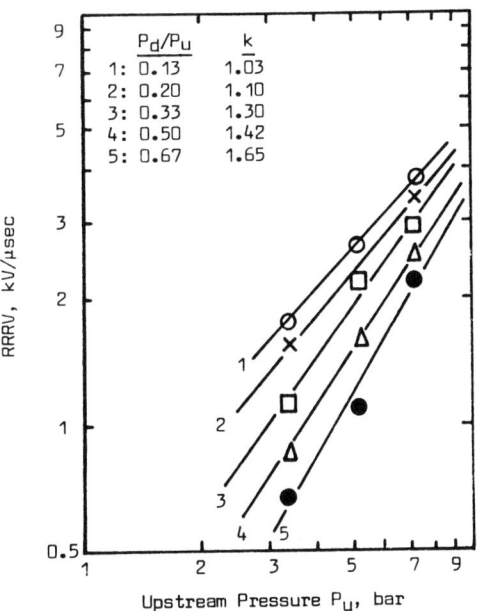

Fig. 22. Dependence of recovery speed on the pressure for different ratios of exit to reservoir pressure, P_d/P_u, in SF_6 for a dI/dt of 9 A/μs.[34]

4. DISCUSSION

4.1. Modeling Technique

In the experiments on the modeling technique no strong sensitivities were found, which would make the use of small-scale gasblast interrupter models questionable. The measured flow relaxation time of approximately 30 μs for a 1/2 inch air nozzle simply sets an upper frequency limit for the power current. Variations of power frequency in the 200 to 4000 Hz range did affect thermal recovery speed by electrode vapor contamination, but only within a factor of 2 and in a controllable manner. Moreover, the effect of vapor on rrrv could be eliminated by the choice of electrode material and/or the use of a high frequency of the power current.

Also, variations in the nozzle throat diameter did affect thermal recovery speed only moderately. If the measured dependency for air (rrrv $\sim 1/D^{0.4}$) is extrapolated to nozzles of 2 inches, a 1/2 inch nozzle is found to be slightly less than twice as fast as a 2-inch nozzle.

Important work has still to be done, however, before modeling of gas-blast interrupters can be considered secure. The experiments discussed above have to be extended both to full-size nozzles and to full-size currents.

4.2. Limiting Curves

4.2.1. Effect of dI/dt

All experiments reported show clearly the sensitivity of thermal recovery speed against high values of dI/dt. The effect is strongest in SF_6 where the slope m is about 2 for high dI/dt values and turns in some results to even higher values for small dI/dt, Fig. 12. The absolute values of rrrv of the curves in Figs. 12 and 13, determined by different experimenters, do not agree too well. We know, however, that double flow interrupters such as used by Toshiba and Fuji are approximately twice as fast as single flow devices. Also, the experiments marked Frind in Fig. 12 and GE in Fig. 13 were made at a larger-than-optimal upstream electrode distance, resulting in about one-half the optimal recovery speed.

Therefore, for the purpose of this discussion if we multiply arbitrarily the double flow recovery speeds with 1/2 and the non-optimal distance measurements with 2, we obtain the revised curves of Figs. 23 and 24. The experimental SF_6 curves, except that of Marchwood, are now grouped very closely around the Westinghouse curve. The agreement is as close as can be expected - in view of the remaining differences and uncertainties among the experiments with respect to nozzle diameter and shape, pressure ratio, and details of the test circuit.

In the revised Fig. 24, the experimental air curves are in good agreement, with a slope m close to 2 in the whole range covered. In contrast, the theoretical curve shows a significantly flatter dI/dt dependence. This difference between experiment and theory appears clearly beyond the uncertainties associated with the measurements.

4.2.2. Effect of Pressure

Perhaps the most important result, with respect to the effect of pressure on thermal recovery speed, is Garzon's measurement up to 1500 psi (102 bar), showing that

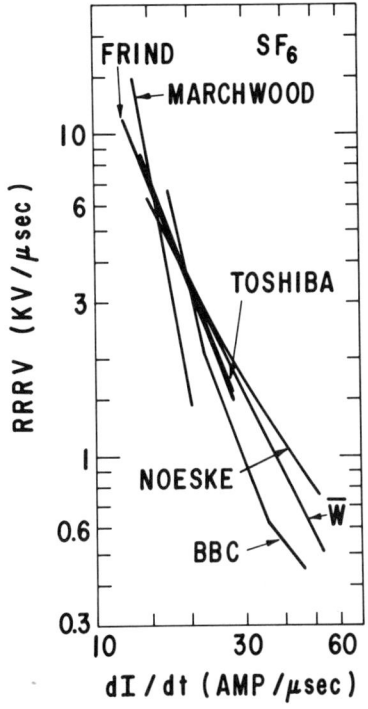

Fig. 23
Repeat of Fig. 12, except that rrrv in the Toshiba curve was multiplied with a factor 1/2 and in the Frind curve with a factor 2.

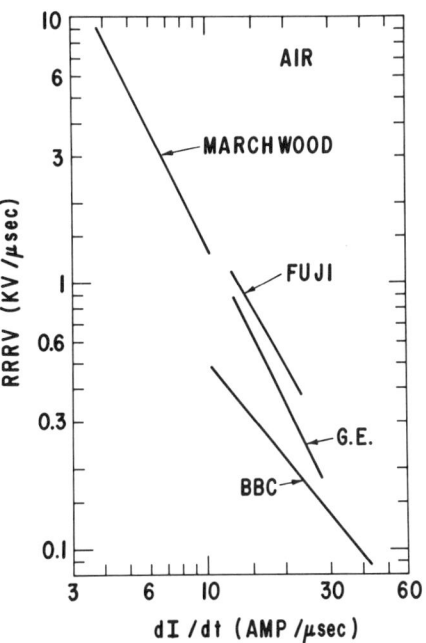

Fig. 24
Repeat of Fig. 13, except that rrrv in the Fuji curve was multiplied with a factor 1/2 and in the GE curve with a factor 2.

even for that high pressure value the curve rises still with the same slope m = 1.46.

For the curves presented in Figs. 16 and 17, we make the same arbitrary adjustments with respect to double flow and non-optimal electrode distance as discussed previously (in Figs. 23 and 24), and we obtain a new set of curves in Figs. 25 and 26 for the adjusted pressure dependency of the thermal recovery speed in SF_6 and in air. For SF_6, all curves, theory, and experiment are close, except that the curve denoted Noeske, which was taken with a 1/4-inch nozzle, has a higher pressure dependence, somewhat out of line with measurements on larger nozzles. The adjusted air curves show excellent agreement with respect to slope. However, the theoretical curve is somewhat low in its absolute values of rrrv.

4.2.3. Effect of Different Gases

It is interesting to see that all gases investigated exhibit a very similar, strong dI/dt dependency of the thermal recovery speed. Therefore this effect appears to be

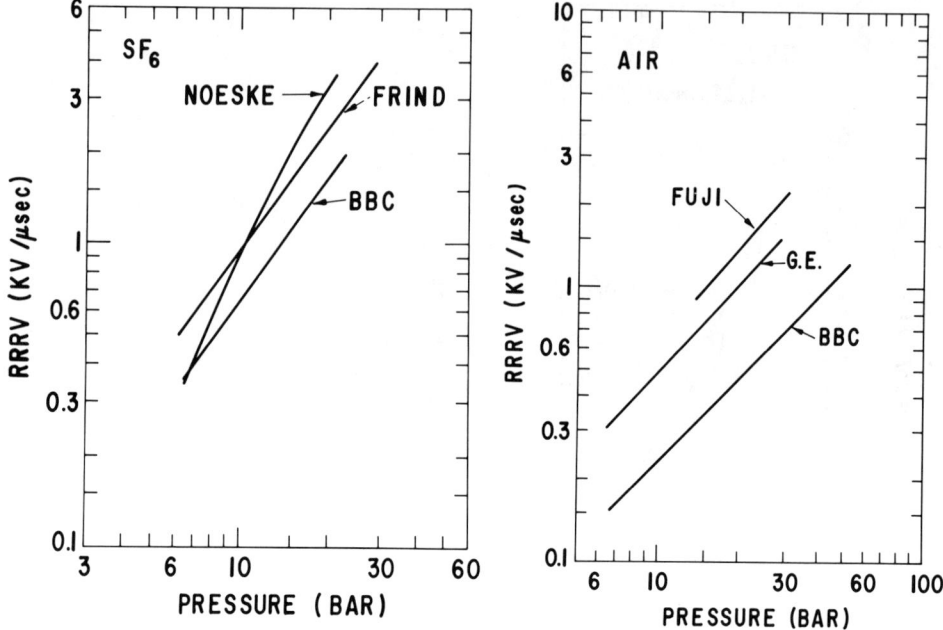

Fig. 25
Repeat of Fig. 16, except that rrrv in the Frind curve was multiplied with a factor 2.

Fig. 26
Repeat of Fig. 17, except that rrrv in the Fuji curve was multiplied with a factor 1/2 and in the GE curve with a factor 2.

of a rather general nature in flow interrupters, independent of the chemical structure of the gas. The measurements have also identified a group of gases and gas mixtures with recovery speeds in the SF_6 rather than in the air class. Some of these gases can be operated at considerably higher pressures than SF_6 without confronting a liquefaction problem. Therefore, they hold promise for achieving very high recovery speeds.

The good agreement of the theory proposed by Kinsinger,[27] with relative thermal recovery speeds measured for different gases, is also encouraging. Candidate gases can be 'screened' theoretically, based on their molecular properties, before an experimental program is started.

The higher pressure sensitivity of the 1/4 inch nozzle and the low nitrogen recovery speed measured with this small interrupter are of some concern for the quantitative evaluation of recovery speed data. These points have to be cleared in future experiments.

4.2.4. Effect of Flow Field Structure

The flow field in the upstream nozzle section before the throat and as a consequence the thermal interruption speed are sensitive to the electrode throat distance. The resulting maximum of rrrv can be quite sharp as in the case of SF_6, Fig. 21. Test results can therefore be affected by small geometric changes, such as a widening of the throat diameter in successive experiments or by misadjustments of the electrode distance. We have observed also that the optimum distance shifts slightly with the pressure level. All these sensitivities make thermal recovery speed measurements prone to error.

The very interesting dependence of thermal recovery speed on pressure ratio found by Campbell et al.[34] deserves our attention. It could reflect well the effect of a changing flow field structure in the expanding nozzle part with a variation in pressure ratio. It seems, however, that the authors' straightline interpretation of their data cannot be extrapolated to significantly higher upstream pressures. The lines would interesect at a pressure somewhat below 20 bar, which is hard to understand.

5. CONCLUSIONS

In the work reviewed in this contribution, the use of reduced size models for thermal interruption testing of gas blast interrupters has not run into serious obstacles. Model tests of thermal recovery speed appear to be a workable alternative to full-size testing.

Thermal interruption speed has been found dependent on several first order parameters - it depends very sensitively on the value of dI/dt at current zero and on the kind of gas used in the blast. It also depends strongly on the pressure level, up to pressures of 100 bar.

In general, theory describes these experimental results well. A few disagreements should be critically evaluated by both theory and experiment.

As for the experiments discussed here, one of their limitations is the lack of a proper definition of the precise waveshapes of current and voltage very close to current zero (microstructure). These should be determined to allow for a more refined comparison. In addition, errors in curve interpretation could be avoided if authors would offer results besides as curves also in the form of tables.

Some information has also been collected on the effect of flow field structure on thermal recovery speed. It is felt, however, that a more detailed definition of important features of that structure, such as flow velocities, shock waves etc., is still needed, before a meaningful assessment of its importance for interruption speed can be made.

6. FUTURE WORK

Future work appears to be promising along the following lines:

- Explore effect of the detailed flow field structure on interruption speed
- Determine effect of current deformation at current zero on recovery speed
- Complete investigation of experimental modeling technique by comparison with full-size tests.

REFERENCES

1. J. Kopainsky, paper in this volume
2. M. B. Humphries, paper in this volume
3. M. Dubanton, paper in this volume
4. H. Nishikawa, A. Kobayashi, T. Okazaki and S. Yamashita, IEEE Trans. PAS 95 (1976) 1834
5. G. Frind and J. A. Rich, IEEE Trans. PAS 93 (1974) 1675
6. T. E. Browne, Jr., Discussion of Rf. 5 and Practical Modeling of the Circuit Breaker Arc as a Short Line Fault Interrupter, IEEE Trans. paper F-77626-5, Mexico City, Mexico, July 17-22, 1977
7. J. C. Henry, J. Passaquin and E. Thuries, 'Improved Performance of Gas-Blast Circuit Breakers', CIGRE Report No. 13-09 (1972)
8. B. W. Swanson, R. M. Roidt and T. E. Browne, Jr., IEEE Trans. PAS 90 (1971) 1094
9. W. Hermann and K. Ragaller, IEEE Trans. PAS 96 (1977) 1546
10. Fathi R. ElAkkari and David T. Tuma, 'Simulation of Transient and Zero Current Behaviour of Arcs Stabilized by Forced Convection', IEEE Trans. Paper F 77 126-6, New York, N.Y., Jan. 30-Feb. 4, 1977
11. Richard R. Kinsinger, IEEE Trans. PAS 93 (1974) 1143
12. G. Frind, R. E. Kinsinger, R. D. Miller, H. T. Nagamatsu and H. O. Noeske, 'Fundamental Investigations of Arc Interruption in Gas Flows', EPRI Final Report EL-284 (January 1977). Also: G. Frind, R. E. Kinsinger and R. D. Miller, 'Power Frequency Scaling and Electrode Vapor Effects in Gas Blast Interrupters', to be published: IEEE Trans. on Plasma Science
13. J. Slepian, AIEE Transact. 60 (1941) 162
14. L. S. Frost, 'Dynamic Analysis of Short-Line Fault Tests for Accurate Circuit Breaker Performance Specification', IEEE Trans. Paper F-77 627-3, Mexico City, Mexico, July 17-22, 1977
15. J. F. Perkins and L. S. Frost, IEEE Trans. PAS 92 (1973) 961
16. R. D. Garzon, IEEE Trans. PAS 95 (1976) 1681
17. D. Birthwhistle, G. E. Gardner, B. Jones and R. J. Urwin, Proc. IEE 120 (1973) 994
18. D. R. Airey, G. E. Gardner and R. J. Urwin, 'Development of SF_6 Arc Interrupters', 11th University Power Engineering Conference, Portsmouth, England (1976)
19. T. Morita, M. Iwashita and Y. Nitta, 'A Theoretical Analysis of Dynamic Arcs and Test Results of Model Synchronous Air Blast Circuit Breakers', IEEE Trans. Paper F-77 703-2, Mexico City, Mexico, July 17-22, 1977
20. H. O. Noeske, Ref. 12 of this paper, Section III
21. A. Eidinger and M. Sanders, Brown Boveri Review 60 (1973) 173
22. R. B. Shores, J. W. Beatty, H. T. Seeley and W. R. Wilson, AIEE Transact. Pow. App. Syst. Part III 78 (1959) 673
23. M. Hudis, 'Arc Recovery Measurements in Sonic Fluorocarbon Gas Arcs', IEEE Conf. Paper C 74-090-7, New York, N.Y., Jan. 1974
24. R. E. Kinsinger and H. O. Noeske, 'Relative Arc Thermal Recovery Speeds in Different Gases', 4th Intern. Conf. on Gas Discharges, Swansea, U.K., 7-10 Sept. 1976, IEE Conf. Publ. 143, pp 2428
25. R. D. Garzon, IEEE Trans. PAS 95 (1976) 140

26. D. M. Grant, J. F. Perkins, L. C. Campbell, O. E. Ibrahim and O. Farish, 'Comparative Interruption Studies of Gas Blasted Arcs in SF_6-He Mixtures', Proc. 4th Int. Conf. Gas Discharges, Swansea, U.K., 7-10 Sept. 1976, IEE Conf. Publ. 143
27. R. E. Kinsinger and H. O. Noeske, 'Arc Thermal Recovery in Different Gases', Symposium Proceedings, New Concepts in Fault Current Limiters and Power Circuit Breakers, EPRI Report EL-276-SR (April 1977)
28. J. F. Perkins and L. S. Frost 'Current Interruption Properties of Gas Blasted Air and SF_6 Arcs', IEEE Conf. Paper C 72-530-4, San Francisco, Cal. (July, 1972)
29. J. Kopeliowitsch, Bulletin SEV 23 (1932) 565 (In German)
30. H. Kopplin, K. P. Rolff and K. Zückler, Proc. IEEE 59 (1971) 518
31. Z. Vostracky, 'Arc Characteristics of Air Blast Circuit Breakers', Third Int. Conf. on Gas Discharges, London, England (1974)
32. H. T. Nagamatsu, R. E. Sheer and R. C. Bigelow, 'Flow Properties of Air and SF_6 in Supersonic Circuit Breaker Nozzles', IEEE Conf. Paper C 74 184-8 (1974)
33. W. Hermann, U. Kogelschatz, L. Niemeyer, K. Ragaller and E. Schade, IEEE Trans. PAS 95 (1976) 1165
34. L. C. Campbell, J. F. Perkins and J. L. Dallachy, 'Effect of Nozzle Pressure Ratio on SF_6 Arc Interruption', Proc. 4th Int. Conf. on Gas Discharges (Swansea), IEE Conf. Publication 143 (Sept. 1974)

DISCUSSION
(Chairman: W. Rieder, Technical University, Vienna)

K. Zückler (Siemens)

In your experiments you avoided clogging effects. We found that in the region of clogging the exponent m in your formula (1) increases to values above 2 and up to 3. Real breakers usually work in the clogging regime, otherwise the nozzle diameters would have to be uneconomically large.

G. Frind

You're right. We studied the asymptotic case without clogging. This, of course, is an approximation but we feel a useful one. At current zero, the effects of a clogging condition around current peak may be forgotten. Take for instance the piston interrupter which can be declogged very closely to current zero.

K. Zückler

We included this effect in our investigations on a double-nozzle breaker. If we go to high current values we get exponents which are larger than 2. Therefore, I think this is very important for a real circuit breaker.

G. Frind

But don't forget that there are additional effects: if you have ablation from the walls you have a different plasma composition, for example.

E. von Bonin (AEG)

In air breakers it is a well-known phenomenon that a parallel resistor of 300 Ω causes a very strong interaction. Can you tell us to what extent you and other authors have used such resistors?

G. Frind
The parallel resistor is used to control the rate of rise of recovery voltage and is therefore different for different parts of the curve. In our own experiments we did measure the deformation time (time delay), though, and found it no more than about 1/3 of the current zero arc time constant.

G. Catenacci (CESI)
In your proposal for future work you mention as one parameter the time delay of the current when going to zero. This current deformation is not an independent parameter (such as the time delay of the inherent voltage) but is caused by the arc voltage.

G. Frind
I agree. I only wanted to stress the need for a better definition of this effect which is important for thermal recovery speed. At the present we are comparing the results of theoretical calculations based on linear waveforms for both current and voltage with measured data of rrrv determined with somewhat deformed waveshapes. Future experiments should therefore be made with the deformations of current voltage quantitatively defined.

D. M. Benenson (State University of New York at Buffalo)
In your Fig. 9 you showed the rrrv as function of nozzle throat. Could the change in mass flow with nozzle diameter be the explanation for the observed dependency?

T. E. Browne (Westinghouse)
With regard to the same curve, it is my impression that the pressure gradient, which is the 'forcing function' for turbulence, may explain this. If you have the same pressure and different nozzle sizes, you have an inverse relation to the pressure gradient.

G. Frind
To Professor Benenson: it has usually been found that recovery speed increases with flow rate. In such experiments the flow rate was often increased by increasing the upstream pressure. This result agrees with the pressure dependence of rrrv discussed in this contribution. An increase in nozzle size, at constant pressure, resulted in our measurements in a decrease of rrrv, however, although flow rate increased.

I agree, therefore, with Dr. Browne's suggestion of the importance of the axial pressure gradient. We are investigating this effect at the present time, in more detail and hope to report on our results in the near future.

THE INFLUENCE OF TURBULENCE ON CURRENT INTERRUPTION

G. R. JONES
Department of Electrical Engineering, University of Liverpool
Liverpool, England

SUMMARY

Optical measurements have convincingly established the existence of turbulence in a gas blast arc downstream of the nozzle throat, during the current zero period. The turbulence appears to be produced in the shear layer separating the high velocity plasma flow from the surrounding, low velocity, imposed flow field. These vortex shear layers have similar properties to those of free jets and wakes, about which much information exists in the literature.

Recent measurements of local voltages along the axial extent of a gas blast arc have shown a rapid electrical decay before current zero along the arc region downstream of the nozzle throat where pronounced turbulence occurs. The plasma decay rate in this region is not explicable in terms of convection and radial thermal diffusion, suggesting that turbulent diffusion has a significant effect. Approximate calculations indicate that the properties of the observed turbulence are consistent with the enhanced recovery measured electrically. However, it is also clear that other power loss processes occurring upstream of the nozzle throat are important even close to current zero in conditioning the arc column for final extinction.

A direct theoretical prediction of the arc electrical behaviour from measured turbulent properties would clearly provide indisputable evidence of the importance of turbulence for arc quenching. Unfortunately such an approach is still plagued by uncertainties in the values of the turbulent parameters required theoretically. Nonetheless encouraging progress is currently being made in correctly predicting electrical recovery characteristics with a semi-empirical arc model involving turbulence.

1. INTRODUCTION

The dominant processes responsible for arc quenching in a gas blast circuit breaker near current zero have for many years defied identification because of the complexity of the physical environment and the difficulty of obtaining relevant and accurate experimental data. This has not deterred many authors from proposing various possibilities from the scant information available. One such proposal has been the enhanced deionisation rates which might be caused by the diffusion mixing of ionised and deionised gases.[1] Improvements of several orders of magnitudes over the recovery rates associated with more conventional processes were anticipated. Recently more accurate comparisons of various loss processes on a purely theoretical basis have suggested the possible dominant effect of turbulent heat transfer during the current zero period.[2]

Modern technology has now enabled sophisticated experimental investigations of the gas blast arc to be undertaken, allowing a more realistic assessment to be made of the existence, nature and influence of turbulence during the current zero period. The purpose of this paper is to review briefly the results of these investigations and their significance.

2. THE NATURE OF TURBULENCE

Turbulence refers to a fluid flow with superimposed small scale rotation of the fluid known as eddies. The fluid rotation is measured in terms of vorticity:

$$\Omega = \partial v/\partial r + v/r$$

where v = rotational component of fluid velocity, r = radial coordinate.

The eddies may be distorted by straining effects orthogonal to the plane of rotation leading to a reduction in the cross-section of eddies. The kinetic energy of rotation simultaneously increases at the expense of the flow kinetic energy in the direction of straining.

In this manner flow energy may be cascaded down from larger to smaller vortices to be ultimately dissipated viscously. In general this dissipation of flow kinetic energy requires a finite time, the rate decreasing with the size of eddy. Also the smaller eddies developed successively in this manner are subjected, in free flow, to approximately equal amounts of straining in all directions leading to an isotropic distribution of rotation. The flow is even more complex since vortices of different sizes may co-exist at a given point. As a result, turbulence will cause a fluid parameter to have a complex fluctuating component superimposed upon a steady value:

$$\text{e.g. } v_{total} = v_{steady} + v_{turb}$$

Clearly the measurement of turbulent effects is difficult on account of their apparent randomness so that recourse needs to be made to statistical methods. Thus in order to unravel the temporal and spatial structure of turbulence correlation functions need to be evaluated which may be written:[3]

$$\phi(\Delta_z,\tau) = \frac{1}{T} \frac{\int_0^T i(z,t)\, i(z+\Delta z, t+\tau)\, dt}{\sqrt{\langle i(z,t)^2\rangle \langle i(z+\Delta z,t)^2\rangle}} \qquad (1)$$

where i = amplitude of a turbulent induced variation observed at different positions z, (z+Δz) and times t, (t+τ). T = sampling time. Such correlation functions yield values, for the eddies, of mean propagation velocity, their coherence time (time over which their identity is retained) and their mixing length (distance over which their identity is retained.

Another important property is the frequency spectrum, i.e. the intensity and occurrence of fluctuations at different frequencies. This may be determined from the spectral power density:

$$S(f) = F^*[i(t)] \cdot F[i(t)] \qquad (2)$$

where F[i(t)] is the Fourier transform of the time signal i(t).

Turbulence is of interest in gas blast circuit breaker arcs as a possible thermal energy transfer process. Although a precise treatment of turbulent heat transfer is difficult, it is normally assumed that the transfer occurs by diffusion of, and between, small scale eddies across a temperature gradient. The process is analogous to molecular diffusion so that a coefficient of turbulent thermal conductivity (κ_t) may be defined[2,4]

$$\kappa_t = \rho c_p \varepsilon_T \qquad (3)$$

where ρ = mass density
 c_p = specific heat at constant pressure
 ε_T = $C_1 \ell (v_1 - v_2)$ = turbulent thermal diffusivity (4)
 C_1 = unspecified dimensionless constant
 ℓ = mixing length of the eddies (determined from $\phi(\Delta z, \tau)$)
 v_1, v_2 = mean fluid velocities on either side of the turbulent zone.

In a similar manner a coefficient of turbulent viscosity may be defined:

$$\eta_t = \rho \varepsilon$$

where ε = turbulent momentum diffusivity related to ε_T by the turbulent Prandtl number
 P_r = $\varepsilon_T / \varepsilon$
For free jets $P_r \sim 0.5$.

The difficulties involved with applying this treatment to real turbulent conditions are:

(i) the mixing length concept is known to fail for some practical pressure gradient flows.
(ii) the coexistence of different eddy scales requires a weighted value of ℓ (the smallest eddies, which are difficult to measure experimentally, may well dominate).
(iii) average values of ρ, c_p are required over steep temperature gradients.

Turbulence may be produced in a gas blast circuit breaker by velocity shear layers between the imposed flow field and the confining walls, and also between the imposed flow and the arc column. Vortex production in the boundary layer between the arc and the imposed flow may occur if a non zero angle exists between the pressure gradient sustaining the flow and the mass density gradient produced by the arc heating.[3] The condition would be satisfied, for instance, by an accelerating flow coaxial with a cylindrical arc column (Fig. 1). The arc core would then be separated from the surrounding flow field by azimuthal vorticity lines.

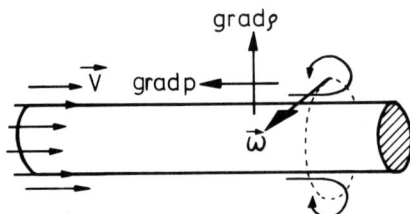

Fig. 1. Production of vorticity at the boundary of an arc in an axial pressure gradient.[3]

The vortex layers formed in this manner are closely related to those of free jets and wakes, which have been extensively studied.[5,6,7] A particularly characteristic feature of these jets is the formation of a well defined axisymmetric instability[7] transforming to a helical instability further downstream (Fig. 2). The axisymmetric instabilities occur periodically and have a well defined wavelength (λ) which is proportional to the thickness of the shear layer (δ). The dominant frequency initially excited in this axisymmetric mode is related to the jet flow velocity v by:

$$f\lambda = v/2. \qquad (5)$$

The formation length of the axisymmetric disturbances is $3\lambda/2$ so that regular disturbances are only apparent beyond this distance from the nozzle exit.

The free jet differs significantly from the gas blast arc in being a constant pressure and temperature flow whereas the gas blast arc is at a considerably elevated temperature to the surrounding flow and is subjected to steep axial pressure gradi-

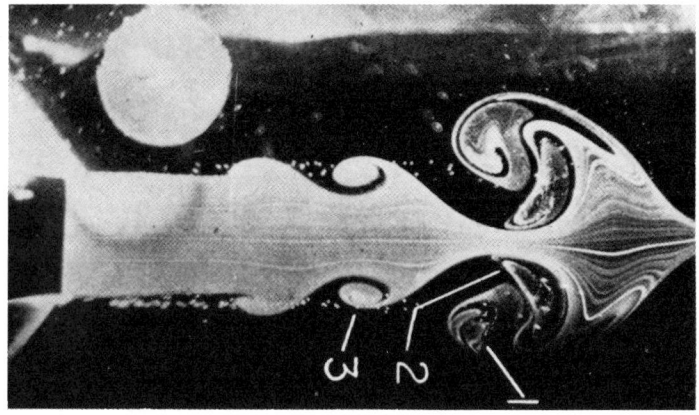

Fig. 2. Visualisation of the instability of a free jet boundary layer.[5]

ents. Consequently there is no unique value for the flow velocity v (equation (5)) for the gas blast arc.

3. EXPERIMENTAL EVIDENCE FOR TURBULENCE IN GAS BLAST ARCS

Experimental evidence for the existence of shear boundary layer turbulence in 20-500 A, 100 Hz arcs subjected to a coaxial flow of gas has been obtained by Howatson and Topham.[8] Their high speed framing photographs (16,000 fps, 2 μs exposure) indicated the formation of axisymmetric instabilities transferring to helical instabilities further downstream (Fig. 3) - a behaviour clearly analogous to free jets (section 2). The arcs were of considerably smaller cross-section than the nozzle throat so that wall turbulence effects could be neglected. The instabilities occurred only in an accelerating coaxial flow (consistent with the requirement of an axial grad p at a non zero angle with the arc grad ρ^3) and were less pronounced at higher currents (up to 500 A), only becoming substantial when the current reduced to about 40 A.

The axisymmetric arc column disturbances are anticipated to change the local energy balance of the arc column leading, under quasi steady conditions to sympathetic changes in the electric power dissipated in the arc. Such fluctuations manifest themselves as regular oscillations of the overall arc voltage with a unique frequency corresponding to the frequency of formation of the axisymmetric disturbances,[8] (Fig. 3).

A similar arc behaviour has also been observed under more realistic circuit breaking conditions (peak arc currents up to 10 kA) in both a relatively slowly accelerating flow through a large nozzle[9] and a rapidly accelerated flow through a short orifice.[10] Under both conditions axisymmetric disturbances accompanied by sympathetic voltage oscillations occurred, but only under specific conditions. A dependence was observed on upstream electrode material and polarity, and the upstream electrode -

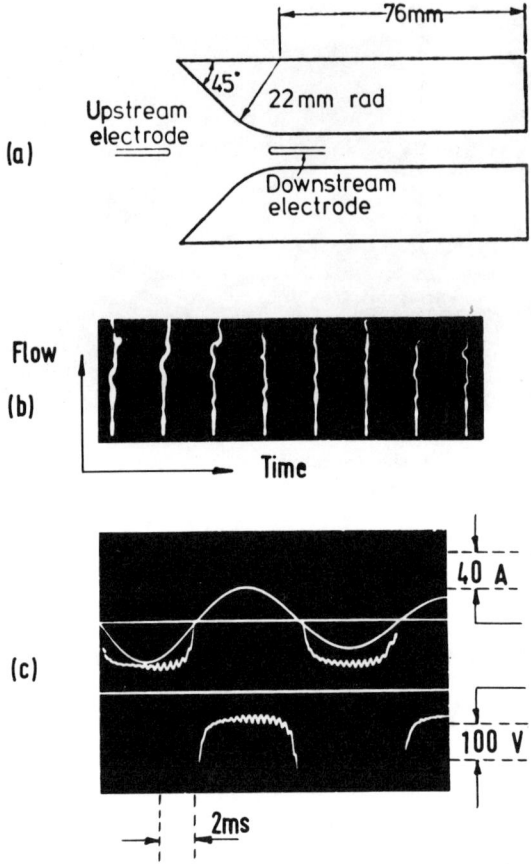

Fig. 3. Axisymmetric and helical arc instabilities produced in an accelerating flow.[8]
a) Experimental configuration.
b) High speed film of arc instabilities in nozzle (I = 40 A, p = 5.2 bar) photographed at 84 μs intervals.
c) Oscillations in arc voltage.

throat separation. Consistent with Howatson and Topham's observations,[8] the axisymmetric instabilities were only detected as the arc current was decreased towards zero. However, when the test head failed to clear, the instabilities persisted throughout the second half cycle of current. The occurrence of these regular instabilities was accompanied by, not only sympathetic arc voltage oscillations, but also regular oscillations in the net radiative emission from the arc column.

In order to examine the extent to which these arc instabilities are similar to those in free jets (section 2) a modified form of the free jet analysis may be used.[8,10] For instance it may be shown that for a linear pressure gradient flow:

$$v = v_o + az \tag{6}$$

the frequency of the dominant intability produced, is given by[10]

$$f \cong \frac{1}{2} a [(T_c/T_\infty)^{0.5} + 1] \tag{7}$$

where T_o = temperature of the arc core
       T_∞ = temperature of the surrounding flow.

The frequencies calculated in this manner are in good agreement with those measured experimentally under the three quite different experimental conditions,[8,9,10] (cf. Fig. 4). The analogous behaviour of gas blast arcs and free jets is therefore convincingly confirmed.

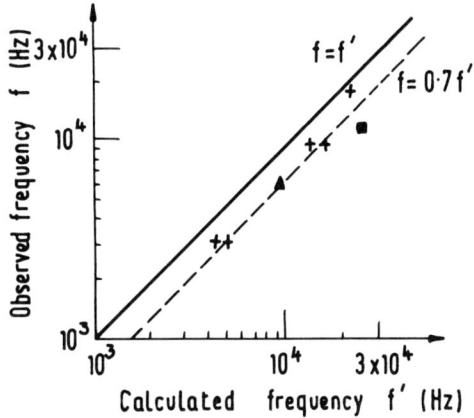

Fig. 4. Comparison of measured and calculated frequency.
+   nozzle flow[8]
□   Nozzle flow[9]
△   Orifice flow[10].

Convincing evidence of less regular, smaller scale turbulence in gas blast arcs has been obtained photographically. Figure 5 shows a shadowgram of an arc column in an orifice air flow at the current zero following a 3 kA current peak at a 100 Hz. The irregular thermal envelope surrounding the arc is obviously turbulent. The manner in which the detailed properties of turbulence varies along the axial extent of a gas blast arc has been investigated by Niemeyer and Ragaller.[3] Figure 6 shows typical fast streak records of an arc core in a Laval nozzle (throat diameter = 12 mm, divergence angle = 8°, downstream length = 90 mm, upstream pressure = 20 atmo-

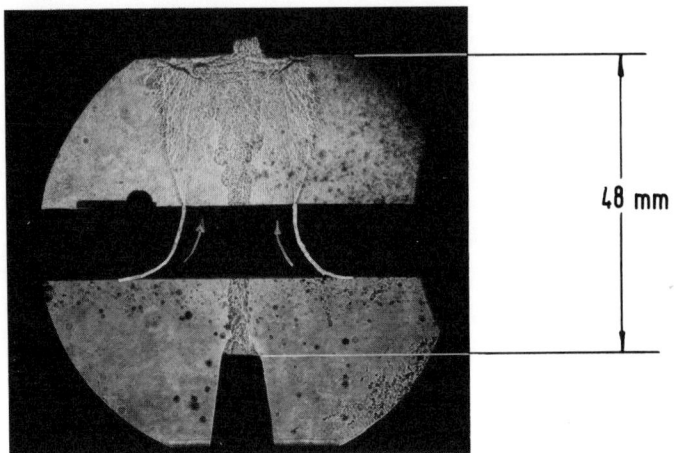

Fig. 5. Shadowgram of orifice arc at current zero (I_{max} = 3 kA, f = 100 Hz, p ~ 7 bar air, exposure time ~ 200 ns).

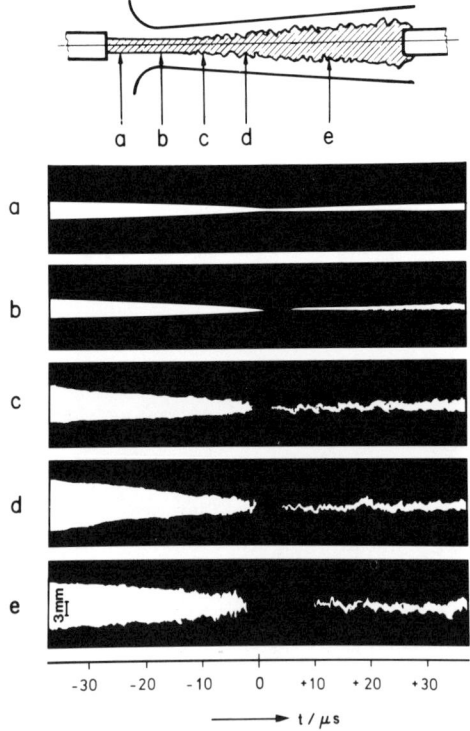

Fig. 6. Streak record of the arc before and after current zero (t = 0) for a re-igniting arc at different axial positions.[11]

spheres) taken during the current zero period as a 2 kA current was ramped to zero at ~ 30 A/µs.[3,11] The records were taken through slotted windows at various axial positions along the length of the nozzle. Clearly, the arc column downstream of the nozzle throat changes from being slightly turbulent during the 30 µs before i = 0, to being strongly turbulent at i = 0, whereas at, and upstream of, the throat, the arc remains fairly laminar throughout the current zero period. The turbulent length scale varies from fine scale distortions of a fraction of a mm to gross transverse drifts of ~ 2-4 mm. Rupture of the luminous arc core appears to occur further downstream at progressively earlier times before i = 0, and the luminous core is reestablished at progressively later times further downstream.

These photographs are therefore particularly important in establishing the onset of turbulence near current zero and its persistence throughout the early post zero recovery period.

More precise information about the current zero turbulence of this arc has been obtained by measuring the space-time correlation function (equation 11) and the spectral power density (equation 12) of light fluctuations from the arc column.[3,11] Such analysis involves synchronous photoelectric measurements of light emission at two axial positions, a small distance, Δz, apart.

Typical results of such measurements are given in Fig. 7[3] from which the turbulent properties may be derived. For instance the coherence time may be evaluated directly from the correlation function (Fig. 7a) whilst the mixing length may be estimated from the spatial decay rate of the correlation function (Fig. 7b). The fluctuations have a mixing length (~ 6 mm) of the same order as the local nozzle radius. Results for the spectral power density (Fig. 7c) indicate that the fluctuations have a low frequency cut off at ~ 500 kHz. The high frequency extreme, which lay beyond the experimental spatial and temporal resolution, has been estimated as 600 MHz.[3] The spectral power density results are consistent with the vortex formation being produced by the shear boundary layer between the arc and external flow, in that the frequency spectrum at onset (near the nozzle throat) is strongly concentrated about a single frequency (Fig. 7c).

The results are particularly important in that for the first time they provide a quantitative estimate of the turbulence properties of gas blast arcs.

4. THE INFLUENCE OF TURBULENCE ON THE ELECTRICAL BEHAVIOUR OF AN ARC NEAR CURRENT ZERO

4.1. Experimental Evidence from Wall Stabilised Arcs

The electrical response of an arc column to rapid changes in electrical power input, such as those occurring near current zero in a gas-blast interruptor, is conveniently described in terms of the variation of electrical conductance along the length of the arc column. Extensive studies of such local conductance variations have been made with geometrically symmetrical arc columns stabilised in a hollow bore with cylindrical

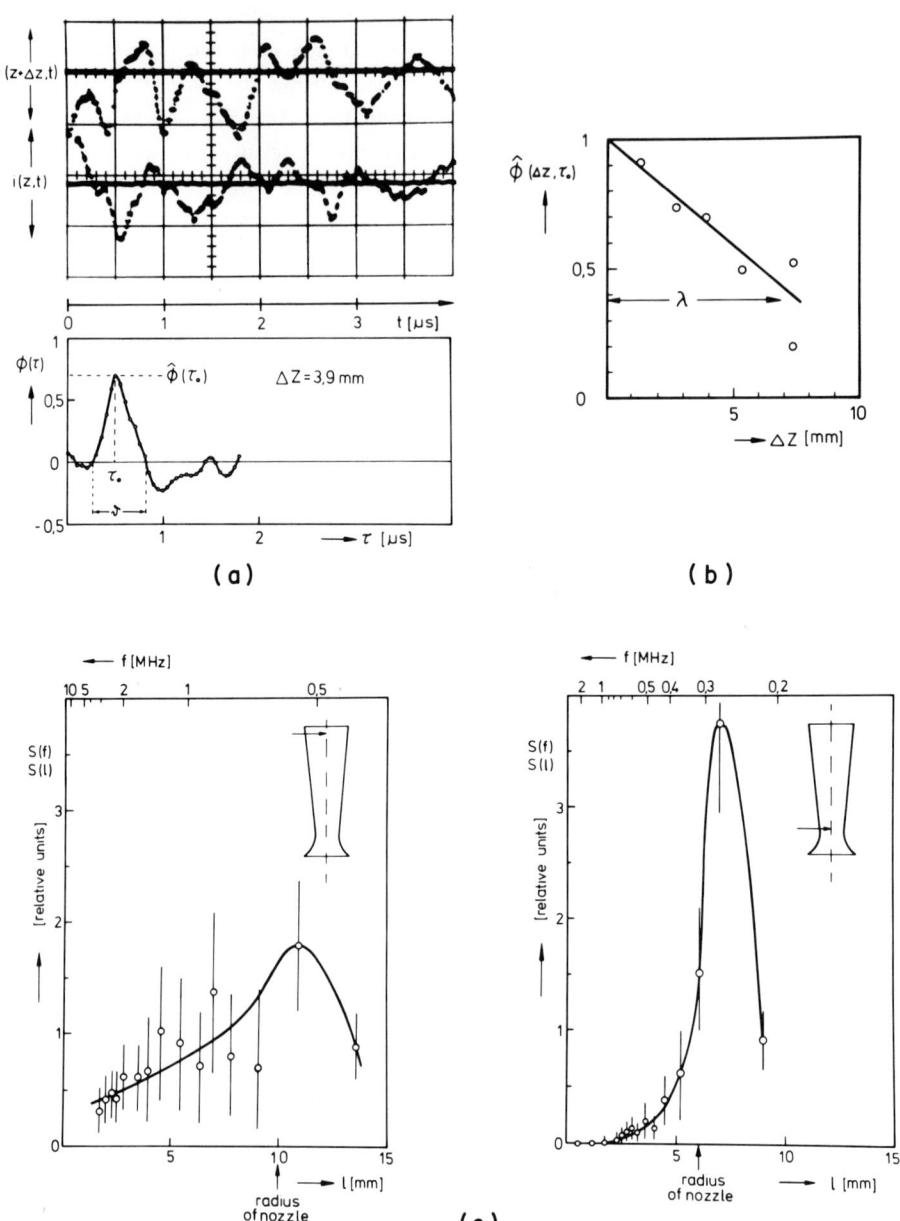

Fig. 7. Typical results of statistical analysis of turbulence in a gas blast arc.[3]
a) The cross correlation function $\phi(\tau)$
b) The spatial correlation function $\phi(\Delta z, \tau_o)$ along the nozzle
c) The spectral power density at two different axial positions.

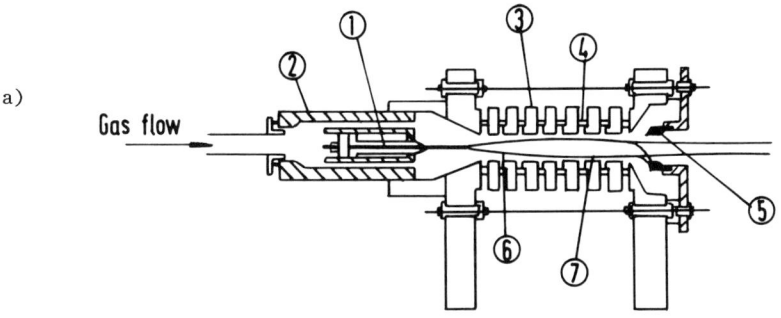

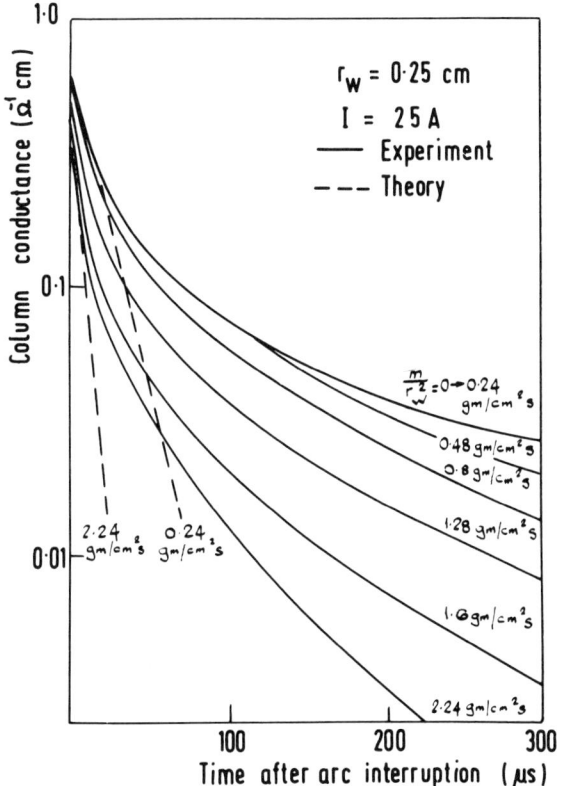

Fig. 8. Wall stabilised arcs.
a) Geometrical arrangement
   1 Cathode
   2 Coaxial cathode holder
   3 Copper probes
   4 Boron nitride insulators
   5 Anode
   6 Arc growing region
   7 Arc fully developed region
b) Conductance decay during free recovery for various mass flow rates.

solid walls (wall stabilised arcs), (Fig. 8a). Segments of the stabilising walls have been used to measure local voltages from which the conductance may be determined. These arcs represent well behaved and highly reproducible arcing conditions amenable to theoretical analysis. One aspect of such studies which is relevant to the circuit interruption problem is the manner in which the arc's electrical conductance decays following the sudden interruption of the sustaining current (free recovery). This conductance decay is governed by the thermal properties of the arc plasma which vary as the plasma temperature decreases and often leads to complicated forms of conductance variations (e.g. Fig. 8b). However, if the radius of the arc core (R) and the plasma thermal diffusivity, α, remain independent of time, such as is the situation immediately after current interruption, the conductance decay may be characterised by a time constant[12]

$$\tau = R^2/(2\alpha B_1^2) \tag{8}$$

where B_1 = first zero value of the Bessel function,
       α = $\kappa/\rho c_p$,
       ρ = gas density,
       κ, c_p = thermal conductivity and specific heat respectively.

Clearly a small arc core accompanied by a high thermal diffusivity encourages a rapid recovery.

Recent work has shown that the electrical behaviour of these wall stabilised arcs may be significantly changed when the arcs are subjected to turbulent gas flows along the axis of the stabilising bore. For instance Frind[13] has shown experimentally that increased power losses occur under steady arcing conditions as a result of turbulence generated along the wall of the stabilising bore whilst Hill et al.[14] have experimentally demonstrated that an improved electrical recovery rate is obtained during free recovery. Hill's results indicated that turbulence of such intensity as to have an insignificant effect upon the gross steady state electrical arc properties still produced an improvement in the conductance decay rates (Fig. 8b). Although the turbulence in these experiments was of a different nature to the 'free jet' turbulence of the circuit-breaker arc, the experiments highlight the favourable influence of micro turbulence on arc recovery. This enhanced recovery may be described analytically by a time constant of the form given in equation (8) except that the thermal diffusivity, α, is increased by the additional effect of the turbulent thermal diffusivity (ε_t) defined by equation (3). Equation (8) therefore assumes the mathematical form given by Swanson and Roidt[15]

$$\tau = R^2/(2\alpha B_1^2 \beta) \tag{9}$$

where $\beta \geq 1$ is the turbulent diffusion coefficient whose value increases with the intensity of the turbulence.

4.2. The Similarity between the Electrical Cores of Wall Stabilised and Flow Stabilised Arcs

Relating the behaviour of the electrical core of a gas flow arc to that of a wall stabilised arc has been justified experimentally using one of the most ideal flow fields that can be realised in practice.[16,17] This is the uniform flow field which is characterised by spatially and temporally constant flow velocity and gas pressure (Fig. 9a). Since there is no pressure or velocity gradients, vortex shear layers are not formed at the arc boundary thus eliminating pronounced 'free jet' turbulence and enabling a comparison to be made with the non-turbulent wall stabilised arc.

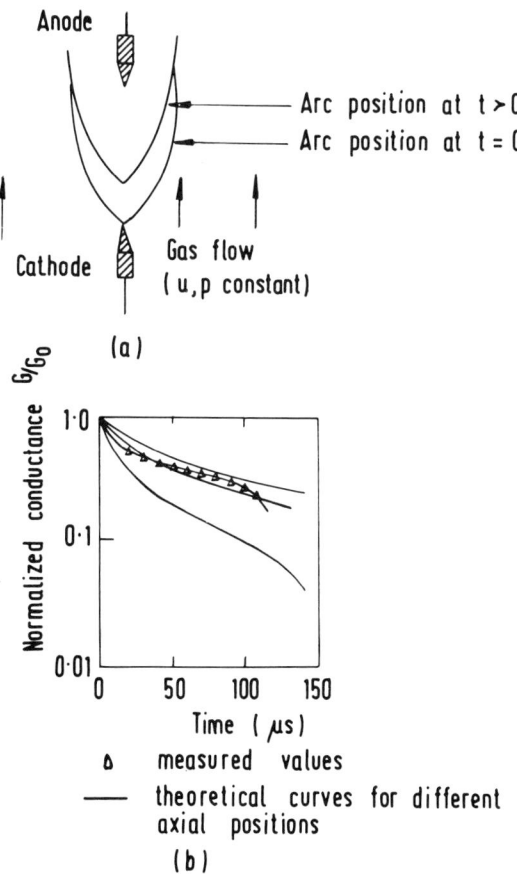

Fig. 9. Uniform flow stabilised arc.
a) Geometrical arrangement.
b) Conductance decay during free recovery of the uniform flow arc for various mass flow rates.

The similarity between the behaviour of the electrical core of the uniform flow and wall stabilised arcs at low currents ($\leq$ 100 A) has been established both under steady state conditions[16] and during free recovery.[17] This similarity exists because the electrical properties of the arc are initially more sensitive to changes in cross-sectional area rather than to detailed changes in thermal profiles of the arc core caused by external influence.[16]

During free recovery the influence of the flow field in axially displacing plasma elements needs to be taken into account. Experiments then show that the decay of each element as it is transported by the flow is identical to that of a wall stabilised arc of the same radius[17] for appreciable periods of the recovery, although the change in thermal plasma properties with time leads to complicated forms of conductance decay curves (Fig. 9b). This forms convincing evidence that the arc conductance decay in a uniform flow field is governed by radial thermal diffusion superimposed upon the convective displacements caused by the flow field..

If arc recovery occurs due to a contraction in cross-sectional area, rather than a decay in axial temperature, the thermal diffusion time constant (equation (8)) is progressively shortened as recovery progresses, whilst the convective rate remains constant on account of the uniformity of the flow field. This means that ultimate arc extinction will be governed by thermal diffusion rather than axial convection effects.

4.3. Experimental Evidence for Gas Blast Circuit Breaker Arcs

It is clear from Figs. 4 and 5 that the overall arc voltage of a gas blast arc reacts sympathetically to the occurrence of axisymmetric column instabilities. Experimental observations have shown that such instabilities on a large scale have a deleterious effect upon the recovery characteristics. However, it is clear from the wall stabilised arc experiments that smaller scale turbulence should enhance arc recovery, since it provides an additional power dissipation process.

An attempt to identify directly the influence of micro-turbulence upon the electrical decay of a gas blast circuit breaker arc has been made by Hermann et al.[11] using spectroscopic measurements of the arc temperature profile and deriving from that the division of the overall arc voltage between the arc sections upstream of the nozzle throat, where there is no evidence of turbulence, and the section downstream of the throat, where turbulence is pronounced near current zero. Their results show that the electrical resistance of the turbulent downstream section increasingly dominates the overall arc value during the current zero period (Fig. 10), particularly during the immediate post zero time.

More direct voltage measurements may be obtained using ablation protected probes.[18] Although such measurements involve inserting metallic probes into the arc plasma, experimental evidence shows surprisingly little effect upon the arc behaviour. Experimental measurements with such probes are available for arcs in both orifice and nozzle air flows at upstream pressures of 7 atmospheres.[18] Typical axial

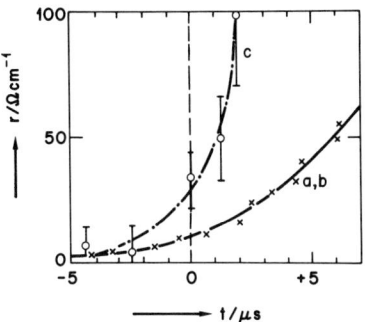

Fig. 10. Measured values of the time dependent specific arc resistance upstream (x) and downstream (o) of the nozzle throat.[11]

variations of electric field strength obtained with such probes along an orifice arc length at various times during a half sinusoid current waveform are shown Figs. 11a and 11b for peak currents of 3 and 8 kA respectively. The 3 and 8 kA arcs represent zero and unity failure rate conditions respectively for the particular test head. The 3 kA results are consistent with those of Hermann et al.[11] in showing a significant shift in axial position of the high electric field strengths from the converging orifice inlet, during the peak current phase, to the throat and turbulent downstream region, during the current zero phase. The considerable increase in electric field strength at the orifice exit near current zero is particularly noteworthy.

For the 8 kA, unity failure condition similar changes occur in the electric field strength values in the convergent orifice region, but the changes of electric field in the throat and downstream are considerably less than for the 3 kA zero failure rate condition. Clearly the throat and turbulent downstream arc regions determine the recovery behaviour of the overall arc gap for this particular test head geometry.

The electric field strength variations shown on Fig. 11b yield the local conductance decays shown on Fig. 12.

4.4. Interpretation of the Electrical Behaviour of the Gas Blast Circuit Breaker Arc

The experimental evidence from wall stabilised arc studies[12,13,14] suggests that the pronounced electrical recovery observed in the turbulent downstream region of the gas blast arcs[11,18] may be a direct result of turbulent heat transfer effects. However, an extension of the approach discussed in section 4.2 for identifying diffusive effects under realistic gas flow conditions is complicated for the gas blast circuit breaker conditions by axial pressure gradients and the high velocity and accelerating nature of the flow. On the other hand the current levels involved near current zero in the gas blast arcs are similar to those used in the wall stabilied arc studies.

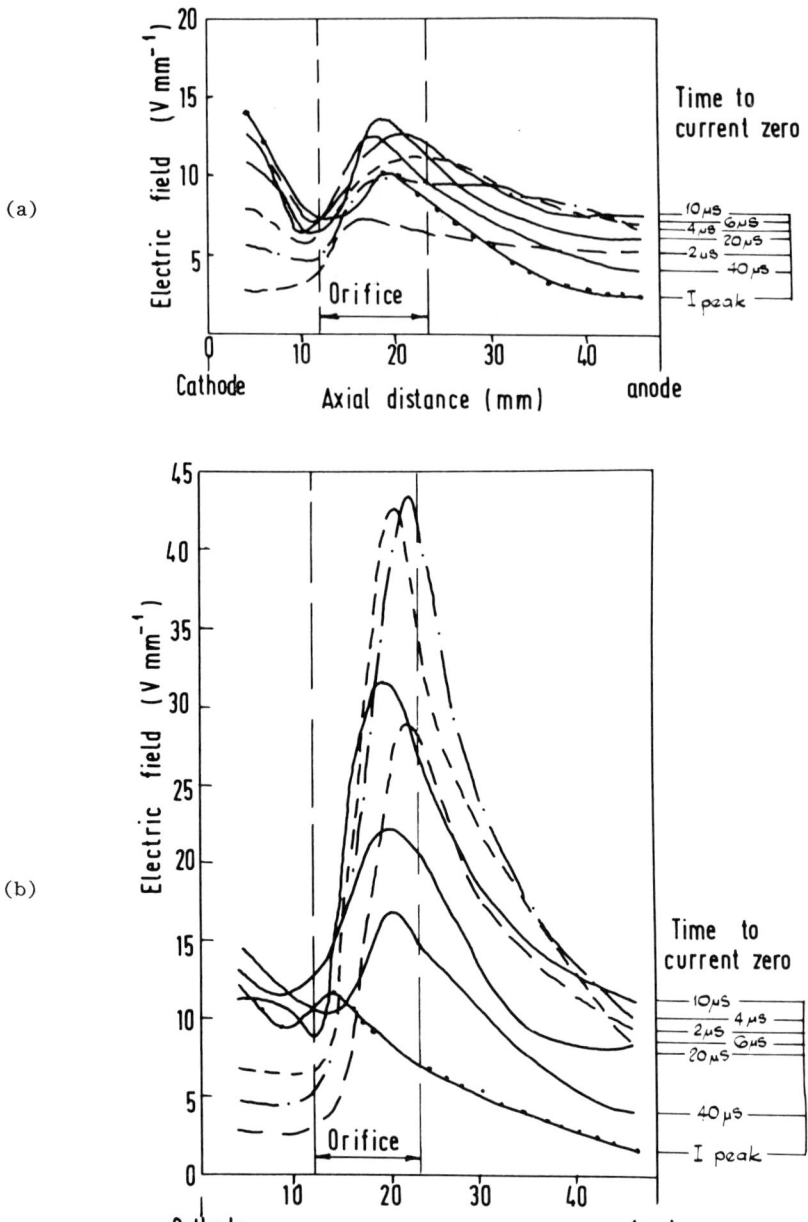

Fig. 11. Measurements of the electric field at various axial positions and at different times before current zero.
a) Orifice arc, unity failure rate[18].
b) Orifice arc, zero failure rate.

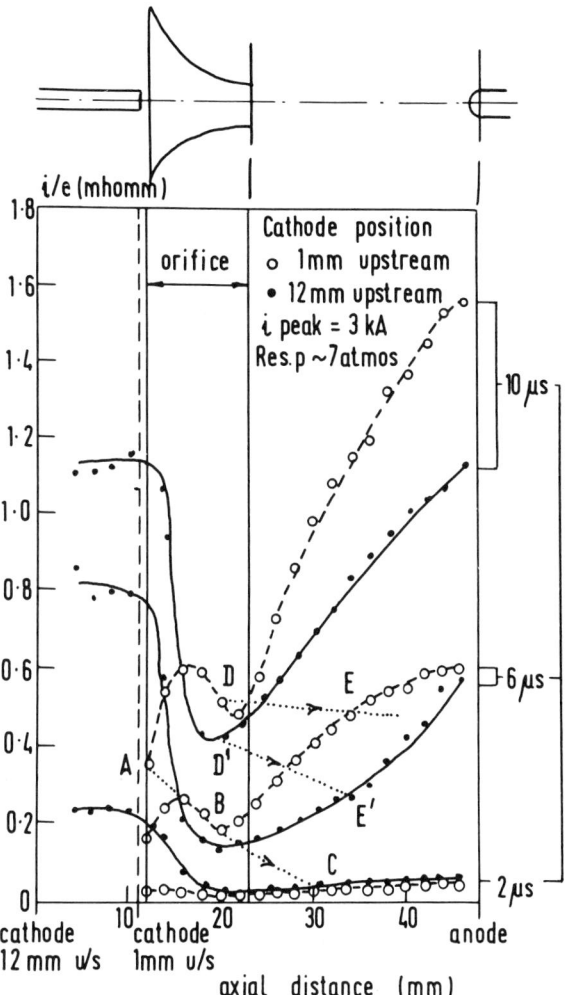

Fig. 12. Electrical Conductance variations of different axial regions of gas blast arcs near current zero for different upstream electrode positions (········ plasma trajectories).

The flow acceleration in the gas blast arc leads to the hot plasma, which forms the arc core, being convected axially at velocities ($\leq 8\cdot10^3 \mathrm{ms}^{-1}$) which are at least an order of magnitude greater than the surrounding cold flow. Not only does this produce turbulent shear layers at the core boundary, but also transports a cylindrical core element across the major portion of the downstream nozzle section within the final few micro-seconds before current zero. The importance of this effect is that any power dissipating processes occurring within the convergent nozzle entry close to current zero may exert their influence even before current zero in conditioning the downstream plasma for final quenching. Typical trajectories of representative plasma elements are shown superimposed upon the conductance curves of Fig. 12.

Another complicating effect of the accelerating flow is the production of plasma straining.[19] This manifests itself as an elongation of an elemental plasma volume at the expense of its cross-sectional area due to the velocity differential across the leading and lagging edges ($\partial v_z/\partial z$) (Fig. 13). The power dissipation associated with

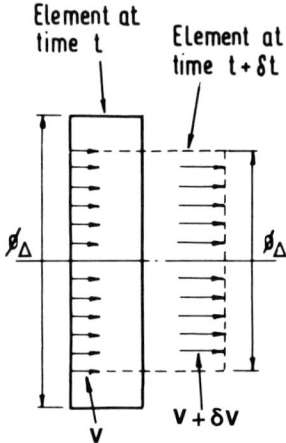

Fig. 13. The physical nature of convective straining

this process is mainly located within the converging throat section of the nozzle and reduces considerably downstream (Fig. 14). Although the straining effect is not located downstream of the throat it does condition plasma elements prior to their entry into the turbulent downstream region. However, within the downstream region itself radial diffusion losses may be isolated if only convective losses are taken into account.

The conductance decay suffered by individual plasma elements as they traverse the arc gap (Fig. 15) may, therefore, be determined in a similar manner to the uniform flow case (section 4.2) by following the trajectories on Fig. 12. An approximate indication of whether laminar thermal diffusion alone can account for the downstream conductance decay may then be obtained by comparing the conductance decay time constants determined from these results with those calculated from the wall stabilised arc expressions (8) and (9).

The most severe theoretical test is provided by a channel arc model with a typical plateau temperature of 7000 K and using published values for the thermal diffusivity α.[20] The channel radius, R (equation (8)) is determined from the measured value of the local conductance and the published value of electrical conductivity at 7000 K.

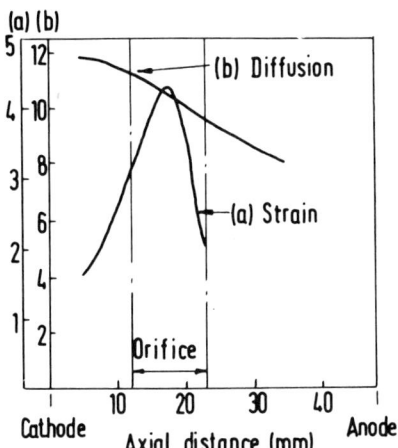

Fig. 14. Comparison of the axial dependence of straining and radial diffusion losses for the orifice arc.

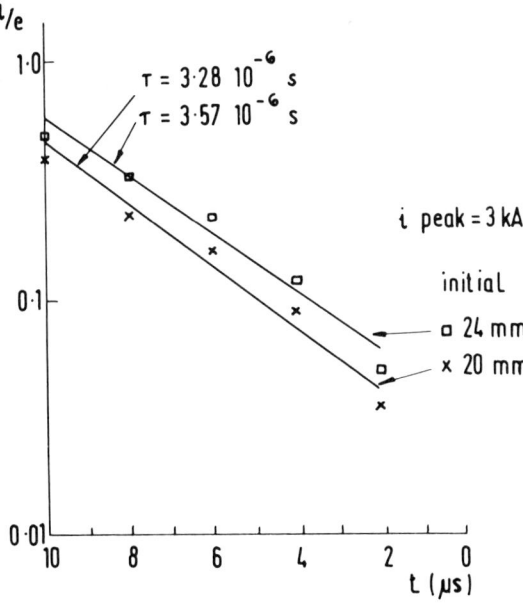

Fig. 15. Measured conductance decay curves for the orifice arc. Derivation of time constants.

The core radius based on this channel model represents the minimum radius consistent with the measured electrical conductance and plateau temperature. The resulting time constant is consequently the minimum consistent with the measured conductance. For a plasma element just downstream of the orifice throat 10 µs before current zero the time constant according to the channel model is typically 10 µs. The corresponding time constant calculated from the measured conductance decays (Fig. 15) are typically 3 µs. The measured conductance decay is more rapid than anticipated from conventional thermal diffusion, thus indicating the importance of the turbulent diffusion. For the measured and calculated decay rates to agree, a turbulent diffusion parameter (equation (9)) has to be invoked with a value $\beta = 3$. This implies through equation (3) that the eddy mixing length ℓ is of the order of a centimeter which is similar to the radius of the orifice throat (~ 0.95 cm) as required by turbulence theory.

5. THEORETICAL PREDICTION OF ELECTRICAL ARC BEHAVIOUR FROM THE FUNDAMENTAL PROPERTIES OF TURBULENCE

The most conclusive proof of the role of turbulence in enhancing arc extinction would be to predict theoretically the complete electrical behaviour of the arc column near current zero from the optically measured turbulent properties. Theoretical calculations performed to date, do indeed, predict a pronounced influence on the arc recovery characteristics due to turbulence.

The most significant purely theoretical approach is due to Swanson and Roidt,[2] whilst more recently Hermann and Ragaller[21] have developed a semi-empirical model to predict rrrv. Unfortunately, such theoretical treatments are plagued by uncertainties in evaluating the necessary turbulent parameters. Swanson and Roidt[2] rely on averaged values of free jet theory parameters; Hermann and Ragaller's model[21] relies also upon adjusting turbulent exchange parameters. Clearly additional work is required before detailed predictive calculations become possible.

6. CONCLUSIONS

The existence of turbulence in the downstream section of some types of gas blast circuit breaker arcs during the current zero period has been convincingly established from optical investigations. The origin of the turbulence lies in the vortex shear layer separating the higher velocity plasma core from the lower velocity surrounding flow field. The vortex formation process has many features analogous to that associated with free jets so that a direct application of free jet theory to the study of such arcs seems possible.

Although direct local electrical measurements indicate a substantial conductance decay downstream of the nozzle throat where turbulence appears to be pronounced, it is difficult to identify the extent of the turbulent effect on account of the superimposition of other effects such as convective displacement and straining, radial thermal diffusion and radiation transport. Straining effects within the nozzle convergence appear to be essential for conditioning the plasma being convectively transferred to the downstream region in preparation for final extinction by radial diffusion losses enhanced by turbulence.

Although turbulence is important for arc quenching there is experimental evidence[22] that additional effects are also important for achieving optimum circuit breaking performance. For instance careful optimisation of the upstream electrode - orifice entry geometry, leads to an improved circuit breaker performance.[22,23] Voltage probe measurements with such an optimised electrode-orifice gap show that the decay rate of the arc column upstream of the nozzle throat is reduced to values comparable with downstream decay rates (Fig. 12). There is also experimental evidence that vapourised electrode material in the arc column can have a deleterious effect on the arc recovery. Clearly further experimental investigations are required before a complete understanding of all the processes contributing to arc extinction will be achieved.

ACKNOWLEDGEMENTS

Stimulating discussions with Professor Dr. K. Ragaller (Brown, Boveri and Co. Ltd.), Drs. Frind (General Electric Research and Development) and Swanson (Westinghouse Research), and colleagues at Liverpool University are gratefully acknowledged.

REFERENCES

1. J. Slepian, Trans. AIEE 60 (1941) 162
2. B. W. Swanson and R. M. Roidt, Proc. IEEE 59 (1971) 493
3. L. Niemeyer and K. Ragaller, Z. Naturforsch. 28a (1973) 1281
4. W. Hermann, U. Kogelschatz, L. Niemeyer, K. Ragaller, and E. Schade, J. Phys. D: Appl. Phys. 7 (1974) 1703
5. D. O. Rockwell, J. Appl. Mech. 39 (1972) 883
6. D. O. Rockwell and W. O. Niccolls, J. Basic Eng. 94 (2972) 720
7. H. A. Becker and T. A. Massaro, J. Fluid Mech. 31 (1968) 435
8. A. M. Howatson and D. R. Topham, J. Phys. D: Appl. Phys. 9 (1976) 1101
9. I. R. Bothwell, K. O. Goodwin and B. Grycz, Univ. Liverpool Arc Res. Rep. ULAP-T 29 (1974)
10. H. L. Walmsley, G. R. Jones, F. Haji and D. C. Strachan, J. Phys. D: Appl. Phys. 10 (1977) 383
11. W. Hermann, U. Kogelschatz, L. Niemeyer, K. Ragaller, and E. Schade, IEEE Trans. PAS 95 (1976) 1165
12. G. R. Jones and H. Edels, Z. Phys. 229 (1969) 14
13. G. Frind and B. L. Damsky, ARL report 68-0067 April (1968)
14. R. J. Hill, G. R. Jones and H. Edels, Proc. 3rd Int. Conf. on Gas Discharges, IEE Conf. Public No. 118 p. 506-511
15. B. W. Swanson, R. M. Roidt and T. E. Browne, IEEE Trans. PAS 89 (1970) 1094
16. D. R. Topham, J. Phys. D: Appl. Phys. 5 (1972) 1837
17. G. R. Jones and S. R. Naidu, J. Phys. D: Appl. Phys. 7 (1974) 2254
18. A. Chapman, G. R. Jones and D. C. Strachan, Proc. 4th Int. Conf. on Gas Discharges, IEE Conf. Public No. 143 (1976) p. 52-55

19. S. K. Chan, M. D. Cowley and M. T. C. Fang, Univ. Liverpool Arc Res. Rep. ULAP-T 26 (1974)
20. J. M. Yos, AVCO Report RAD-TM-63-7 (1967)
21. W. Hermann and K. Ragaller, IEEE Trans PAS 96 (1977) 1546
22. G. Frind, R. E. Kinsinger, R. D. Miller, H. T. Nagamusu and H. O. Noeske, Final report (Research project 246-1) EPRI EL-284 (1977)
23. A. A. Hudson, The Engineer, 200 (1955) 249-252, 288-290

DISCUSSION
(Chairman: D. M. Benenson, State University of New York at Buffalo)

K. Ragaller (Brown Boveri)

You describe in your paper straining and turbulence as two effects which act in the same way and therefore cannot easily be separated. I want to mention here that in our experiment the effects of straining and turbulence could be separated (Fig. 1).

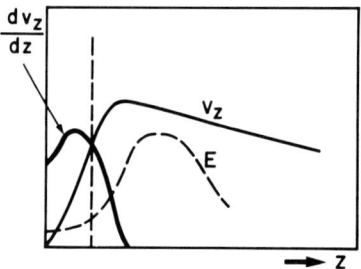

Fig. 1. Axial variation of flow speed, 'straining' (dv_z/dz) and electric field strength near current zero for the Brown Boveri experiment.[1]

The measured and calculated velocity profile shows that in our geometry, straining (i.e. the axial increase of velocity) occurs up to the nozzle throat at an approximately constant rate in the upstream part. However, the increase of the electrical field around current zero and the appearance of turbulence both occur downstream of the nozzle throat. This is direct proof that turbulence, and not straining, is the main effect in interrupting the current. Straining is, of course, important in producing a thin arc column in the turbulent flow section.

G. R. Jones

Yes, I think that is correct. I merely wished to indicate that under many circumstances it may not be so easy to separate these two effects.

D. M. Benenson

Do you see a way to determine characteristic parameters of the turbulence, such as mixing length or exchange factors, perhaps by investigating a variety of configurations or something like a general configuration?

G. R. Jones

With the aid of equations (6) and (3) we tried to derive the mixing length from the arc diameter and the conductance decay time constant.

G. Frind (General Electric)

How did you measure the radius that you have to put into this formula?

G. R. Jones

One knows the electrical conductance at each point along the axis at various times and therefore by going back to cylindrical-column theory one can get the radius. If you assume a square temperature profile this gives the smallest radius and therefore the fastest time constant.

W. Hermann (Brown Boveri)

Referring to the question of the type of turbulence, it is important to realize that one knows more than can be explained just by a time constant. As the turbulence is produced in a free shear layer, the extensive results available for free-shear-layer turbulence can be used, although one still has to adjust an exchange constant to the arc flow situation. This empirical constant is a fairly basic property and therefore this is a much better approximation than adjusting the total heat flux, for example, or even the power loss as is done in simpler arc models.

G. R. Jones

A deficiency in our understanding of turbulent arcs at present is that no obvious predictive criteria have yet been established. It is therefore not clear what the influence of many different configurations would be on the turbulent effect. However, in the case of turbulent, wall stabilised arcs which are free recovering,[2] an extension of the Swanson and Roidt analysis[3] shows that the turbulent diffusion coefficient, depends upon the radius of the stabilising bore, R. Consequently equation (9) predicts that the variation of time constant, τ, with bore radius, R, should be different for turbulent and laminar, wall stabilised arcs. Hill[4] has confirmed experimentally that this is so for mildly turbulent arcs, the dependence of τ upon R being in accordance with theoretical predictions.

REFERENCES

1. W. Hermann, U. Kogelschatz, L. Niemeyer, K. Ragaller, E. Schade, IEEE Trans. PAS 95 (1976) 1165
2. R. J. Hill, G. R. Jones and H. Edels, Proc. 3rd Int. Conf. on Gas Discharges, IEE Conf. Public No. 118, p. 506-511
3. B. W. Swanson and R. M. Roidt, Proc. IEEE 59 (1971) 493
4. R. J. Hill, 'Wall stabilised axial flow arcs', Ph.D. Thesis, University of Liverpool, University of Liverpool, 1973

See also Discussion of Swanson's paper.

THE INFLUENCE OF ARC ROOTS ON CURRENT INTERRUPTION

J. MENTEL
Ruhr University Bochum, Germany*

SUMMARY

Until now it has not been possible to determine a simple relationship between the interruption capability of a circuit breaker and the arc-root behavior. Observation of the roots of a high-current arc on WCu-electrodes in a model circuit breaker reveals a strong vaporization of electrode material. Since contamination of the quenching medium by several percent of metal vapor increases its electrical conductivity at low temperatures (2000-6000 K) by at least an order of magnitude, this metal vapor should decrease the quenching ability. It is also observed that frequent separation of the arc and the jet of electrode vapor occurs, accompanied by removal of the arc roots from the ignition point by the vapor. Accordingly mixing of the arc plasma and the vapor does not take place. To clarify this behavior, spectroscopic measurements were made of the plasma- and vapor temperature in front of the graphite cathode of a high-current dc-arc. The measurements show that in the case of strong vaporization, the electrode vapor in the vicinity of the electrode is much colder than the arc plasma. By using an appropriate electrode material, it is possible to avoid the contamination of the quenching medium and uncontrolled movement of the arc roots.

1. INTRODUCTION

For a given current strength and gas atmosphere, the arc-root behavior is determined by the electrode material and the electrode configuration. Only a few investigations[1,2] have been carried out under the conditions which exist in a circuit break-

---

*) This work is based upon material in the final report 'Basic research program for plasmatechnology, high-pressure arcs in SF_6,' presented to the Federal Department of Research and Technology of the F.R.G. The work was carried out by the author whilst he was with Siemens AG., Erlangen, Forschungslaboratorien.

er. It is to be expected that arc root behavior is irreproducible and probably not determined uniquely by the experimental conditions. Therefore, it seems pointless to search for a quantitative correlation between the interruption capability and the arc-root behavior. Nevertheless it seems possible to clarify some qualitative aspects from which conclusions can be drawn relevant to circuit-breaker design.

2. ARC ROOTS UNDER CIRCUIT-BREAKER CONDITIONS

To identify important effects we have used a high-speed camera operating at 10^4 frames per second to photograph the arc roots in a model circuit breaker provided with windows. The electrode system of the model circuit breaker comprises a solid rod with a flat end face on which the arc root can easily be observed and a nozzle-shaped counter-electrode a distance of 15 mm away. The breaker is operated with SF_6-gas at a pressure of 8 bar. As the arc burns the SF_6 flows through the nozzle-shaped electrode into a reservoir where the pressure is 3 bar. The arc was supplied with a current half cycle having a duration of 10 ms and an amplitude I = 10 kA.

A high-speed film of an arc root on a copper/tungsten cathode shows a green jet of copper vapor escaping from the electrode. At the beginning of the half wave, the green jet is enveloped by the arc which emits light of a dirty red color. Toward the end of the half wave, particularly at the cathode, the jet separates from the arc and the arc root moves toward the edge of the electrode face. It seems as if the root is driven out of the center of the electrode by the vapor and that the position of the arc is determined only partially by the cold gas flow. Initially, the most disturbing finding is the strong vaporization of copper. The ionization energy of copper (7.72 eV) is much lower than that of the usual quenching media, for example atomic nitrogen from dissociated N_2 with 14.54 eV and atomic sulfur from dissociated SF_6 with 10.36 eV.

Therefore the admixture of copper vapor to the quenching media increases the electron density considerably, especially at lower temperatures, and with it the electrical conductivity. The metal vapor apparently contaminates the quenching gas.

An estimate showing the increase in the electrical conductivity σ of the contaminated gas in comparison with that of the pure quenching media is presented in Fig. 1. The relationship between σ-contaminated and σ-pure is plotted against temperature. The two upper curves show the effect of 10 % and 1 % copper-contamination on nitrogen at a pressure of 10 bar [3] and the lower curve of 5 % contamination on SF_6.[4]

It is now known[5,6] that immediately after current zero the arc channel in the nozzle throat is the critical part of the breaking distance. The arc temperature is relatively low in the boundary zones, about 6000 K in N_2 and 4000 K in SF_6,[7] so that an admixture of metal vapor considerably increases the conductance of the residual channel. It is therefore to be expected that the metal vapor in the nozzle throat during the current zero period reduces the value of du/dt which can be withstood (see also paper by Frind).

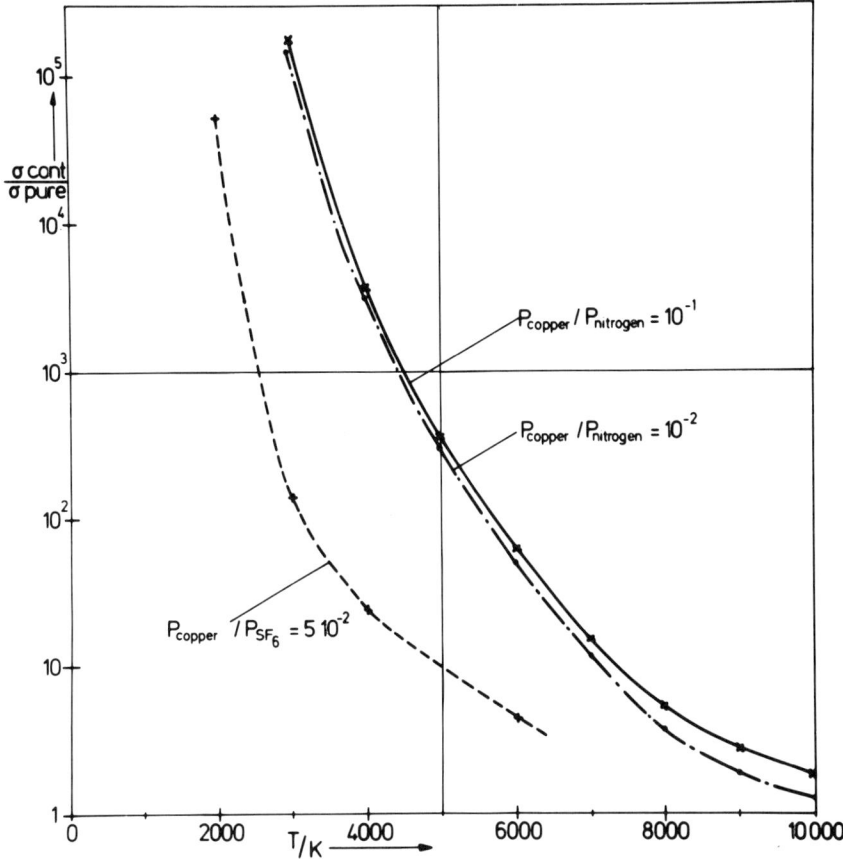

Fig. 1. The ratio of the electrical conductivity of the quenching medium contaminated by copper σ_{cont} to that of the pure quenching medium. The values for nitrogen are taken from Ref.[3], that for SF_6 from Ref.[4].

It is also the case that if gas contaminated with metal vapor remains upstream of the nozzle for a period of 100 µs after current zero, the dielectric strength of the gap will be lowered. It is further decreased by protrusions left behind on the electrode surface by resolidification of the metal melted by the arc roots.

What happens in a particular case depends upon the arc-root behavior before current zero and the properties of the electrode vapor. Since the detailed investigation of these effects is difficult under the experimental conditions present in a circuit breaker, it has been carried out under more reproducible conditions in a special arrangement which is operated with dc-current.[8]

3. INVESTIGATION OF THE VAPOR IN FRONT OF THE ROOTS OF A HIGH CURRENT DC-ARC

3.1. Experimental Arrangement

Attempts to observe the arc roots under stationary conditions led to the arrangement shown in Fig. 2. It consists essentially of a rod electrode mounted in a water-cooled collar. The opposing electrode is split into four parts arranged in the form of a ring above the rod. The rod is separated from the ring of electrodes by an electrically floating copper plate with a circular aperture. This arrangement prevents interactions between the mass flows from the anode and cathode. The single electrode vaporizes very rapidly, so that its tip can only be held at a nominally fixed position by a system which continually monitors and moves the electrode. For this purpose a laser beam is directed onto the tip of the electrode; as the electrode is eroded, the laser beam falls on a photodiode which then generates the signal used to move the electrode.

To avoid complications due to electrode melting, graphite rather than WCu was chosen as the electrode material for the detailed investigations. The measurements were carried out on the cathode since on thin graphite anodes at high currents the arc roots have been shown to be highly unstable.

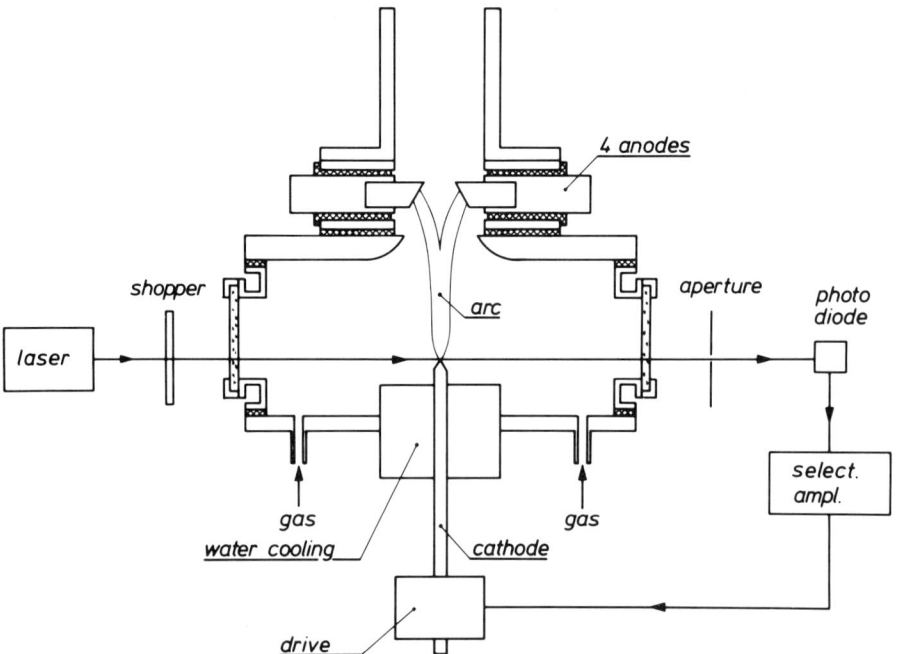

Fig. 2. Experimental arrangement consisting of the discharge vessel and a system which continually monitors and moves the electrode. A detailed description is given in[8].

The experiments were performed in SF_6, N_2 and Ar. The last was chosen since it is easier to achieve a stationary arc root with argon. However, in the case of strong vaporization, the root of the dc-arc in argon is also unstable unless special precautions are taken.

3.2. Unstable Arc Root

Figure 3 is taken from a section of film obtained using high-speed photography and shows the unstable root of a dc-arc burning on a graphite cathode in argon. The arc current was 1000 A and the exposure rate was 10,000 frames per second. The area from which current is extracted can be seen on the cathode surface. In addition a broad trail of green vapor spatially separated from the arc is observed. The area of current transfer and the vapor trail move around together on the surface of the cathode with a frequency of 1.5 kHz. Also visible in the photographs is a needle-shaped jet (also green) which is ejected from a small crater.

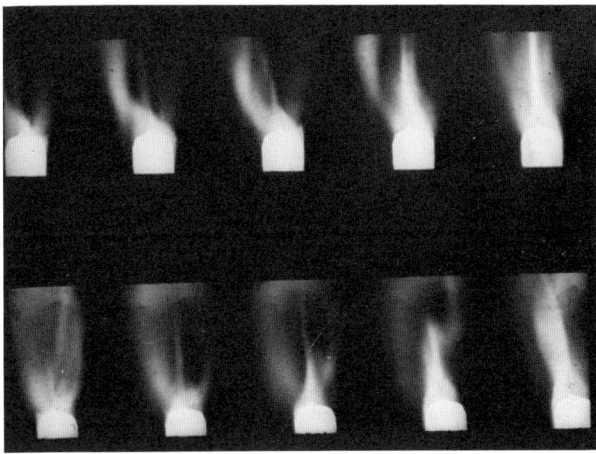

Fig. 3. Short-time exposures of a cathodic arc root on graphite. The current is 1000 A. Time between frames 100 µs.[8]

By taking care to achieve a symmetrical arrangement, it is possible to investigate various phenomena on the cathode under stationary conditions. The form of the arc root can, moreover, be influenced by the degree of electrode cooling which depends upon the length of the cathode projecting from the cooled collar.

3.3. Arc Root with Strong Cooling

An example of a strongly cooled arc root is shown in Fig. 4. A cathode 6.15 mm in diameter carrying a current of 1000 A extends 2 mm above the cooled collar. From the photograph it can be seen that with intensified cooling the green vapor vanishes

Fig. 4. Strongly cooled root of a 1000 A arc on a graphite cathode 6.15 mm in diameter extending 2 mm beyond the cooled collar.[8]

and is replaced by a darker core consisting of atomic carbon, as is confirmed by the UV-spectrum of the arc root.

We have measured the radial temperature distribution using two atomic and two ionic carbon lines. In addition, we have determined the densities of atomic and ionic carbon from the absolute line intensities. A detailed description of the method is given elsewhere.[9] The results of the temperature measurements are shown in Fig. 5 for

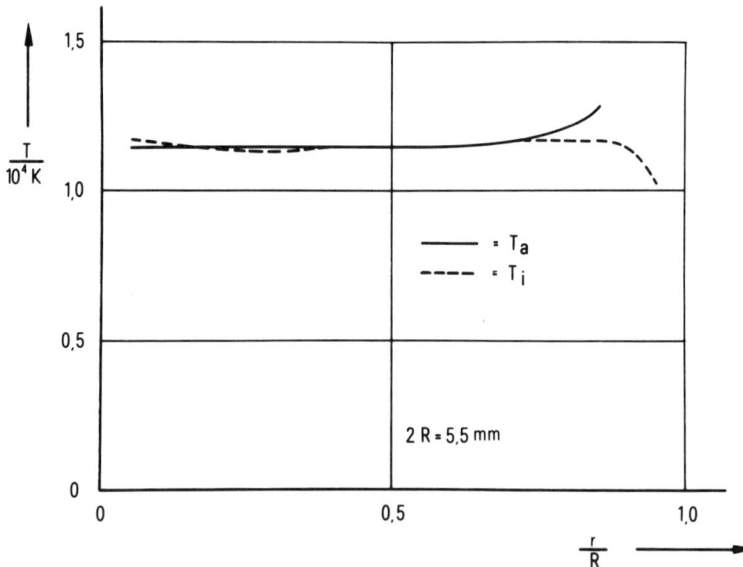

Fig. 5. Excitation temperature of the C-atoms $T_a(r)$ and of the C-ions $T_i(r)$ 1 mm from the cathode surface.[9]

a section 1 mm from the electrode tip. Values determined from the atomic lines are indicated as $T_a(r)$ and from the ionic lines by $T_i(r)$. The temperature of the carbon plasma immediately in front of the cathode lies between 11,000 and 12,000 K. Particle densities are given in Fig. 6 for the same section, as determined from the spectroscopic measurements. The difference between n_i and n_e must be attributed to errors of measurement. The curve for $n_{i1} = n_e$ represents a corrected value of the charge carrier density.

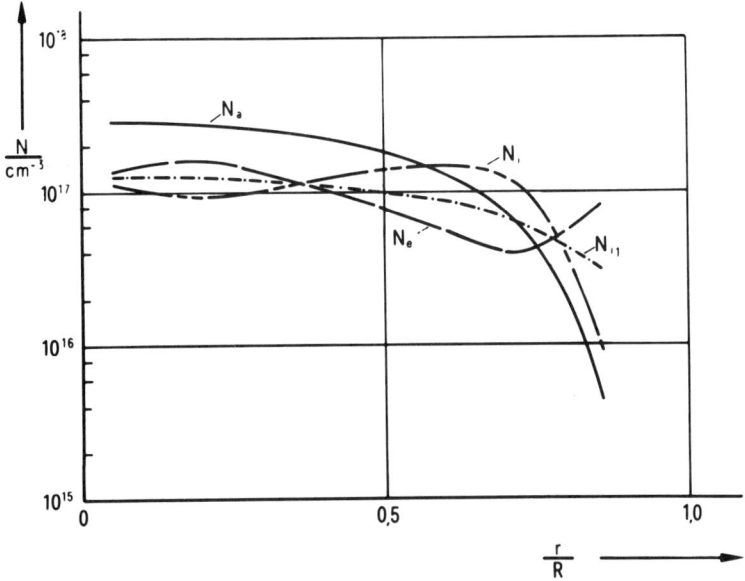

Fig. 6. Densities of the C-atoms $N_a(r)$, the C-ions $N_i(r)$ and the electrons $N_e(r)$ 1 mm from the cathode. N_{i1} is calculated from N_a and T_a.[9] (In the text particle densities denoted by n).

The magnitudes of the particle densities in front of the cathode are found to be:

$$n_a = 3 \cdot 10^{17} \text{ cm}^{-3}$$
$$n_{i1} = 1.25 \cdot 10^{17} \text{ cm}^{-3}$$

It thus follows that the partial pressure of the carbon plasma is

$$p = (n_a + n_i + n_e) kT \simeq 1 \text{ bar},$$

which is comparable with the pressure of the filling gas within the discharge vessel.

The measurements demonstrate that in front of the arc root, the plasma is composed almost entirely of atomic carbon with a degree of ionization of 30 %. The plasma extends up to the cathode surface. Thus the arc root corresponds to the area of current transfer mentioned above in the discussion of the unstable arc root.

The measured erosion rate and particle densities were inserted into the equation for the mass balance and used to estimate an average plasma velocity in the vicinity of the cathode. As the lower curve in Fig. 7 shows, the velocity initially increases with increasing distance from the cathode, but quickly reaches a constant value of approximately 300 m/s. The increase of velocity in the region in front of the cathode can be attributed to magnetic acceleration. In the divergent current-density distribution in this region the Lorentz force has an axial component $j_r B_\phi$. The acceleration is evident from the upper curve which was determined using the momentum balance given by Maecker.[10] The result for the plasma velocity is supported by the form of the dark carbon jet which originates at the cathode. Over the first few millimeters the cross section of the jet contracts and then becomes constant.

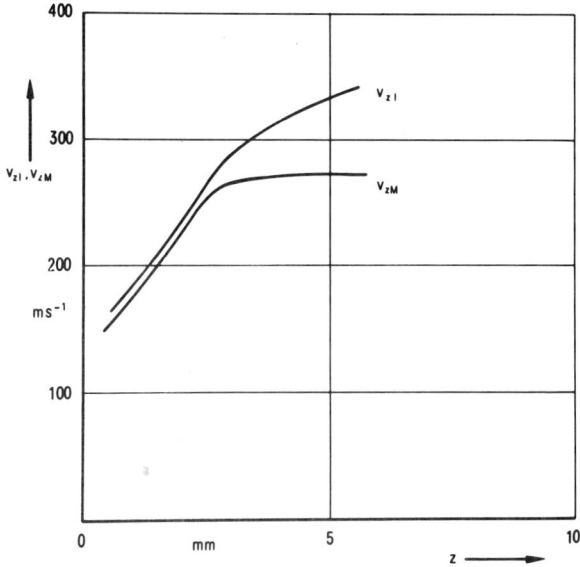

Fig. 7. Dependence of the plasma velocity upon the distance from the cathode. v_{zM} is obtained from a mass balance, v_{zI} from a momentum balance.[9]

3.4. Arc Root with Strong Vaporization from the Surface

With the appropriate choice of experimental conditions, it is possible to produce an arc root with strong vaporization from the electrode surface under stationary conditions. In Fig. 8 a cathode of 6.15 mm diameter carrying 1000 A is again shown, but now the electrode tip extending 10 mm beyond the cooled collar, i.e. it is only weak-

Fig. 8. Weakly cooled arc root of a 1000 A arc with a green axially symmetric vapor jet on a graphite cathode 6.15 mm in diameter extending 10 mm beyond the cooled collar.[8]

ly cooled. In the center of the arc a broad green jet enveloping the electrode tip can be seen. The radiation from the jet is much brighter than that of the surrounding arc.

However, an arc root with an axially symmetric vapor jet is not very stable. Even small disturbances to the rotational symmetry cause the separation of the arc and the vapor. The moment of separation is illustrated in Fig. 9. Subsequently the arc root moves around the cathode accompanied by a vapor trail as has already been illustrated in Fig. 3.

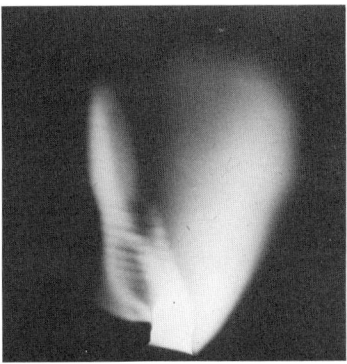

Fig. 9. Separation of arc and vapor jet in front of a graphite cathode.[8]

We have also recorded spectra from the arc root with strong vaporization under stationary conditions. The spectrogram shows that in the visible region the radiation from the carbon vapor originates for the most part from the C_2-molecular band spectrum which is also responsible for the green color. On one of the C_2-bands we have measured the (rotational) temperature as a function of the distance from the cathode.[11]

Figure 10 shows a vapor jet together with the temperature as a function of the distance from the cathode. As expected the temperature immediately in front of the cathode is slightly above the sublimation point of carbon. Surprisingly, however, with increasing distance from the cathode there is only a slight increase in temperature. The relatively weak heating of the vapor shows that only a small fraction of the total current flows through the vapor jet. It may be concluded from the measurements that the vapor is relatively cold, almost current free and, in the rotationally symmetrical case, enveloped by the current-carrying arc plasma.

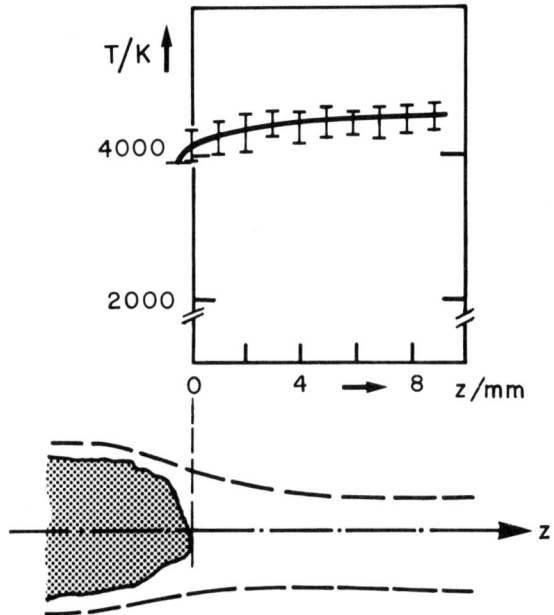

Fig. 10. Vapor temperature as a function of the distance from the cathode, together with a picture of the vapor jet.[11]

From the erosion rate per unit area A and the measured vapor temperature ~4500 K, a mean velocity for the vapor within the jet can be estimated as

$$v = \frac{A}{\rho_{C_2}} = 45 \text{ m/s} \tag{1}$$

This is in reasonable agreement with the value 35 m/s measured for carbon particles which had become detached from the electrode surface. Although the values vary somewhat from experiment to experiment they have the expected order of magnitude. For metal electrodes the vapor velocities would probably be higher.

3.5. Vapor Eruptions from the Arc Root

On a weakly cooled cathode it is frequently possible to observe an arc root resembling that on a strongly cooled cathode, i.e. with very little vapor emission from the surface. Instead, sudden eruptions of green vapor occur from within the interior of the electrode: an example is shown in Fig. 11. These eruptions can last for several seconds and bring about the breakup of the electrode tip after which they cease.

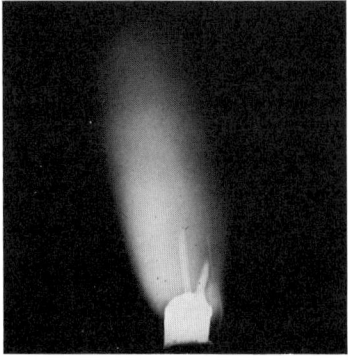

Fig. 11. Needle-shaped jets originating from vapor eruptions from within the interior of the graphite electrode.[8]

In Fig. 12 a magnified photograph of the surface and breaking area of such an electrode tip is reproduced. Isolated holes can be seen in the rough surface crust

Fig. 12. Microscopic exposure of a broken-up graphite cathode tip showing electrode surface with holes.

which remains on such electrode tips. These holes lead into the interior of the electrode and end in cavities which are laid bare by the breaking area shown in Fig. 12. The eruptions of vapor are clearly supplied from these cavities.

Figure 13 also shows an eruption of vapor, this time from the electrode shaft outside the current-transfer region. This picture clearly demonstrates that the vapor eruptions are not directly correlated with the current transfer to the electrode, but are only a consequence of the heating of the electrode material and vaporization from within the bulk of the material.

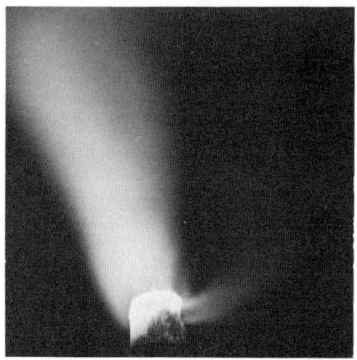

Fig. 13. Vapor eruption from the electrode shaft outside the arc root.[8]

The green eruptions also emit the band spectrum of the C_2 molecule. The occurrence of the molecular spectrum leads to the conclusion that the vapor eruptions are cold in comparison with the arc plasma and thus the electrical conductivity small. This is to be expected considering the origin of the eruptions.

Except to say that it probably varies over a wide range, it is difficult to make any definite statements about the velocity of the vapor ejected by the vapor eruptions.

4. INTERPRETATION OF ARC ROOT BEHAVIOR

Very effective heating of the arc roots is caused by the charge carriers. At the anode, energy is released equivalent to the work-function of the electrons which transport the current to the positive electrode. At the cathode the ionization energy of the ions becomes free: these ions carry a remarkable part of the total current immediately in front of the negative electrode. The ion current is generated in a plasma sheet at the cathodic end of the arc column.

For the special case investigated here in detail both the charge carrier density and the power balance results show that the ion current increases from a negligible amount in the arc column to 40 % immediately in front of the cathode. The ion current consists of vaporized electrode material, which is partially brought back to the cathode by the current. This is the case for a strongly cooled cathode with a hot plasma in front of it and a relative weak vaporization.

The vapor cools the plasma in front of the electrodes. The electrical conductivity is decreased by the vapor, and the ion emission of the plasma in front of the cathode is suppressed. The local vaporization, and hence the cooling of the plasma layer, can become so great that the ion current at that point is reduced to zero. The back-scattering of the vapor to the cathode by the ion current ceases with the result that the effective vaporization increases momentarily. This process is characterized by the transition from the first arc-root form shown in Fig. 4 to the second form in Fig. 8. The cold vapor is observed as a trail or jet originating from the arc root. The transition between the two forms is accompanied by a considerable increase of the vaporization rate, as can be seen by comparing the lower and upper curves in Fig. 14.

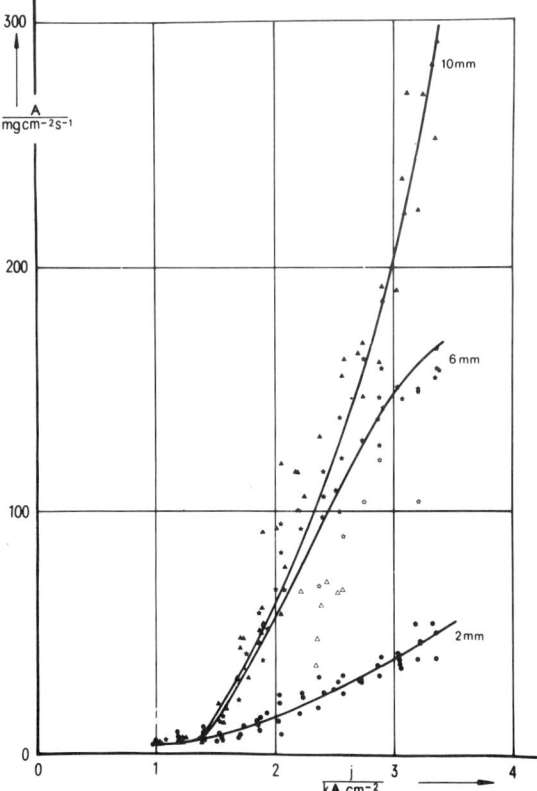

Fig. 14. Erosion of graphite cathodes with a diameter of 6.15 mm for different heights h of the electrode tip above the cooled collar. The open symbols for 6 and 10 mm height indicate measurements of arc roots without vapor jet.[8]

As a result of the low temperature of the vapor, almost no current flows in the region in front of the electrode area from which the pronounced vaporization occurs. To enable current transfer between the arc column and the electrode, the arc root must therefore shift to a part of the electrode where less vaporization occurs. In general this destroys the rotational symmetry of the arc which gives rise to magnetic forces, the effects of which combine with those of the vapor in driving the root from its original site. The net effect is that the arc root is driven away from its own vapor. On a rod shaped electrode this results in a rotational movement of the arc root.

A high-speed film was made of this behavior for an arc produced in nitrogen on a rod-shaped graphite cathode 6 mm in diameter carrying a current of 1000 A. From the film it could be seen that the vapor destabilizes the arc root: cold vapor arising on the edge of the arc root grasps like a hand under the root and pushes it aside.

The arc root behavior observed on graphite cathodes is even more pronounced on metal electrodes. The reason is the lower enthalpy of vaporization which is 4.8 kJ/g for copper at its boiling point at atmospheric pressure, compared with 30 kJ/g for graphite at its sublimation point.

Two single exposures taken from a film of the behavior of the arc roots on copper tungsten are shown in Fig. 15, one for the early phase of the current half wave and

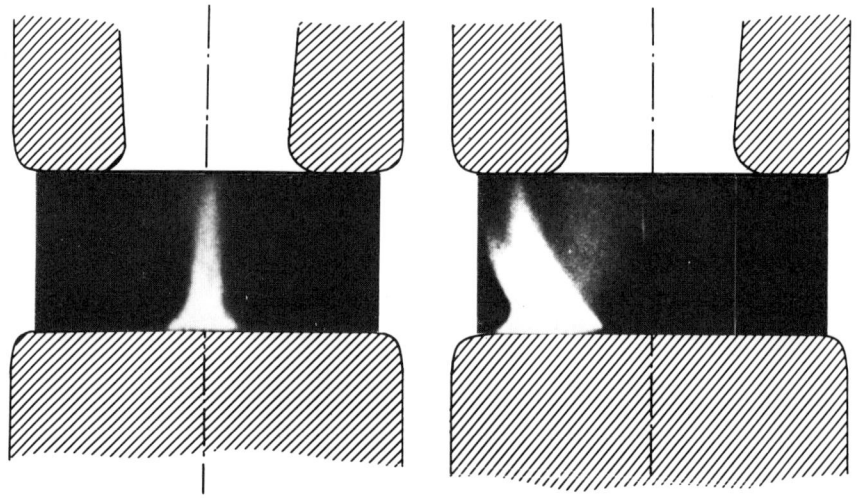

Fig. 15. Arc root of an ac-discharge on a WCu cathode in a SF_6 flow with a peak current of 10 kA. Distance between the electrodes is 15 mm.[8]
Left: the arc envelopes the green vapor jet.
Right: arc and vapor jet have separated from one another.

the other for a later phase. In the picture on the left, for the early phase, the vapor is still enveloped by the SF_6-arc. In the other picture, vapor and arc have already been separated from another. Figure 16 shows measurements made in Liverpool[1] carried out at a cross section of a 10 kA arc 10 mm from the tip of a copper

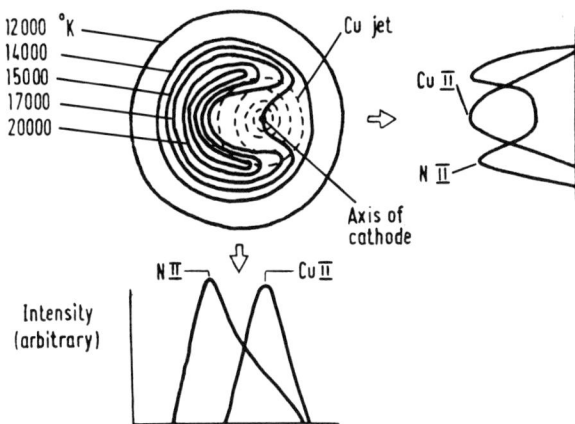

Fig. 16. Isotherms 1 cm above the cathode of a 10 kA Cu-air arc. Case of non-coincident vapor and plasma.[1]

cathode. From these measurements it is seen that even at this distance from the cathode, the copper vapor is colder than the nitrogen plasma, which is pushed aside by the vapor.

CONCLUSIONS

Contamination of the quenching medium by metal vapor produces a strong increase in electrical conductivity. Accordingly, a lowering is to be expected of the steepness of the recovery voltage which can be held. The dielectric strength of the gap can also be reduced as a result of the lower ionization potential of the vapor and the protrusions left on the electrode surfaces by melting in the arc roots. Finally, the vapor can result in the arc root moving against the external forces which should drive it as far as possible into the nozzles before current zero.

Our detailed investigations have shown that the vapor is relatively cold and not necessarily mixed with the arc plasma. In addition, contamination can be avoided, if an electrode material is used with an ionization energy which in the case of SF_6, for example, is higher than that of sulfur.

Finally, if a material is selected with a high enthalpy of vaporization an arc root on the cathode can be achieved with a vaporization considerably reduced by the ion current. In cases where metal cannot be substituted, for example for contact fingers, the problem must be overcome by appropriate design.

REFERENCES

1. H. Edels, Properties of the high pressure ultra-high-current arc, in Proc. XI Int. Conf. Ion. Gases, Invited papers Prague (1973)
2. D. R. Airy, R. E. Kinsinger, P. H. Richards and H. D. Swift, IEEE Trans. PAS 95 (1976) 1
3. P. J. Shayler and M. T. C. Fang, The transport and thermodynamic properties of a copper-nitrogen mixture. Report ULAP-T 45 of the University of Liverpool, Department of Electrical Engineering and Electronics (December 1976)
4. J. Ruffer, Siemens AG, Forschungslaboratorien Erlangen, Germany, private communication
5. W. Hermann, U. Kogelschatz, L. Niemeyer, K. Ragaller and E. Schade, IEEE Trans. PAS 95 (1976) 1165
6. D. W. Branston and J. Mentel, XIII Int. Conf. Phen. Ionized Gases, Berlin (1977) 525-526
7. W. Hermann and K. Ragaller, IEEE Trans. PAS 96 (1977) 1546
8. J. Mentel, Appl. Phys. 14 (1977) 269
9. J. Mentel, Appl. Phys. 14 (1977) 361
10. H. Maecker, Z. Physik 141 (1955) 198
11. J. Mentel, Appl. Phys., in press

DISCUSSION
(Chairman: D. M. Benenson, State University of New York at Buffalo)

K. Ragaller (Brown Boveri)

I would like to relate these results to an important question for circuit-breaker applications: Is the interruption a statistical process? We have found in our experiments that under well-defined, reproducible conditions the scatter in the interrupting capability is very small. Perhaps Mr. Frind can comment on this problem too. Also, from our theoretical understanding, there is no effect which could produce a statistical variation. Of course, in a real circuit breaker we don't have the ideal conditions of a model experiment. A major difference between a real breaker and a model experiment is the arc roots which are formed during contact separation. As was shown in this lecture, the behavior of these arc roots is not at all reproducible. One way to avoid statistical effects due to the arc roots, such as result from metal-vapor jets, is to drive them downstream into the two nozzles of a double-nozzle configuration as quickly as possible. The arc roots can then no longer influence the current-interruption zone. In order to achieve this effectively it is, of course, necessary to understand the physical phenomena associated with arc roots.

I want to mention here another effect which can also produce a large scatter of the interrupting capability of a real circuit breaker. Figure 1 shows a Schlieren picture of the cold-gas flow in an SF_6-circuit breaker in the region between the two nozzles.[1] The dark helically-shaped filament is the visualization of a vortex in the flow which is produced by a asymmetry of the inflow. The vortex is unstable and therefore changes the flow and pressure field in a non-reproducible way. It is my impression that people who develop statistical theories to explain the behavior of circuit breakers actually have such effects as these occurring in their circuit breakers. If one carefully avoids these effects, the performance of a circuit breaker is very

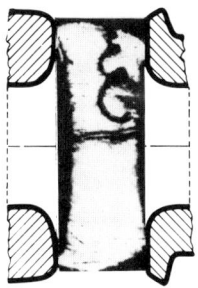

Fig. 1. Schlieren-picture of the region between the two nozzles of a double-nozzle SF_6 breaker showing strong vortex filaments.[1]

reproducible. Although a statistical theory would be necessary if effects existed which obeyed certain statistical laws, there is in fact no evidence for their existence.

J. Mentel
In addition to this remark one should add that the choice of materials is also of importance. As the ionization energy of carbon is higher than that of SF_6, the carbon vapor does not deteriorate the quenching performance of SF_6. Since contact fingers must be made of copper, one must seek a design where the metal vapor does not enter the arc plasma.

G. Frind (General Electric)
I have not yet investigated the double-nozzle configuration, although we are currently building one and will be able to comment on this later. Another comment is that the single-flow device is, of course, more economical than a double-nozzle breaker. For an application where you don't need the double flow, you would make a single flow and then, of course, the metal vapor produced at the upstream electrode is transported into the gap, giving you some statistic variations. It's a matter of cost.

REFERENCES

1. U. Kogelschatz, E. Schade and K. D. Schmidt, Brown Boveri Review 61 (1974)

THEORETICAL MODELS FOR THE ARC IN THE CURRENT ZERO REGIME

B. W. SWANSON
Westinghouse Research and Development Center
Pittsburgh, USA

SUMMARY

This paper describes the current state-of-the-art of nozzle arc modeling. It begins with a discussion of arc turbulence and radiation followed by a review of three arc models that have the most potential for design application. The merits and limitations of each model are discussed and recommendations are made for further model development. The discussion shows the importance of arc turbulence and the practicality of modeling the arc as a circuit element. A comparison of arc models gives insight into the influence of gas properties on arc interruption and the explanation of observed arc phenomena. The potential for design application is demonstrated by analyzing the effects of normal shock waves on arc interruption to explain the known variation of interrupting ability with nozzle pressure ratio.

INTRODUCTION

For over thirty years the classical equations of Cassie and Mayr have been used in circuit breaker work. In 1939 Cassie[1] developed a differential equation for arc resistance by assuming an arc column in which the arc temperature is constant and the arc area varies with the current. In this arc model, energy is carried away by the gas blast and this convection cooling is proportional to the arc area. In 1943 Mayr[2] developed a differential equation for arc resistance by assuming an arc column in which the arc radius is constant and the arc temperature varies with time. In this arc model energy is transferred only by radial conduction.

In 1948 Browne[3] developed a composite arc model in which the Cassie and Mayr equations are used before and after current zero, respectively. In 1958 Browne[4] extended the application of his Cassie-Mayr model to the analysis of thermal reignition in the critical post-current zero energy balance period. Since then Browne has used this composite model extensively in circuit breaker design and analysis. In using it to analyze near-critical tests on model interrupters, Browne found that interrupting ability varies inversely with the square of the current, as also shown by Frind and Rich.[5] In a recent paper,[6] Browne shows the utility of Cassie-Mayr theory by analyzing the effects of capacitance and resistance shunts, and gives practical equations for estimating the amount of shunt capacitance required for short-line fault protection. Frost[7] has extended Browne's work by developing a Cassie-Mayr-Cassie theory and has used it to develop computer programs for getting complete performance and circuit information quickly from high power lab tests.

There is no question that Cassie-Mayr arc models have practical application in circuit breaker development. However, since these models are described by ordinary differential equations, there is a limit to the amount of information that can be obtained from them. In particular they cannot shed light on the physics of arc interruption where such physics is described by the partial differential equations of gas dynamics. Moreover, there are good reasons today which compel us to understand nozzle arc interruption.

The problems of inflation and recession of the past decade and their effects on costs make it imperative that we apply analysis wherever possible to reduce empiricism in design and development.

The growth in short circuit currents of power systems over the past two decades has emphasized the short line fault limitations of gas blast breakers. As is well known, the rate of rise of recovery voltage caused by a line fault is proportional to the fault current. On the other hand, the interrupting ability of gas breakers at high currents varies inversely with the square of the fault current.[5] Equating interrupting ability with circuit stress gives a performance equation involving the cube of the current. Thus, to double the interrupting current of an air breaker from 40 to 80 kA requires an eightfold increase in either the number of breaks or the gas pressure. To uprate an SF_6 breaker over the same current range requires a 4.4 increase in gas pressure. Such breaker modification and redesign is expensive. Although shunt capacitance and resistance may also be used to update performance, these options are also expensive especially at higher voltages. Therefore, to reduce costs, we must optimize breaker designs and this requires a clear understanding of nozzle arc physics.

Another reason for developing advanced arc models involves nozzle clogging. Clogging occurs when the arc expands to fill the nozzle throat. Because of the high plasma temperature and low density, clogging greatly reduces nozzle mass flow and limits the amount of energy carried away by convection. Clogging can blow back hot gas into upstream plenums which can adversely affect bus fault and short-line fault interrupting ability. Clogging is particularly important in the design of puffer breakers which may operate under heavily clogged conditions. To understand this phenomenon we must have arc models which can account for nozzle geometry.

The design engineer today asks questions which cannot be answered by the Cassie-Mayr models. How do we optimize a breaker nozzle? What are the optimum length, diameter and expansion angle? How does clogging affect interruption? What are the effects of shock waves and pressure ratio? To answer such questions we are compelled to develop nozzle arc models based on the partial differential equations of gas dynamics, and this task is not easy.

One difficulty arises from the complex nature of the equations describing nozzle arcs. We must solve the continuity, energy and momentum equations of gas dynamics along with Ohm's Law and an equation of state. A second difficulty arises from the nonlinear character of the thermodynamic and transport properties of interrupting gases. A third difficulty arises from the complex flow fields generated within Laval nozzles with supersonic flow and shock waves. A fourth difficulty arises from stability problems encountered in numerical methods. Another difficulty arises from the uncertainties in modeling arc turbulence and radiation. We can name more. In view of these difficulties, the main objective of nozzle arc modeling is to reduce the problem to the simplest form that still contains the essential physics of arc interruption. It is this simplification process that makes arc modeling an art. In this paper we shall discuss the current state-of-the-art, its potential for design application, and what insight we now have into the interruption process. We begin with the modeling of arc turbulence and radiation.

ARC TURBULENCE

Although much is known about aerodynamic turbulence, relatively little is known about turbulence in high pressure arcs. Until the recent development of nozzle arc models requiring turbulence information, there has not been a need for extensive arc turbulence studies. There is now a real need for turbulence information, since it appears that arc turbulence controls short-line fault interruption in gas blast interrupters. Until we have better information, we must let our knowledge of aerodynamic turbulence guide us in formulating preliminary arc turbulence models.

In the momentum equation, the transfer of momentum in the y direction normal to the flow is given by the term

$$\partial/\partial y \,[(\eta + \eta_t) \,\partial V_z/\partial y] \tag{1}$$

where η is the molecular viscosity and η_t is the turbulent viscosity defined by the equation

$$\eta_t = \rho \varepsilon_m \tag{2}$$

where ρ is the gas density, and ε_m is the eddy diffusivity of momentum.

Similarly, in the energy equation the transfer of heat in the y direction is given by the term

$$\partial/\partial y \,[(\kappa + \kappa_t) \,\partial T/\partial y] \tag{3}$$

where κ is the thermal conductivity and κ_t is the turbulent thermal conductivity defined by the equation

$$\kappa_t = \rho C_p \varepsilon_H \tag{4}$$

where ε_H is the eddy diffusivity of heat and C_p is the specific heat. The eddy diffusivities of heat and momentum are related by the equation

$$\varepsilon_H = \varepsilon_m / Pr_t \tag{5}$$

where Pr_{rt} is the turbulent Prandtl number generally assumed to equal 1/2. Therefore, to estimate the effects of arc turbulence we must estimate the eddy diffusivity of momentum ε_m. Before doing this we first consider eddy diffusivity equations for the turbulent boundary layer.

According to experimental data,[8] a turbulent boundary layer may be regarded as a composite layer made up of inner and outer regions. In the inner region the eddy diffusivity varies according to the equation

$$(\varepsilon_m)_i = \ell^2 \, \partial u/\partial y \text{ for } y_0 \leq y \leq y_c \tag{6}$$

where y_0 is a small distance from the surface, and ℓ is the Prandtl mixing length which varies linearly with y in this region.

In the outer region of the boundary layer, the velocity is fairly uniform, and the eddy diffusivity varies according to the equation

$$(\varepsilon_m)_0 = \alpha_1 V_e \delta^* \text{ for } y_c \leq y \leq \delta \tag{7}$$

where α_1 is a constant, δ is the boundary layer thickness, V_e is the free stream velocity outside the boundary layer, and δ^* is the displacement thickness which is proportional to δ and varies in the z direction.

From Equations (6) and (7) we see that in the inner region of a turbulent boundary layer, the eddy diffusivity is proportional to a velocity gradient while in the outer region it is proportional to a velocity and some function of z. Furthermore, experimental evidence[9] exists indicating that a fine scale vorticity structure in the viscous wall layer may influence the turbulent inner region.

With this brief background in aerodynamic turbulence, let us now infer some possible properties of arc turbulence. Within the core of a radiation dominated arc the flat temperature profile produces a fairly uniform core velocity and core turbulence may well resemble turbulence in the outer region of the boundary layer where the velocity is also fairly uniform.

However, at the arc boundary, a steep temperature gradient creates steep gradients in density and velocity, and arc boundary turbulence may well resemble turbulence in the inner region of the turbulent boundary layer. The steep density gradient in the arc boundary layer may in itself induce turbulence. Niemeyer and Ragaller [10] have shown that the generation of vorticity at the arc boundary is proportional to $\nabla \rho \times \nabla p$ giving the nozzle arc a unique mechanism for generating vortex layers to produce turbulence.

Thus, it is quite possible that there may exist two regions of arc turbulence similar to the inner and outer regions in a turbulent boundary layer. If this is the case then we can expect arc turbulence effects to vary with current as the flat radiation dominated temperature profile at high current gives way to the parabolic conduction dominated profile at low current.

We now consider an arc turbulence model based on aerodynamic turbulence. From their arc experiments, Niemeyer and Ragaller[10] observed that disturbances at the arc boundary gradually penetrate into the arc until they finally disturb the total arc cross-section. This led Hermann, et al.[13] to assume that arc turbulence resembles free shear turbulence. The results of free shear turbulence are assumed valid, since the low density gas in the arc column will be accelerated to considerably higher velocities than the denser cool gas surrounding the arc and will therefore induce free turbulent mixing of the two streams. In free turbulent shear flows, Prandtl assumed that the eddy diffusivity of momentum can be represented by the equation

$$\varepsilon_m = Cb(z)(V_1 - V_2) \tag{8}$$

where C is a constant, b(z) is the width of the turbulent mixing zone and V_1 and V_2 are the different shear velocities. In turbulent jet mixing the width b of the mixing zone varies linearly with the distance from the first point of contact of the jets. [11,12] Furthermore, Niemeyer and Ragaller found their nozzle arc to be laminar up to the nozzle throat but highly turbulent downstream. Consequently, following Prandtl's hypothesis, Hermann, et al.[13] assume that in a nozzle arc the eddy diffusivity of momentum is given by the equation

$$\varepsilon_m = C_t (z - z_t) \left| V_1 - V_2 \right| \tag{9}$$

where z_t is the distance to the throat, V_1 and V_2 are the core and cool gas velocities and C_t is a turbulence constant. Equation (9) could have been obtained from equation (7) by replacing the displacement thickness δ^* by the mixing zone width b(z) and by replacing the free stream velocity V_e by the difference in gas velocities $V_1 - V_2$.

The turbulence model of Hermann, et al.[13] is used in three nozzle arc models to be discussed later. In two of these models C_t is a constant while in the third model C_t varies with the current. Equation (9) is only one model of arc turbulence. Swanson and Roidt[14] considered a mixing length model based on Equation (6), but this area of mixing length models still remains to be developed. As our knowledge of arc turbulence increases, these first models will be replaced by more advanced models.

Judging from engineering experience, relatively simple turbulence models should be adequate for assessing the importance of arc turbulence in the interruption process and for ultimately making design calculations.

ARC RADIATION

Radiation is the principal energy loss mechanism at the arc center when the current exceeds 1000 A. Tuma and Lowke[15] show that nozzle arc models can be developed using two emission coefficients of radiation, U and U_t, as functions of temperature, pressure and arc radius, assuming the arcs are approximately isothermal. The coefficent U gives the radiation loss at the arc center and is dominated by ultraviolet radiation which is reabsorbed at the arc boundary. The coefficient U_t gives the radiation which is completely lost from the arc and which is dominated by the visible region of the spectrum. Figure 1[16] shows the net emission coefficient ε_N[16] of SF_6

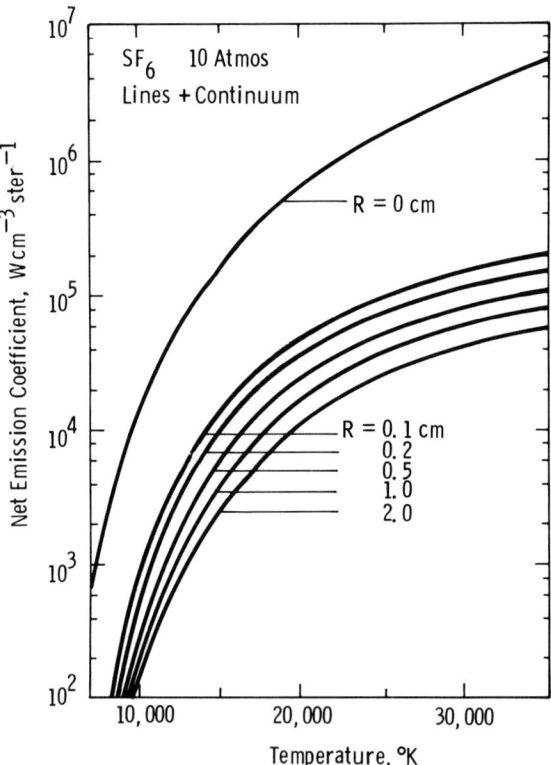

Fig. 1. Calculated net emission coefficient for radiation from SF_6 arc plasmas of various radii at 10 atmospheres.

for line and continuum radiation as a function of temperature and radius for a pressure of 10 atm. From radiation calculations at 1 atm it was found[16] that ε_N is approximately proportional to pressure, consistent with the radiation being dominated by line radiation. Knowing ε_N, the emission coefficient is given by $U = 4\pi\varepsilon_N$. Liebermann and Lowke[16] have compared line and continuum radiation with continuum radiation only. They found that for an arc radius of 2 mm and a temperature of 20.000° K, over 80 % of the radiation is from line radiation, while at 30.000° K, this percentage is over 90 %. Thus line radiation is quite significant in high current arcs.

Hermann and Ragaller[17] use a radiation model that also accounts for the effects of pressure and arc radius. Their average emission coefficient $\bar{U}$ is given by the equation

$$\bar{U} = U_r(T) f_r f_p C_1 \tag{10}$$

where C_1 is a constant, $U_r(T)$ accounts for the effects of temperature, and f_r and f_p account for the effect of radius and pressure, respectively.

The important feature of these radiation models is that they can include the effects of lines and arc radius as well as pressure. Swanson[21] based his nozzle arc model on the radiation information in Fig. 1. At low currents the radiation model can have a significant effect on the nozzle arc temperature distribution as shown in Fig. 2. Here, Swanson compares nozzle arc temperature profiles for a 150 A SF_6 arc for the cases of no radiation, continuum radiation and line and continuum radiation. Note that the difference between no radiation and line and continuum radiation accounts

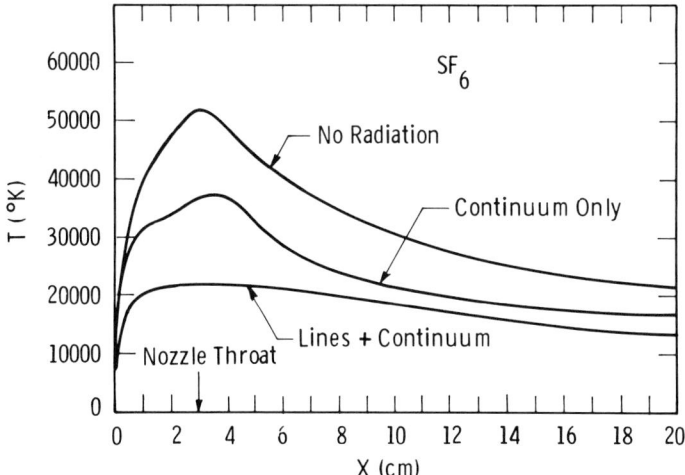

Fig. 2. Effect of radiation on nozzle arc temperature distribution.

for a temperature difference of 30,000° K at the nozzle throat and 8.000° K at the nozzle exit. Although arc radiation can be quite important, even at a low current of a hundred amperes, Fig. 1 shows that once the arc temperature falls below 10.000° K radiation ceases to be important. Thus we would not expect radiation to play a direct role during the critical post-zero energy balance period when the current zero arc temperature is below 10.000° K. Rather we find that radiation can play an indirect role by influencing the arc temperature profile in the initial steady state. Having considered radiation and turbulence models, we now consider their application in nozzle arc models.

NOZZLE ARC MODELS

We now review several nozzle arc models that have potential for design application. This arc model research may be broadly classified into (a) the work of Hermann, Kogelschatz, Niemeyer, Ragaller and Schade;[13,17,18] (b) the work of Lowke, Ludwig, Tuma and El-Akkari;[15,19,20] and (c) the work of Swanson.[21] Each group of researchers has made significant contributions to the development of nozzle arc models. We will first discuss the merits and limitations of each approach and then compare these models and recommend the best features of each model for further model development. We will also consider what insight these models now give us into understanding the interruption process. We begin by first considering some general arc equations which will be useful in this discussion.

Arc Integral Equations

The continuity, momentum and energy equations of gas dynamics and Ohm's Law form the basis of nozzle arc modeling. The following equations apply to a rotationally symmetric arc.

Continuity

$$\frac{\partial \rho}{\partial t} + \frac{\partial}{\partial z}(\rho V_z) + \frac{1}{r}\frac{\partial}{\partial r}(r\rho V_r) = 0 \tag{11}$$

Axial Momentum

$$\rho \frac{\partial V_z}{\partial t} + \rho V_z \frac{\partial V_z}{\partial z} + \rho V_r \frac{\partial V_z}{\partial r} = -\frac{\partial P}{\partial z} + \frac{1}{r}\frac{\partial}{\partial r}\left[(\eta+\eta_t) r \frac{\partial V_z}{\partial r}\right] \tag{12}$$

Energy

$$\rho \frac{\partial h^\circ}{\partial t} + \rho V_z \frac{\partial h^\circ}{\partial z} + \rho V_r \frac{\partial h^\circ}{\partial r} = \sigma E^2 - U + \frac{1}{r}\frac{\partial}{\partial r}\left[(\kappa+\kappa_t) r \frac{\partial T}{\partial r}\right] \tag{13}$$

Ohm's Law

$$I = E \int_0^{r_1} 2\pi r \sigma \, dr \tag{14}$$

In these equations ρ is the gas density, V_z is the axial velocity and V_r is the radial velocity. In Equation (12), P is the gas pressure, η is the molecular viscosity and η_t is the turbulent viscosity defined by Equation (2). In Equation (13) σ is the electrical conductivity, E is the voltage gradient, U is the net emission coefficient, κ is the thermal conductivity, κ_t is the turbulent thermal conductivity defined by Equation (4) and h° is the total enthalpy. Equations (11) through (13) may be integrated radially from r = a to r = b to give the following integral equations.

Continuity

$$\frac{\partial}{\partial t} \int_a^b 2\pi r \rho \, dr + \frac{\partial}{\partial t} \int_a^b 2\pi r \rho V_z \, dr + q(b) - q(a) - \lambda(b) + \lambda(a) = 0 \tag{15}$$

Axial Momentum

$$\frac{\partial}{\partial t} \int_a^b 2\rho r \rho V_z \, dr + \frac{\partial}{\partial z} \int_a^b 2\pi r \rho V_z^2 \, dr - \phi(b) + \phi(a) + q(b)V_z(b) - q(a)V_z(a) =$$

$$- \frac{dP}{dz} \pi(b^2 - a^2) - 2\pi b S(b) + 2\pi a S(a) \tag{16}$$

Energy

$$\frac{\partial}{\partial t} \int_a^b 2\pi r \rho h° \, dr + \frac{\partial}{\partial z} \int_a^b 2\pi r \rho V_z h° \, dr - \psi(b) + \psi(a) + q(b)h°(b) - q(a)h°(a) =$$

$$\int_a^b 2\pi r [\sigma E^2 - U] \, dr - 2\pi b W(b) + 2\pi a W(a) \tag{17}$$

where q(a) and q(b) give the radial mass flow across boundaries a and b respectively and λ, ϕ and ψ are functions of $\partial r/\partial t$, ρ, V_z and h° at these boundaries. The functions S(b) and W(b) give the radial transport of momentum and energy across r = b and are defined by the equations

$$S(b) = -[(\eta + \eta_t) \frac{\partial V_z}{\partial r}]_{r=b} \tag{18}$$

$$W(b) = -[(\kappa+\kappa_t) \frac{\partial T}{\partial r})]_{r=b} \tag{19}$$

Similar equations define $S(a)$ and $W(a)$. These integral equations give us a common basis for discussing nozzle arc models. Herman, et al.[13,17] base their models on Equations (15) through (17) while Lowke, et al.[15,19,20] and Swanson[21] base their models only on Equation (17). On this basis alone we can expect some significant differences between these models. We now consider the theoretical work of Hermann, et al.[13,15]

ARC MODELS OF HERMANN, KOGELSCHATZ, NIEMEYER, RAGALLER AND SCHADE[13,17]

Hermann and Ragaller[17] used Equations (15) through (17) to define a two zone arc model where zone 1 is the arc core extending from $r = 0$ to $r = r_1$ while zone 2 is the surrounding hot gas region extending from r_1 to r_2. The arc core equations are obtained by setting $a = 0$ and $b = r_1$ while the surrounding gas zone equations are obtained by setting $a = r_1$ and $b = r_2$. By neglecting λ, ψ and ϕ and replacing the integrals by appropriate averages, these equations may be put into the following form.

Continuity

$$\frac{\partial}{\partial t} \overline{\rho}_1 A_1 = - \frac{\partial}{\partial t} \overline{(\rho v)}_1 A_1 - q(r_1) \equiv M_1 \tag{20}$$

Axial Momentum

$$\frac{\partial \overline{(\rho v)}_1 A_1}{\partial t} = - \frac{\partial}{\partial z} \overline{(\rho v^2)}_1 A_1 - q(r_1)V_z(r_1) - \frac{dP}{dz} A_1 - 2\pi r_1 S(r_1) \equiv I_1 \tag{21}$$

Energy

$$\frac{\partial}{\partial t} \overline{(\rho h)}_1 A_1 = - \frac{\partial}{\partial z} \overline{(\rho v h)}_1 A_1 - q(r_1) h^\circ(r_1) + A_1 [\overline{\sigma}_1 E^2 - \overline{U}_1] - 2\pi r_1 W(r_1) \equiv E_1 \tag{22}$$

In these equations, $A_1 = \pi r_1^2$ and the average quantities $\overline{(\rho)}_1$ $\overline{(\rho v)}_1$, $\overline{(\rho v^2)}_1$ $\overline{(\rho h)}_1$ and $\overline{(\rho v h)}_1$ are obtained from the general equation

$$\overline{(\rho \omega)}_1 A_1 = \int_0^{r_1} 2\pi r \rho \omega dr \tag{23}$$

by setting ω equal to 1, v, v², h and vh, respectively. Similar equations define the average conductivity $\bar{\sigma}_1$ and emission coefficient $\bar{U}_1$. Equations (20) through (22) apply to the arc zone. A similar set of equations can be written for the surrounding gas zone where $A_2 = \pi(r_2^2 - r_1^2)$. Equations (20) through (22) in zone 1 and their counterparts in zone 2 define a general dynamic nozzle arc model that contains all three arc models discussed in this paper.

Hermann, et al. developed a steady state arc model from these equations by setting M_1, I_1 and E_1 equal to zero. The steady state equations for the two zones can be transformed into a set of six ordinary differential equations in the six unkown quantities.

$$\frac{d\bar{T}_1}{dz}, \frac{d\bar{V}_1}{dz}, \frac{d\bar{V}_2}{dz}, \frac{r_1}{dz}, \frac{dP}{dz}, q \qquad (24)$$

This set of equations is difficult to solve near the nozzle throat because of the singularity in transition from subsonic to supersonic flow. This difficulty is overcome by an iterative method involving the pressure gradient and nozzle contour. These equations were solved for a Laval nozzle with a pressure ratio large enough to ensure supersonic flow without internal shocks. Typical comparisons of theory and experiment are shown in Figs. 3 and 4.[13] Figure 3 shows the nozzle contour and compares calculated arc radius with measured values obtained by using streak records and spectroscopic measurements. Figure 4 compares measured and calculated field

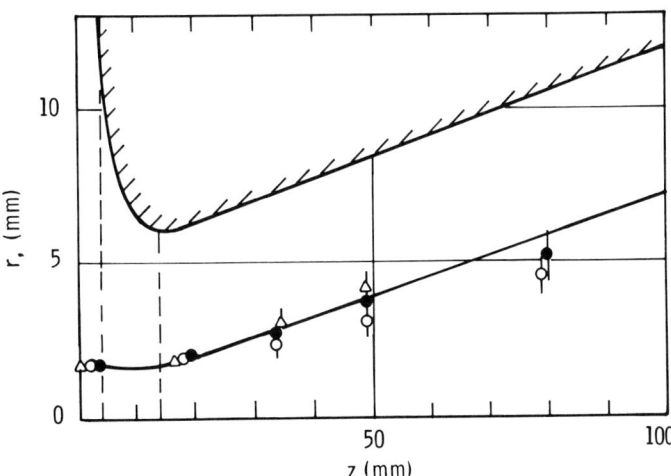

Fig. 3. Nozzle contour R(z) and measured and calculated arc zone radius $r_1(z)$

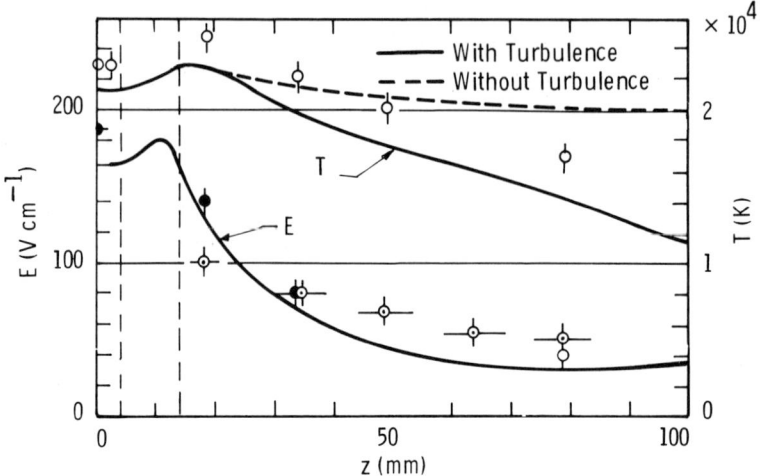

Fig. 4. Measured and calculated electric field strength and nozzle arc temperature for a 2000 A nitrogen arc.

strength, and axial temperature distributions with and without turbulence. The test of any theory lies in its agreement with experiment. The good agreement between theory and experiment illustrated in these figures confirms the validity of the two zone steady arc model.

Hermann and Ragaller[17] developed a dynamic arc model from Equations (22) through (24) and the corresponding equations in the surrounding gas zone. To facilitate solution of these equations, the time derivatives are expanded by differentiating $\overline{(\rho)}$, $\overline{(\rho v)}$, and $\overline{(\rho h)}$ with respect to $\overline{T}$ and solving for $\partial A/\partial t$ and $\partial \overline{T}_1/\partial t$. Equations (20) through (22) then become

$$\frac{\partial \overline{T}_1}{\partial t} = \frac{1}{A_1} [t_1 E_1 + t_2 M_1] \tag{25}$$

$$\frac{\partial A}{\partial t} = a_1 E_1 + a_2 M_1 \tag{26}$$

and

$$\frac{\partial \overline{(\rho v)}_1}{\partial t} = \frac{I_1}{A_1} - \frac{1}{A_1} \overline{(\rho v)}_1 \frac{\partial A_1}{\partial t} \tag{27}$$

where the coefficients t_1, t_2, a_1 and a_2 are functions of average gas properties defined by the equations

$$t_1 = \frac{-\overline{\rho}_1}{D_1} \qquad t_2 = \frac{(\overline{\rho h})_1}{D_1} \tag{28}$$

$$a_1 = \frac{\partial \overline{\rho}_1/\partial T}{D_1} \qquad a_2 = \frac{-\partial(\overline{\rho h})_1/\partial T}{D_1} \tag{29}$$

and

$$D_1 = (\overline{\rho h})_1 \frac{\partial \overline{\rho}_1}{\partial T} - \overline{\rho}_1 \frac{\partial (\overline{\rho h})_1}{\partial T} \tag{30}$$

These equations apply in the arc zone and a similar set of equations can be written for the surrounding gas zone. This two zone arc model is fairly involved and to simplify it, Hermann and Ragaller[17] made several assumptions, the most important being that (a) around current zero the nozzle arc column is cylindrical and the temperature distribution is independent of z; and (b) the average gas velocities in both zones are proportional to z and to time dependent functions $B_1(t)$ and $B_2(t)$. While these assumptions introduce a z dependence into the model, they do not utilize the full potential of the general model to investigate the effects of nozzle geometry.

With these simplifying assumptions, the two zone equations can be reduced to a system of ordinary differential equations in the variables

$$\frac{d\overline{T}_1}{dt}, \frac{dA}{dt}, \frac{d(\overline{B\rho})_1}{dt}, \frac{dA_2}{dt}, \frac{d(\overline{B\rho})_2}{dt} \tag{31}$$

Starting at a current of 1.9 kA, these equations were solved for a linear current ramp to current zero, followed by a linear recovery voltage. Using the same arc model, calculations were made for SF_6 and nitrogen (air).

The calculated variation of interrupting ability with current is shown in Fig. 5. What is significant here is that (a) the variation of interrupting ability with pressure agrees with the data of Frind and Rich;[5] and (b) the model accounts for the superiority of SF_6 over air. This is the first time that a dynamic arc model has been used to quantitatively evaluate the differences in the interrupting abilities of air and SF_6 which is a crucial test for any arc model.

With respect to current dependency, the model predicts that the interrupting ability varies as I^{-n} where the exponent n differs for air and SF_6 and is not equal to the value of 2 obtained by Frind and Rich[5] for both gases. Nevertheless, the results in Fig. 5 confirm the validity of this dynamic arc model.

In summary, the arc models of Hermann, et al.[13,17] have the following properties: (a) the partial differential equations for the average temperature and areas of

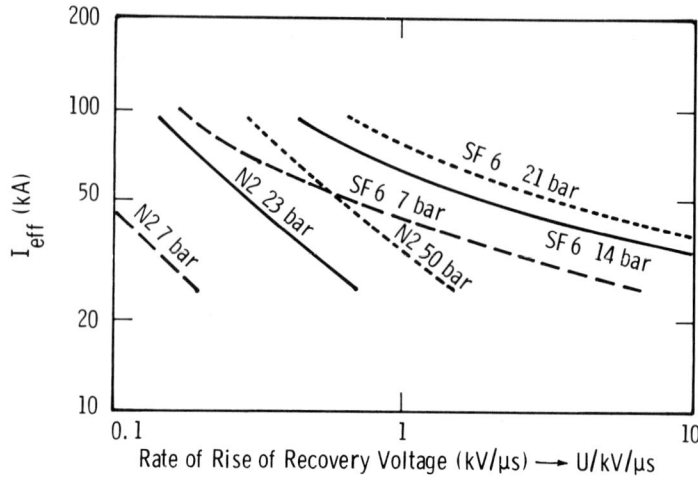

Fig. 5. Thermal extinction limiting curves for air and SF_6 at different pressures.

the arc and surrounding gas zone have been rigorously derived from the integrated continuity and energy equations; (b) in these equations, the coefficients t_1, t_2, a_1 and a_2 are only functions of gas properties for an assumed radial temperature distribution; (c) the steady state arc model agrees well with experimental data; (d) the dynamic arc model accounts for the difference in the interrupting abilities of air and SF_6 and the known variation of interrupting ability with pressure.

The principal limitations of these models are that (a) the present dynamic arc model is described by ordinary differential equations which do not really account for nozzle geometry; (b) the general dynamic arc model that can account for nozzle geometry is involved requiring the solution of six partial differential equations, and (c) an iterative method must be used to handle the transition through Mach 1. We now consider a simpler but less rigorous model by Lowke, et al.[15,19,20]

ARC MODELS OF LOWKE, LUDWIG, TUMA AND EL-AKKARI[15,19,20]

Lowke and Ludwig[19] developed an arc model for convection stabilized arcs that rigorously applies only to a certain current regime. The current must first be large enough so that thermal conduction and turbulence losses do not influence the central arc temperature. The current must also be small enough, or the nozzle large enough, so that the arc area is small compared to the nozzle area and therefore the pressure and velocity distributions are functions only of nozzle shape. The following assumptions are crucial to their model: (a) the arc temperature is isothermal with respect to arc radius; (b) radiation losses are represented by net emission coefficients which are independent of arc radius; and (c) the Mach number of the arc plasma is assumed equal to the Mach number of the surrounding gas and is determined by nozzle geometry.

For the arc temperature Lowke and Ludwig[19] write the energy equation in the form

$$\rho C_p \frac{\partial T}{\partial t} = \sigma E^2 - U - \rho C_p V_z \frac{\partial T}{\partial z} - \frac{4\pi \kappa T}{A} \tag{32}$$

where the last term was added to the isothermal model to simulate the effect of conduction loss by assuming a parabolic temperature profile.

By using a form of Equation (17), Lowke and Ludwig arrive at the following equation for arc area

$$\frac{\partial}{\partial t} (\rho h A) = \rho h \frac{\partial A}{\partial t} = (\sigma E^2 - U_t) A - \frac{\partial}{\partial z} (\rho h V_z A) \tag{33}$$

In developing this equation the assumption is made that ρh is a constant which is a reasonable assumption for the high temperature range for which the model was developed. However, this assumption is invalid in the lower temperature range near current zero, and therefore Equation (33) is not rigorous at current zero. To complete the model, the voltage gradient is given by Ohm's Law and the plasma velocity V_z ist obtained from the equation

$$V_z = M V_s \tag{34}$$

where M is the Mach number of the cold gas and V_s is the sonic velocity.

To avoid the difficulty at Mach 1 encountered by Hermann, et al.,[13] Lowke and Ludwig[19] assume that the nozzle Mach number and pressure distributions are defined by the following isentropic relationships.

$$\left(\frac{Q}{Q^*}\right)^2 = \frac{1}{M^2} \left[\frac{2}{\gamma+1} \left(1 + \frac{\gamma-1}{2} M^2\right)\right]^{\frac{\gamma+1}{\gamma-1}} \tag{35}$$

and

$$\frac{P_0}{P} = \left[1 + \frac{1}{2}(\gamma-1) M^2\right]^{\frac{\gamma}{\gamma-1}} \tag{36}$$

where Q is the nozzle cross-sectional area, Q* is the throat area, and P_0 is the tank pressure. Equations (35) and (36) define their adiabatic approximation. By analyzing the steady state forms of Equations (32) and (33), Lowke and Ludwig establish some interesting properties of nozzle arcs.[19]

Tuma and Lowke[15] applied the Lowke-Ludwig model to analyze the data of Hermann, et al. Besides the adiabatic approximation, they also consider clogging and isothermal approximations for pressure and Mach number. At intermediate currents, the arc area can be a significant fraction of the nozzle area. Since the mass flow through the arc area A is negligible compared to that through the surrounding area Q-A, the Mach number and pressure can be determined from Equations (35) and (36) by replacing Q by Q-A. This modification defines their clogging approximation.

In the adiabatic and clogging approximations it is assumed that there is no heating of the gas surrounding the arc. Therefore, because of adiabatic expansion, the surrounding gas temperature can fall below the freezing temperature at the nozzle exit. In reality the gas surrounding the arc is heated by radiation. It is therefore more realistic to consider an isothermal approximation where the cooling by expansion is offset by the heat received from the arc. Making the isothermal approximation the Mach number and pressure are found from the equations[15]

$$M/M^* \exp(-M^2/2M^{*2}) = 0.61 \ (Q^*-A^*)/(Q-A) \tag{37}$$

and

$$P(z) = P_0 \exp(-M^2/2M^{*2}) \tag{38}$$

where M* is the Mach number in the nozzle throat. Using the adiabatic, clogging and isothermal approximations, Tuma and Lowke calculated the steady state arc radius, temperature, electric field, pressure and velocity for a 2000 A nitrogen arc in the nozzle used in the Brown Boveri experiments. The calculations of all arc parameters are in good agreement with the experimental data as illustrated in Fig. 6 for electric field and temperature.

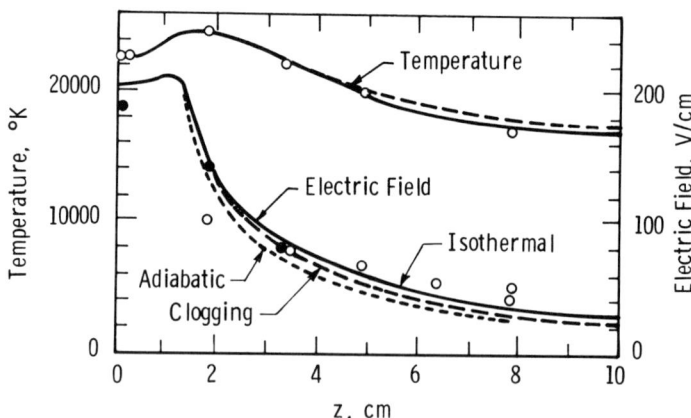

Fig. 6. Theoretical temperature and electric field distributions for a 2000 A arc in nitrogen using different approximations in calculating the pressure distribution.

El-Akkari and Tuma[20] extended the Tuma-Lowke model and applied it to analyze the transient arc behavior. In their model the arc temperature is defined by Equation (32) with κ replaced by $\kappa_{eff} = \kappa + \kappa_t$ where κ_t is defined by the Equations (4), (5) and (9), used by Hermann, et al. To calculate arc area they added a term to Equation (33) giving the equation

$$\rho h \frac{\partial A}{\partial t} = \sigma E^2 A - \frac{\partial}{\partial t}(A\rho h V_z) - U_t A - 4\pi \kappa_{eff} T \tag{39}$$

The last term was added to estimate the effects of turbulent conduction. The arc model equations are solved for the conditions of the Brown Boveri experiment and the model calculations agree well with data. Figure 7 shows the transient variation of

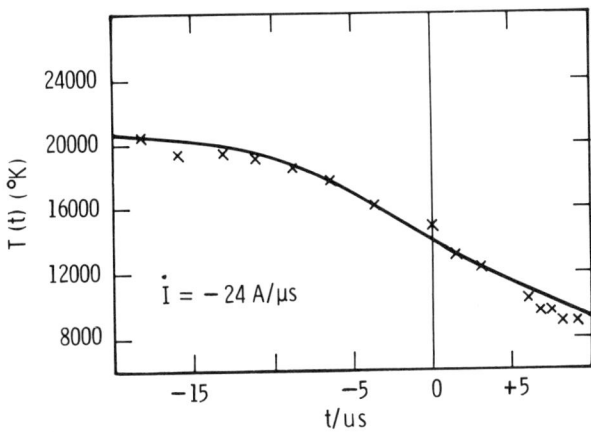

Fig. 7. Comparison of experimental (crosses) and theoretical (solid curve) values for the decay of the arc temperature.

arc temperature for a point before the nozzle entrance showing good agreement with the data crosses. Figure 8[20] shows the calculated arc voltage for $dI/dt = -24$ A/µs which is also in good agreement with experiment. The significance of arc voltage will be discussed later.

A significant result from these model calculations is the importance of arc turbulence on interrupting ability. Figure 9 shows the axial profiles of arc temperature, with and without turbulence, as a function of current.

At 2000 A turbulence has a negligible effect on arc temperature. At 100 A turbulence makes a difference of 4000° K in the nozzle exit arc temperature, while at current zero, turbulence makes a difference of over 6000° K. With turbulence the arc cools down to a minimum temperature of 7000° K approaching the Mayr temperature

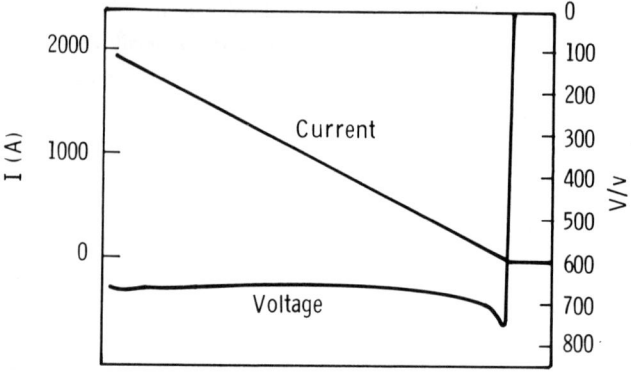

Fig. 8. Calculated arc voltage and arc current when the arc current decays at 24 A/μs from 2000 A to zero current.

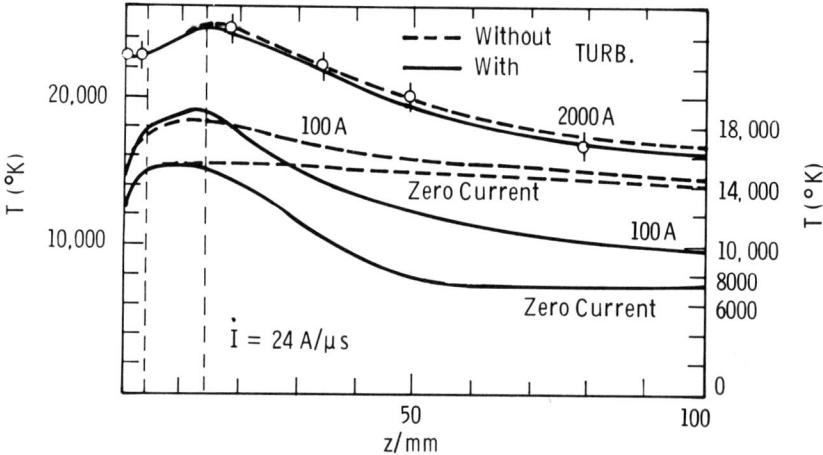

Fig. 9. The axial profiles of the arc temperature as a function of current, with and without turbulence effects.

range where electrical conductivity varies exponentially with temperature. Without turbulence, the minimum arc temperature at current zero is 14,000° K. These model calculations indicate that arc turbulence in supersonic flow is the principal factor that controls arc interruption in a circuit breaker nozzle.

In summary, the arc models of Lowke, Ludwig, Tuma and El-Akkari have the following properties: (a) the models account for nozzle geometry; (b) the models are simpler than the general model of Hermann, et al.;[13,17] (c) the pressure and velocity distributions are calculated from analytic formulas involving nozzle shape thereby avoiding the singularity in the transition from subsonic to supersonic flow; (d) the steady-state Tuma-Lowke[15] model calculations agree with the nitrogen data of Hermann, et al; (e) the dynamic El-Akkari-Tuma[20] model calculations also agree well with Hermann's data. The principal limitations of these models are that (a) the model equation for arc area is not rigorous in the vicinity of current zero and lacks the a_2 term of Hermann's model; and (b) the energy equation is only rigorous for an isothermal model and lacks the t_2 term of Hermann's model. Consequently, the El-Akkari-Tuma[20] model may not be accurate in comparing the interrupting abilities of air and SF_6. This point will be discussed later. We now consider an equivalent arc model developed by Swanson.[21]

SWANSON'S ARC MODEL[21]

Swanson[21] has developed a nozzle arc model using only the energy equation (13). Integrating this equation from $r = 0$ to the arc radius $R(z,t)$ yields the equation

$$\frac{\partial}{\partial t} \int_0^{R(z,t)} [F-F_e] r dr + \frac{\partial}{\partial z} \int_0^{R(z,t)} \rho V(h-h_e) r dr + N(z,t,R)R^2 = I^2/(4\pi R^2 \int_0^1 \sigma \eta d\eta) \qquad (40)$$

where F is the energy density defined as $\int_0^T \rho C_p dT$, F_e and h_e are the energy density and enthalpy at the arc boundary and $N(z,t,R)$ accounts for energy losses by radiation and turbulent conduction. Equation (40) is fully equivalent to Equation (17).

During the high current portion of a half cycle, the arc temperature is essentially constant and the arc radius varies with the current. For this condition, Equation (40) can be reduced to the equation

$$\frac{\partial R^2}{\partial t} = \frac{-R^2}{\theta(z,t)} + \frac{C_1 I^2}{\tilde{\sigma}\psi} \qquad (41)$$

In this equation $\theta(z,t)$ is a total cooling time constant defined as

$$\frac{1}{\theta(z,t)} = \frac{1}{\theta(z,t)} + \frac{1}{\theta_w(z,t)} + \frac{1}{\theta_c(z,t)} \qquad (42)$$

where θ is the diffusion time constant, θ_w is the radiation time constant and θ_c is a convection time constant. These cooling time constants are in turn defined by the equations

$$\theta = R^2(z,t)/(\alpha(1+\beta)\delta_1^2) \tag{43}$$

$$\theta_w = (\tilde{F}-F_e)/\tilde{U} \tag{44}$$

and

$$\theta_c = \frac{U\psi}{\partial\psi/\partial z} \tag{45}$$

where

$$\upsilon(z,t) = (\tilde{F}-F_e)/(\tilde{\rho}\tilde{V}(\tilde{h}-h_e)) \tag{46}$$

and

$$\psi(z,t) = \tilde{\rho}\tilde{V}(\tilde{h}-h_e)R^2 \tag{47}$$

The tilda sign over a variable denotes its value at the arc center. In these equations θ, θ_c, θ_w, N and ψ vary with z and t making θ a function of z and t.

In Equation (42), α is the thermal diffusivity and β is the ratio of turbulent to molecular conduction, given by the equation

$$\beta = \frac{\kappa_t}{\kappa} = \frac{\rho C_p \varepsilon_m}{\kappa P_{rt}} = \frac{C_t(z-z_t)|V_1-V_2|}{\alpha P_{rt}} \tag{48}$$

As the current approaches zero, the radial diffusion losses become controlling and the arc temperature begins to fall. For this condition Equation (40) reduces to the equation

$$\frac{\partial \hat{S}}{\partial t} = -\alpha\tilde{\rho}\tilde{V}\frac{\partial \tilde{h}}{\partial z} - \frac{\tilde{S}}{\theta} - \alpha\tilde{U} + \frac{\alpha C_1 I^2}{\tilde{\sigma}R^4} \tag{49}$$

where $\hat{S} = \tilde{S}-S_e$ and $\tilde{S}$ is the heat flux potential defined as $\int_0^T \kappa dT$. Equations (41) and (49) define Swanson's dynamic nozzle arc model. This model has been derived solely from the integrated energy equation and is based on the assumption that two limiting forms of this equation can be solved simultaneously to define a reasonable arc model around current zero. Using Equations (42) through (46) it is easily shown that Equations (41) and (49) reduce to the forms

$$(F-F_e)\frac{\partial A}{\partial t} = A(\sigma E^2 - U) - \frac{\partial}{\partial z}\rho V_z(h-h_e)A - \delta_1^2 \pi(1+\beta)\tilde{S} \tag{50}$$

and

$$\rho C_p \frac{\partial T}{\partial t} = \sigma E^2 - U - \rho C_p V_z \frac{\partial T}{\partial z} - \delta_1^2 \pi \frac{(1+\beta)\tilde{S}}{A} \tag{51}$$

Comparing Equations (39) and (59) we see that Swanson's area Equation (50) is equivalent to the El-Akkari-Tuma[20] Equation (39) when $\tilde{F}-F_e$ is related to ρh and $\delta_1^2 \pi(1+\beta)\tilde{S}$ is related to $4\pi \kappa_{eff} T$. Similarly a comparison of Equations (32) and (51) shows that Swanson's temperature Equation[21] (51) is equivalent to the Lowke-Ludwig [19] Equation 32. Thus Swanson's arc model is equivalent to that of Lowke, et al. In solving his model Swanson uses the adiabatic approximations for nozzle pressure and Mach number given by Equations (35) and (36).

There are however several important differences between the models of Swanson and Lowke, et al. Swanson assumes a radial Bessel function distribution for $(\tilde{F}-F_e)$, $\tilde{\rho V}(\tilde{h}-h_e)$, $\tilde{\sigma}$, $\tilde{U}$ and $\tilde{S}$ while Lowke[19] assumes an isothermal distribution for all terms except the diffusion term which is evaluated using a parabolic distribution. Swanson also uses the turbulence model of Hermann et al. but permits C_t to vary with current to make the interrupting ability vary as $1/I^2$. A third difference is that Swanson starts his calculations at low currents to reduce computation costs while Lowke, et al.[15,19,20] and Hermann, et al.[13,17] start at 2000 A. Another difference is that Swanson lets θ in Equation (41) be a constant during the linear current ramp. This constraint was necessary in the SF_6 calculations to obtain arc voltage curves that agreed with known experimental data. This point will be discussed later.

Swanson uses his model to study the effects of dI/dt on nozzle arc temperature distribution. He considers various dI/dt's and compares nozzle arc temperature distributions corresponding to critical post arc currents.

A typical case is shown in Fig. 10 where the nozzle arc interrupts a rate of rise of 5.51 kV/μs but fails a rate of rise of 5.55 kV/μs.

Figure 11 shows nozzle arc temperature profiles for thermal interruption at three post-zero times. At 1.05 μs after current zero, the nozzle exit arc temperature has just fallen to 6,000° K. At 1.15 μs the last 6 cm of nozzle arc are below 6,000° K while at 1.25 μs, half of the nozzle arc is below 6,000° K. If, by effective arc length, we mean that section of arc below 6,000° K, then in only 0.2 μs, the effective arc length has grown from 1 to 11 cm. Below 6,000° K the arc resistance varies exponentially with temperature.

Figure 12 compares the arc resistance calculated by Swanson's model with that calculated by the Mayr model for the same interrupting ability of 5.52 kV/μs. Figure 12 indicates that a nozzle arc can be significantly different from the Mayr arc, and the sharp increase in resistance implies a variable time constant. Figure 13 shows the 'Mayr time constant' for Swanson's model indicating that the nozzle arc does have a variable electrical time constant.

An important test of an arc model is its ability to match test data. Figure 14 shows arc voltage and post-arc current for a small scale nozzle tested by Browne and Spindle. It was characteristic of these tests that the precurrent zero arc voltage E_0 was constant up to the last microsecond before current zero. When SF_6 calculations were made with a variable time constant θ in Equation (41) the calculated arc voltage resembled that in Fig. 8. There was a similar extinction peak in contrast with all available SF_6 test data. This extinction peak was moderate for a dI/dt of 25 A/μs

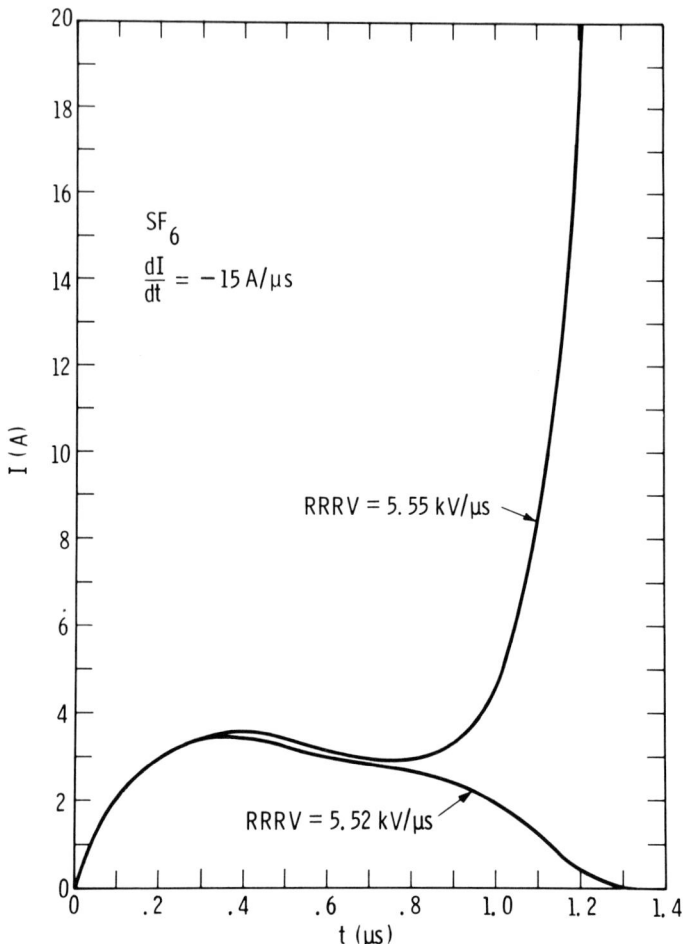

Fig. 10. Nozzle arc post-arc current

and an interrupting ability of 2 kV/μs. The extinction peak increased sharply when dI/dt was decreased to 15 A/μs and the interrupting ability increased to 6 kV/μs.

Now the diffusion time constant θ is the controlling time constant in Equation (42), and the average value of θ varied from 15 μs at 10 μs before zero to 1 μs at current zero. To simulate the precurrent arc voltage in Fig. 15, the total time constant θ in Equation (41) had to be set to 0.3 μs during the linear ramp. However the time constant θ in Equation (49) always varies with z and t. After current zero θ varied with z and t. The calculated precurrent zero arc voltage with θ = 0.3 μs is

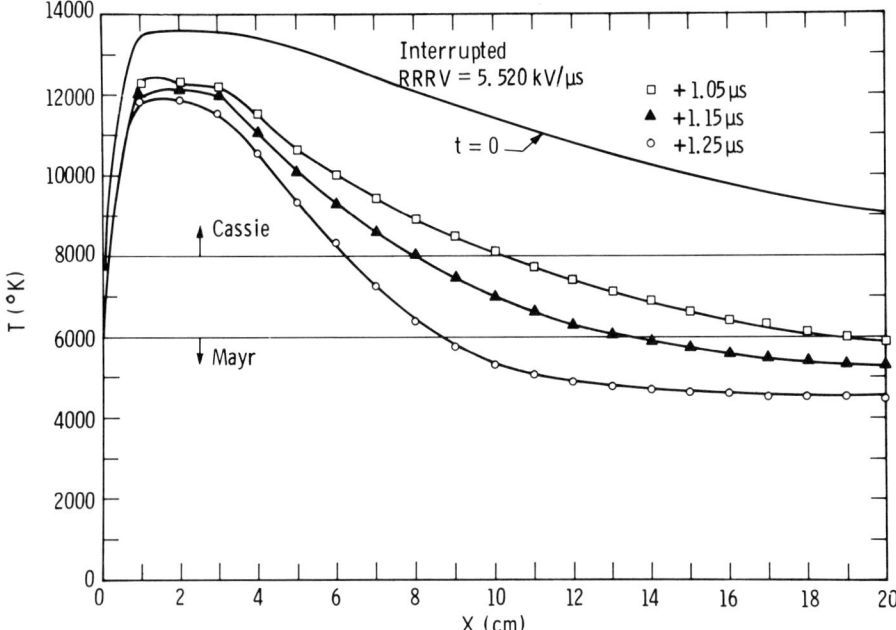

Fig. 11. Post-zero nozzle arc temperature profiles

shown in Fig. 16, showing good simulation of test data. Swanson's nozzle arc model has been used to make calculations with arc-circuit interaction. The calculated arc voltage and post arc current with circuit interaction are shown in Figs. 16 and 17. A comparison of Fig. 14 with Figs. 16 and 17 shows that Swanson's model can simulate realistic test conditions indicating that nozzle arc models can be used as a circuit element.

In developing nozzle arc models it is important that we relate them to the vast amount of data that has been correlated using Cassie-Mayr theory. In this regard Swanson's model shows that (a) a Cassie type arc can exist after current zero because the nozzle arc temperature can exceed 8,000° K; (b) the Cassie-Mayr transition time can exceed 1 µs; (c) thermal interruption requires a Cassie-Mayr transition, and (d) the nozzle arc behaves like a Mayr arc with a variable time constant. Like El-Akkari and Tuma,[20] Swanson[21] has also found that arc interruption is controlled by turbulent cooling in supersonic flow.

In summary Swanson's arc model has the following properties: (a) the model accounts for nozzle geometry; (b) the model is equivalent to that of El-Akkari and Tuma; (c) the nozzle arc temperature can vary significantly in the axial direction;

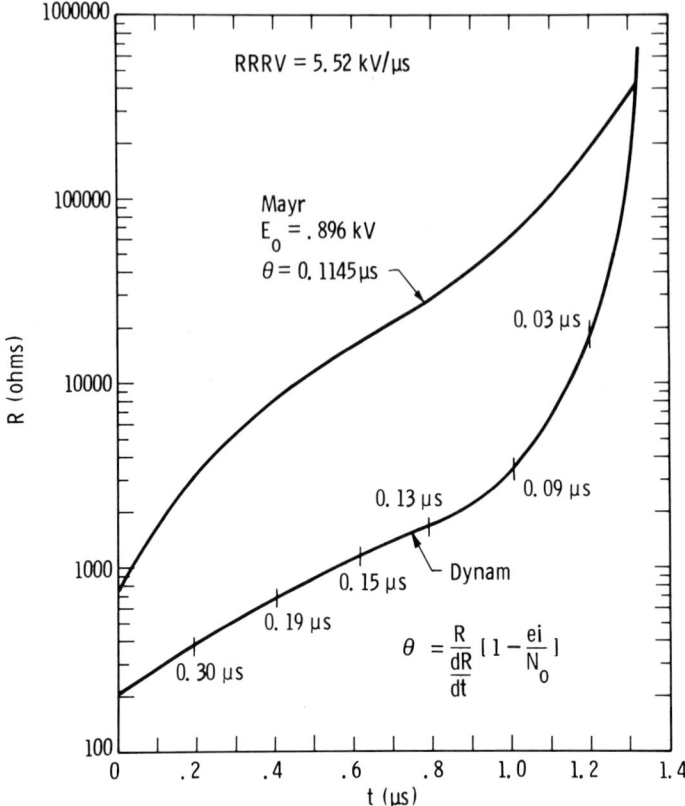

Fig. 12. Comparison of Mayr and Dynam arc resistance

(d) the model can be used as a circuit element to calculate realistic arc voltage and post-arc current; (e) the model can be used to interpret arc phenomena in terms of Cassie-Mayr theory, and (f) the model emphasizes the significance of the cooling time constant in understanding thermal arc physics. The principal limitations to the model are that (a) the arc model equations for area and temperature are approximate and do not include the a_2 and t_2 terms of Hermann's model; (b) for SF_6, the arc radius time constant θ is a model constant prior to current zero; and (c) the turbulence constant C_t is required to vary with current. While the discussion on arc turbulence indicates that turbulence may well vary with current, such variation is only rigorously justified on the basis of a core and boundary turbulence model.

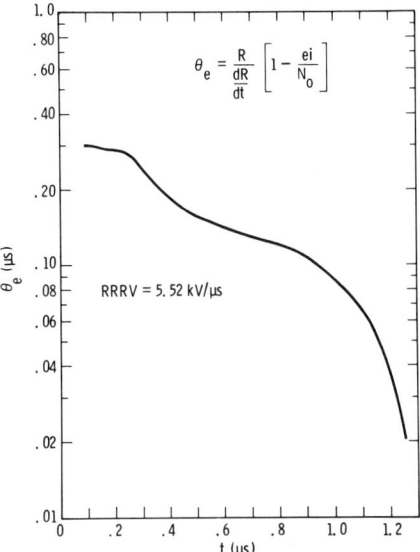

Fig. 13. Mayr 'time constant' for dynamic nozzle arc model.

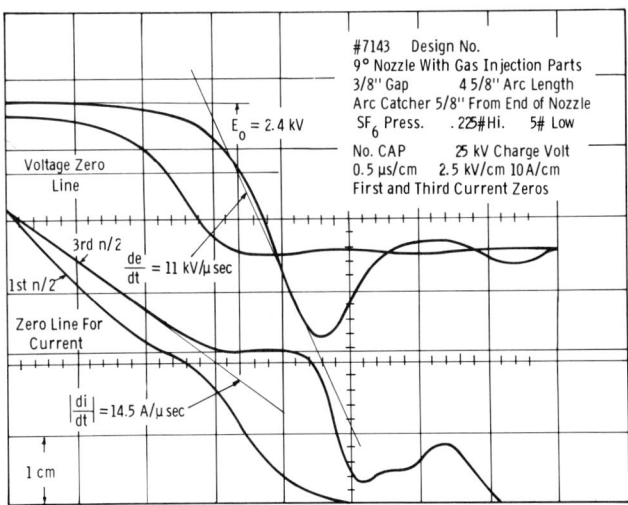

Fig. 14. Current and voltage oscillogram for test nozzle.

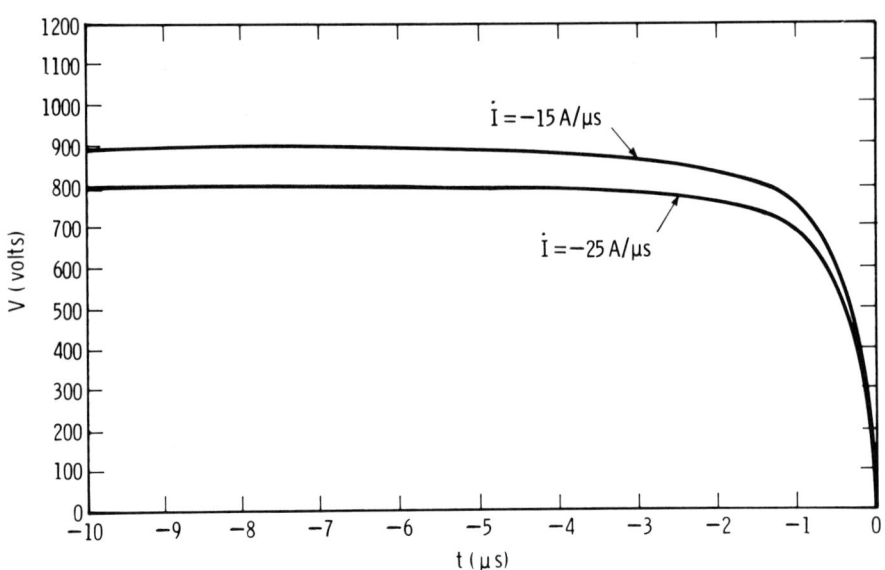

Fig. 15. Pre-current zero arc voltage.

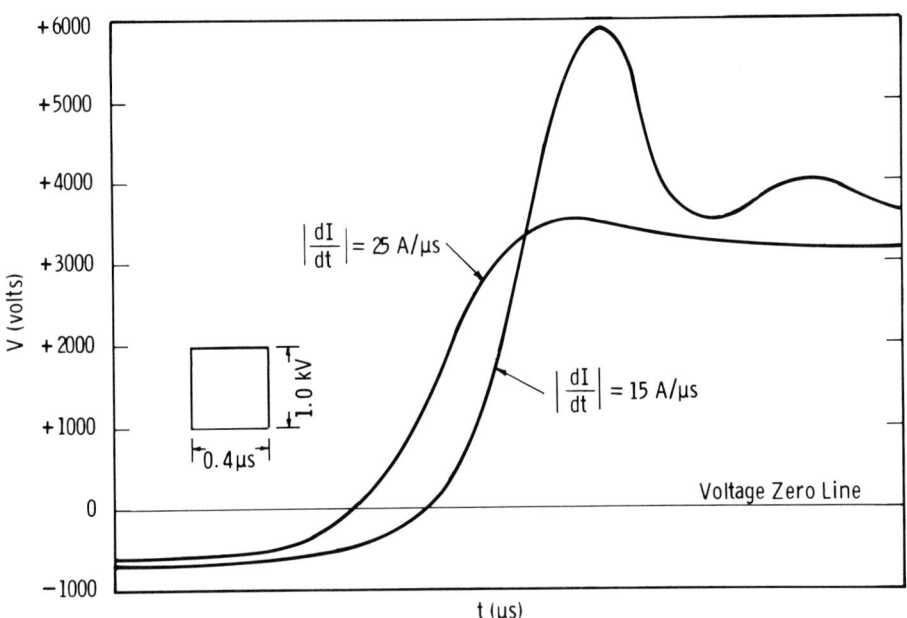

Fig. 16. Nozzle arc voltage with circuit interaction.

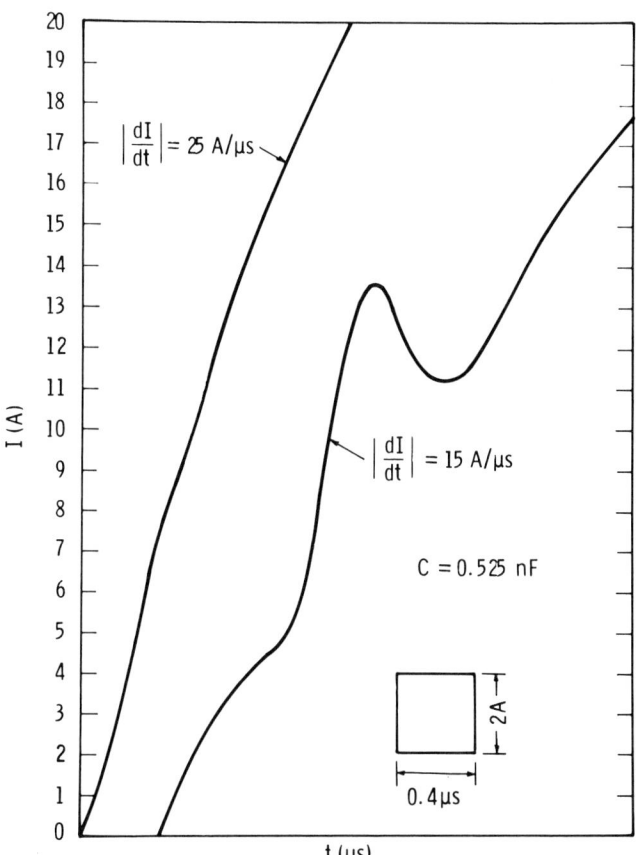

Fig. 17. Post-arc current with circuit interaction.

DISCUSSION

At this time we cannot compare the predictions of these arc models with respect to interrupting ability because Swanson's model[21] has not been applied to air while the El-Akkari-Tuma model[20] has not been applied to SF_6. Even if these calculations had been made, the comparison would not be fair, since the models of El-Akkari-Tuma and Swanson are based on partial differential equations while the present dynamic model of Hermann and Ragaller[17] is based on ordinary differential equations. We can, however, make a limited comparison of these models on the basis of a_1 and t_1 coefficients to highlight their differences. The equivalent a_1 coefficient of the Swanson model is $1/(F-F_e)$ while that of the El-Akkari-Tuma model is $1/\rho h$. Figures 18 and 19 give a comparison of the equivalent a_1 and t_1 coefficients of these models for air and SF_6. On the basis of area coefficient Fig. 18 indicates Swanson's model would overestimate the superiority of SF_6 while the El-Akkari-Tuma model would underestimate it. On the basis of temperature coefficients, Fig. 19 shows that the Swanson and El-Akkari-Tuma models could overestimate the superiority of SF_6.

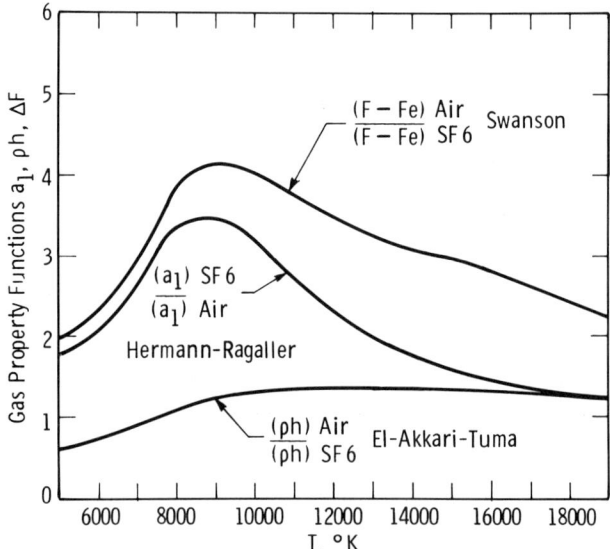

Fig. 18. Gas property functions versus temperature for air and SF_6.

We have considered the merits and limitations of three nozzle arc models. The desirable features of the models of Swanson[21] and Lowke, et al.[15,19,20] are that they

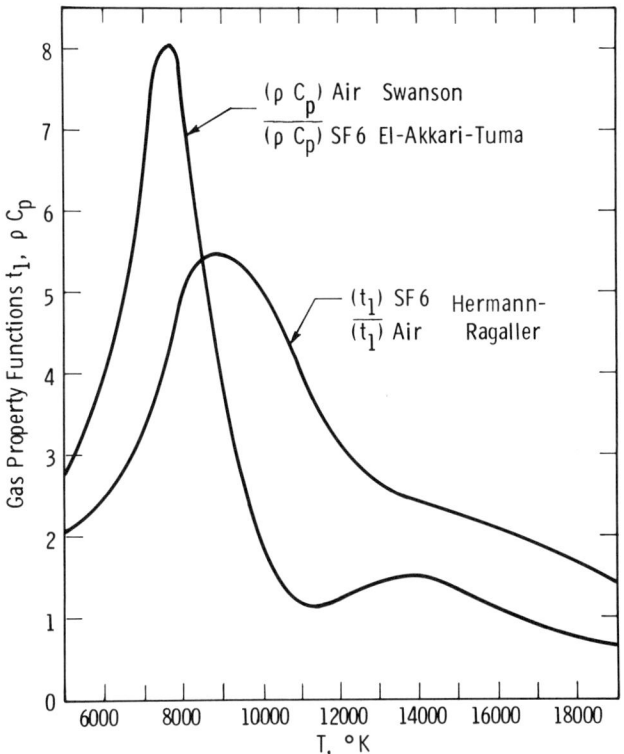

Fig. 19. Gas property functions versus temperature for air and SF_6.

account for nozzle geometry and are relatively simple. Their undesirable feature is that their equations for arc area and temperature are not rigorous near current zero.

The desirable features of the general model of Hermann, et al.[13,17] are that the equations are rigorous at all currents and the general model can account for nozzle geometry. Its undesirable feature is that it is difficult to solve six simultaneous partial differential equations that define a two zone arc model. Therefore, a reasonable compromise would attempt to develop a one zone model using only the three partial differential Equations (25), (26) and (27).

Such a model would retain the desirable features of all three models and should be solved using the clogging and isothermal approximations for pressure and Mach number developed by Lowke, et al.

We now consider what insight these models give into understanding observed physical phenomena. Airey and Abbott[26] have done experiments in SF_6 and N_2 using the same nozzle and gas pressure and have found that (a) the N_2 arc has a higher arc voltage; (b) the N_2 arc is about 2000° K cooler than the SF_6 arc; and (c) the N_2 arc is slightly more constricted. Using Ohm's Law, the smaller electrical conductivity inferred from Item (b) and the smaller arc area explain the differences in arc voltage.

Now Hermann and Ragaller[17] used the same nozzle for calculating the interrupting abilities of N_2 and SF_6 and found that the thermal interrupting ability of SF_6 is about 10 times that of N_2 as shown in Fig. 5. From Cassie-Mayr theory the interrupting ability is proportional to E_0/θ_e where E_0 is the precurrent zero arc voltage and θ_e is the electrical time constant. Since $(E_0)_{N2} > (E_0)_{SF6}$ while $(E_0/\theta_e)_{SF6} >> (E_0/\theta_e)_{N2}$, it follows that $(\theta_e)_{SF6} << (\theta_e)_{N2}$. In making arc model calculations the author has found that the arc cooling time constant θ is proportional to θ_e. Consequently, it follows that $\theta_{SF6} << \theta_{N2}$ and since the cooling time constant is proportional to arc area, then it follows that at current zero, $A_{SF6} < A_{N2}$. Thus we have an interesting paradox. Initially the steady state N_2 arc is more constricted than the SF_6 arc. But at current zero, the SF_6 arc must be more constricted to account for the differences in time constants.

Can these arc models explain this phenomenon? The approximate models of Swanson, El-Akkari and Tuma cannot, but the more rigorous model of Hermann and Ragaller can. To understand why, we now consider Equation (26) for arc area.

Since $\partial \bar{\rho}/\partial T < 0$ we can rewrite the a coefficients as

$$a_1 = \frac{|\partial \overline{\rho_1}/\partial T|}{D} \qquad a_2 = \frac{\partial (\overline{\rho h})_1/\partial T}{D} \tag{52}$$

$$D = (\overline{\rho h})_1 \left|\frac{\partial \overline{\rho_1}}{\partial T}\right| + \overline{\rho_1} \frac{\partial (\overline{\rho h})_1}{\partial T} \tag{53}$$

Assuming a parabolic temperature profile, we can evaluate $\bar{\rho}$, $\overline{\rho h}$ and $\bar{h}$ for air and SF_6. The variations of these functions with temperature are shown in Figs. 20 and 21. An inspection of these figures shows significant differences in the behavior of ρh in that $\partial \overline{\rho h}/\partial T$ is positive for air, making a_2 positive, while $\partial \rho h/\partial T$ is negative for SF_6 making a_2 negative. Furthermore, from Equation (53) this difference is the sign of $\partial \rho h/\partial T$ decreases D for SF_6 and increases D for air, creating differences in a_1 and a_2 for both gases. Figure 22 shows the relative values of a_1 and a_2 for air and SF_6 indicating a significant difference in a_2. Now in Equation (26) E_1 is negative and M_1 is positive during current decay. The negative value of a_2 makes both terms on the right hand side of Equation (26) negative, thereby making the collapse of arc area much faster in SF_6 than in air. Thus, the Hermann-Ragaller model can resolve this paradox in arc area.

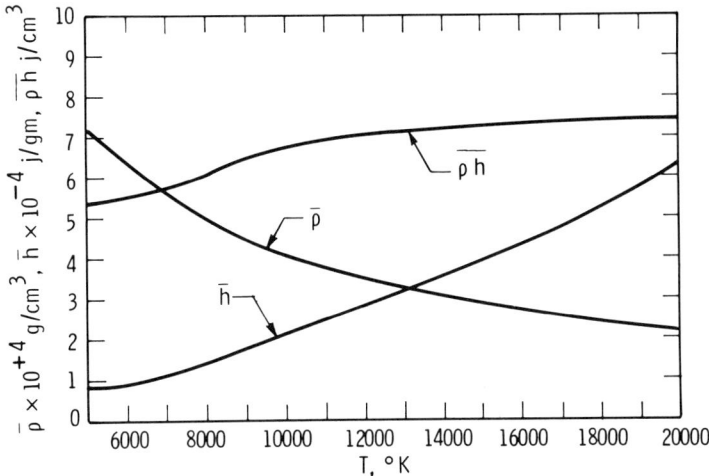

Fig. 20. Average air properties versus arc centerline temperature.

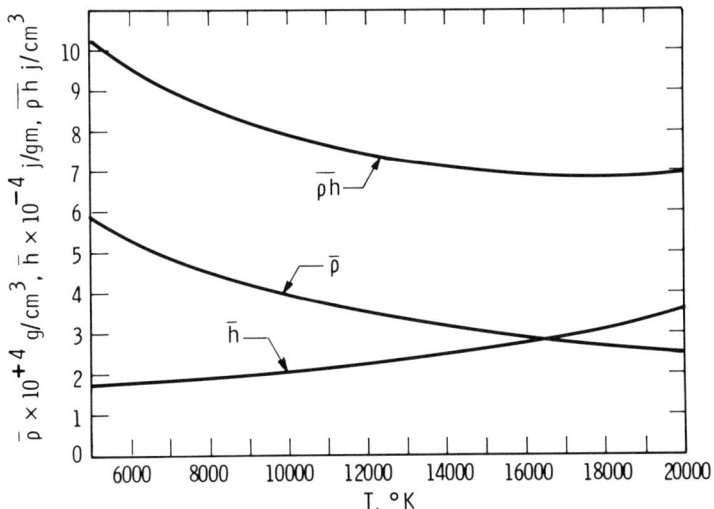

Fig. 21. Average SF_6 properties versus arc centerline temperature.

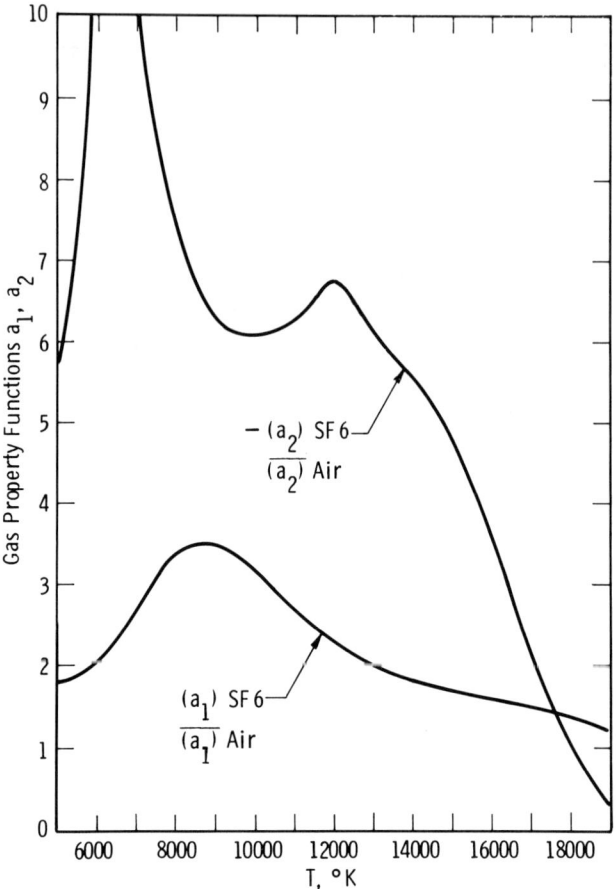

Fig. 22. Gas property functions versus temperature for air and SF_6.

This feature of their model also explains differences in observed arc voltage. The voltage gradient and Joule heating are given by the equations

$$E = \frac{I}{\sigma A} \qquad \sigma E^2 = \frac{I^2}{\sigma A^2} \qquad (54)$$

If the arc area time constant θ is very small, the arc area will closely follow the current keeping the Joule heating constant, the voltage gradient constant, and the arc temperature fairly constant. These conditions lead to a flat voltage characteristic as shown in Fig. 15 for the precurrent zero arc voltage. An SF_6 arc should therefore approach current zero with a relatively high temperature but small area and small cooling time constant. For $dI/dt = -25$ A/μs, the author calculates a current zero nozzle exit arc temperature of 15,000° K, an average arc radius of 0.5 mm, and an average cooling time constant of 1.4 μs.

On the other hand, if the arc area time constant is large, the arc area will lag the current causing a decrease in Joule heating and arc temperature. The electrical conductivity will then decrease with the arc temperature. If the arc voltage remains constant, it is only because the conductivity is decreasing at a rate that compensates for the lagging area. As the arc temperature falls below 10,000° K the conductivity can decrease faster than I/A causing an extinction peak in arc voltage as shown in Fig. 8. Therefore, an air arc should approach current zero with a relatively low temperature but large arc area and cooling time constant. For $dI/dt = -25$ A/μs, El-Akkari and Tuma[20] calculate an average current zero temperature of 7,000° K and an arc radius that increases with z to a value of 1.5 mm at the nozzle exit. At the nozzle exit the author estimates a cooling time constant of 8.2 μs. Thus these arc models can explain differences in observed arc voltage and in thermal conditions at current zero.

What about differences in interrupting ability? In particular, what properties of a gas principally determine its interrupting ability? The a and t coefficients amplify the terms E_1 and M_1 in Equations (25) and (26) so that the greater these coefficients, the greater will be the time rates of change of $\bar{T}_1$ and A_1. For good interrupting ability we want $\bar{T}_1$ and A_1 to be as small as possible at current zero which implies that large values of these coefficients favor arc interruption. Figure 23 shows the relative values of the ratios of t coefficients for air and SF_6. Figures 22 and 23 clearly imply the superiority of SF_6 assuming that both gases can be compared at the same temperature.

The t coefficients corresponding to Equation (52) are given by

$$t_1 = \frac{\overline{\rho}_1}{D} \qquad t_2 = \frac{-\overline{(\rho h)}_1}{D} \tag{55}$$

If we now assume that $\overline{\rho h} = \overline{\rho}\,\overline{h}$ then it is easy to show that Equations (52) and (55) reduce to

$$t_1 = 1/(\overline{\rho}_1 \overline{C_{P_1}}) \qquad t_2 = -\bar{h}_1/(\overline{\rho}_1 \overline{C_{P_1}}) \tag{56}$$

$$a_1 = 1/(\overline{\rho}_1 \overline{C_{P_1}} \bar{T}_1) \qquad a_2 = \overline{C_{P_1}} \bar{T}_1 - \bar{h}_1/(\overline{\rho}_1 \overline{C_{P_1}} \bar{T}_1) \tag{57}$$

where $\overline{C_{P1}} = \partial \bar{h}_1/\partial T$. It thus appears that, as a first approximation, the product ρC_p, which appears in all four coefficients, is that gas property which accounts for its interrupting ability. This conclusion is supported by Kinsinger and Noeske[22] who have correlated interrupting ability with a figure of merit ξ where

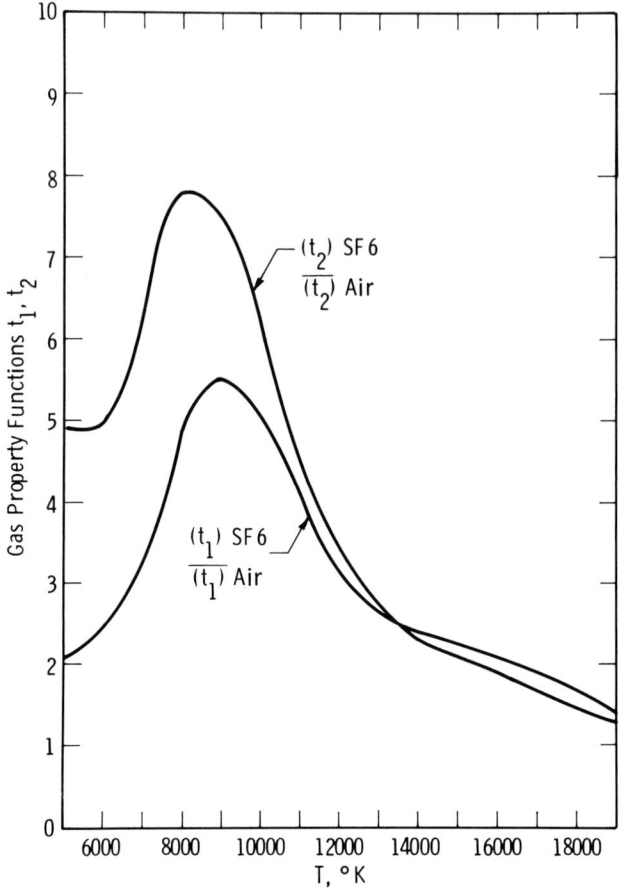

Fig. 23. Gas property functions versus temperature for air and SF_6.

$$\bar{\xi} = \frac{1}{\rho C_p \sigma} \frac{d\sigma}{dt} \tag{58}$$

This correlation is shown in Fig. 24 for 14 different gases and gas mixtures. The correlation is reasonably good for 10 gas mixtures, but SF_6, CF_4 and two other gas mixtures do not fit the correlation. In addition, while the assumption that $\overline{\rho h} = \bar{\rho}\bar{h}$ is reasonable for SF_6 it does not hold for air as shown in Fig. 25 and may not hold for other gases. Therefore, more information about a gas is contained in the four prop-

THEORETICAL MODELS

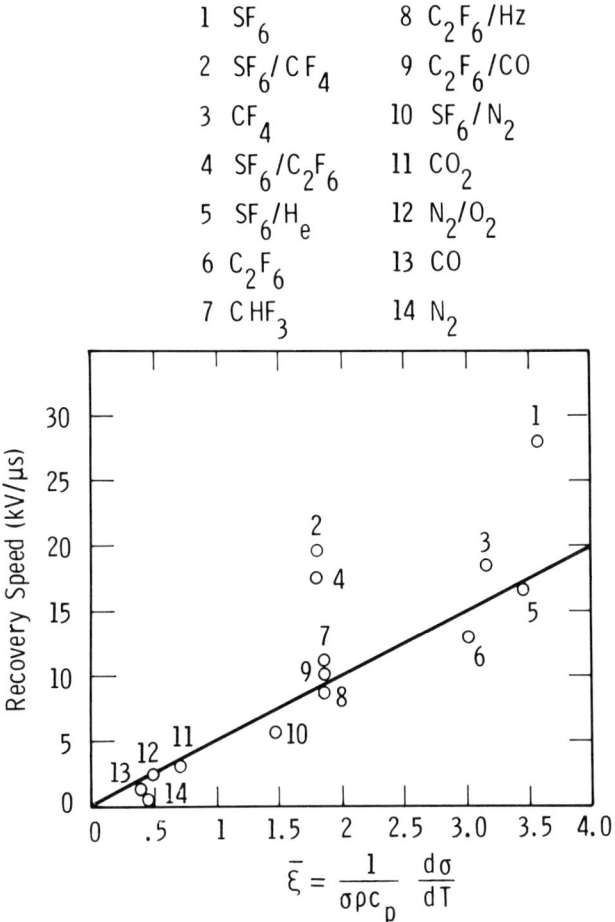

Fig. 24. Recovery speed vs gas property 'figure at merit'.

erty coefficients than is revealed by the simplifying assumption that ρC_p determines interrupting ability. Thus, the property coefficients in the model defined by Equations (25) and (27) are a significant feature of this model.

We may also ask why does interrupting ability decrease with current and increase with pressure? Swanson[21] has shown that increasing the current and therefore dI/dt causes a significant increase in the current zero arc temperature. It will also increase the current zero arc area and cooling time constant. Thus, the combination of increased temperature and cooling time constant at current zero is responsible for the decrease in interrupting ability.

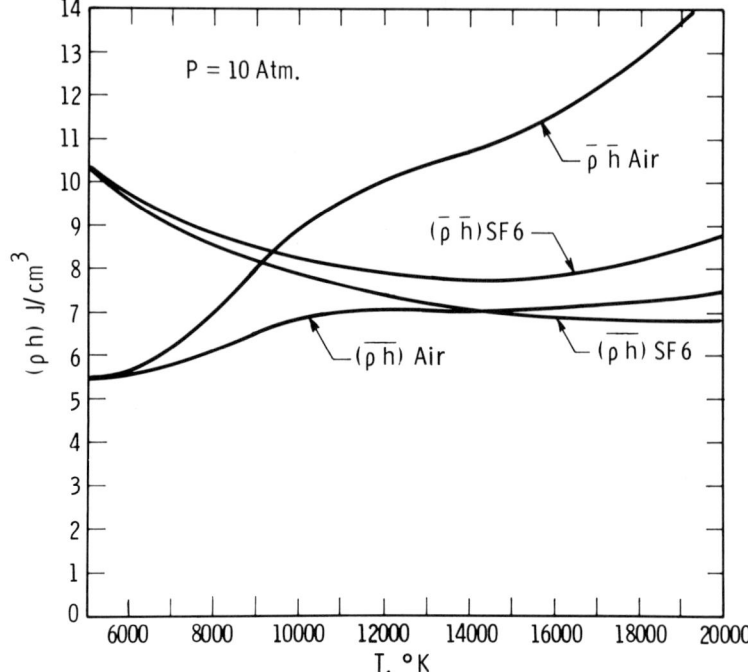

Fig. 25. Comparison of $\overline{\rho h}$ with $\overline{\rho}\,\overline{h}$ for air and SF_6.

With respect to the effect of pressure, Lowke and Ludwig[19] have shown that the arc temperature is independent of the tank pressure P_0 while the arc area (and hence cooling time constant) are inversely proportional to $\sqrt{P_0}$. Thus, increasing tank pressure decreases the cooling time constant which decreases the current zero temperature and increases interrupting ability. In this paper we have seen that nozzle arc models can give quantitative and qualitative answers regarding the effects of gas properties, current and pressure on interrupting ability. This insight cannot be obtained with Cassie-Mayr models.

ARC MODEL APPLICATION

We are ultimately interested in nozzle arc models that can be used in design application. We need answers to nozzle design questions concerning optimum length, diameter and expansion angle. We must understand clogging and how it affects bus fault and short line fault interruption. We must understand the effects of flow separation and shock waves on interrupting ability. We are a long way from this goal, but we can at least indicate the potential of these models for design application.

Recently Campbell, Perkins and Dallachy[23] published experimental data on the effects of nozzle pressure ratio on interrupting ability. Figure 26 shows their experimental results on the variation of interrupting ability with pressure ratio for three upstream pressures. The data indicates that interrupting ability falls off monotonically as the ratio P_d/P_u of downstream to upstream pressure increases from 0.1 to 0.67. Campbell[23] et al. attribute the decrease in interrupting ability to the formation of a compression shock wave that moves upstream towards the nozzle throat as P_d/P_u increases. They note that the formation of this shock wave reduces the extent of supersonic flow having two main effects: (a) the shock reduces the length of arc that is effectively cooled by turbulent supersonic flow; and (b) there is a greater tendency for flow separation in the divergent section of the nozzle producing a progressive deterioration in interrupting ability.

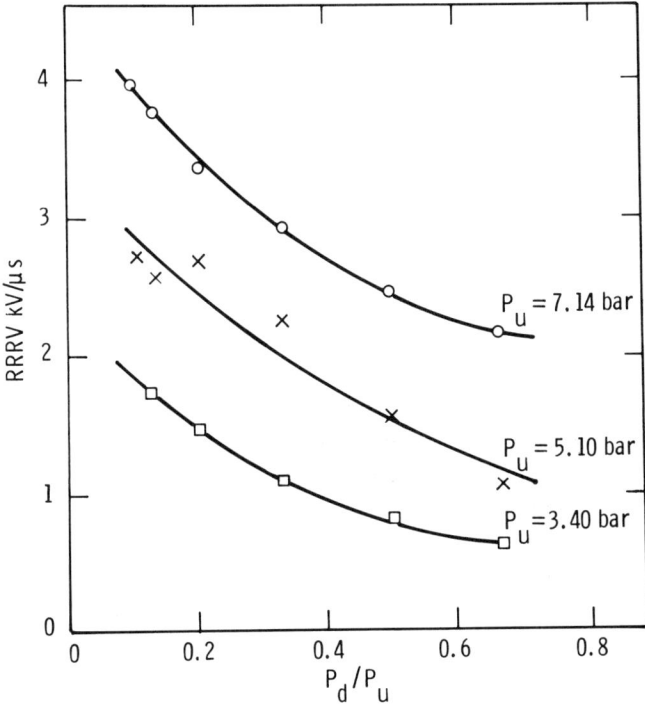

Fig. 26. Plot of variation of rate of rise of recovery voltage rrrv with pressure ratio P_d/P_u for different upstream pressure.

We now want to investigate these conclusions on the basis of nozzle arc theory.

Increasing the downstream pressure will cause a compression shock in the nozzle. When the shock is weak and the boundary layer is thin, as is the case just downstream of the throat in a supersonic nozzle, the nozzle shock extends over most of the nozzle as shown in Fig. 27(a). When the shock is strong and the boundary layer is thick, as is the case well downstream of the throat, a series of forked normal shocks can occur as shown in Fig. 27(b),[24] accompanied by flow separation. Nagamatsu, et al.[25] have shown that nozzle flow separation depends on the nozzle pressure ratio so that the more complex flow picture in Fig. 27(b) is likely to occur in

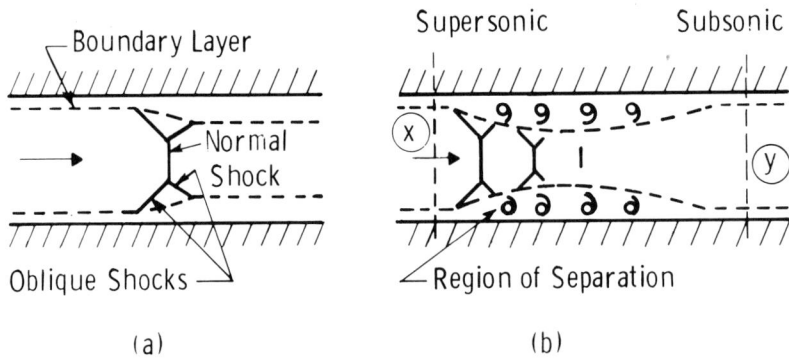

Fig. 27. Typical shock-boundary layer interactions in a duct.[24]
(a) Thin boundary layer and weak shock.
(b) Thick boundary layer and strong shock.

the divergent section of the nozzle. Thus, increasing the downstream pressure can cause a complex flow field in the nozzle and design calculations must take this into account. For our purpose here we shall greatly oversimplify the problem by assuming that the compression shock is a sharp discontinuity without flow separation. Figure 28 shows a simple compression shock located 14 cm from the nozzle inlet. If subscripts x and y denote conditions upstream and downstream of the shock, then the corresponding Mach numbers and pressures are related by the normal shock relations.[24]

$$M_y = \frac{M_x^2 + 2/(\gamma-1)}{2\gamma/(\gamma-1)M_x^2 - 1} \tag{59}$$

and

$$\frac{P_y}{P_x} = \frac{2\gamma}{\gamma+1} M_x^2 - \frac{\gamma-1}{\gamma+1} \tag{60}$$

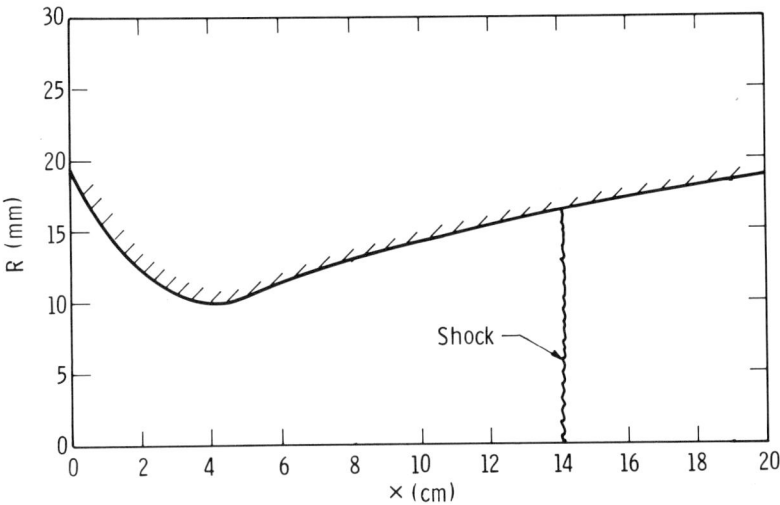

Fig. 28. Laval nozzle with compression shock.

For the shock location in Fig. 28, the nozzle pressure and Mach number distribution are shown in Fig. 29. Upstream of the shock the geometrical Mach number is 2.2 and the pressure is 1.5 atm. Downstream of the shock the Mach number is .48 and the pressure has increased to 7.5 atm.

The effect of the shock on the cooling time constant θ is shown in Fig. 30. This figure shows the axial distribution of cooling time constant at -10 μs, at current zero and at 1.35 μs after current zero. The shock causes a significant increase in the cooling time constant. To understand this effect we substitute Equation (48) for β into Equation (43) for θ giving the equation

$$\theta = \frac{R^2(z,t)}{[\alpha + C_t(z-z_t)(V_1-V_2)]\delta_1^2} \quad (61)$$

Now the shock causes a large decrease in V, and a large increase in P which causes a large decrease in α. Equation (61) shows that decreases in α and V will increase θ, as shown in Fig. 30.

Therefore the section of arc downstream of the shock cannot cool as fast as the upstream section. Figure 31 shows the effect of the shock on the nozzle arc temperature profiles. At 1.35 μs after current zero the arc temperature upstream of the shock has cooled down to 4000° K while that downstream of the shock is still over

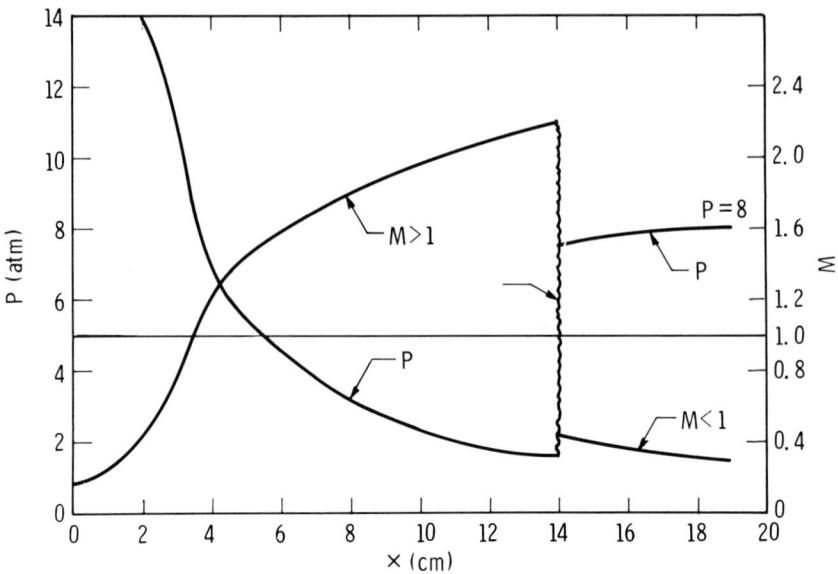

Fig. 29. Pressure and Mach number distributions with compression shock at X=14 cm.

10,000° K. The result is a reduction in interrupting ability from 6.3 to 4.5 kV/μs. The theoretical variation of interrupting ability with downstream pressure is shown in Fig. 32. Thus Swanson's arc model calculations support Campbell's[23] assumption that the compression shock decreases interrupting ability. Of course, realistic nozzle design calculations would also have to account for the effect of flow separation mentioned by Campbell, et al. which is a difficult problem that has not been considered. This example does illustrate the ultimate potential of nozzle arc models in design application.

CONCLUSION

We have seen that advanced arc models are needed to answer today's difficult design questions. We have considered the uncertainties in modeling arc turbulence and radiation. We have compared the merits and limitations of different models and made recommendations for further model development. We have seen the importance of turbulence in supersonic flow and the hazards of shock waves. We have gained insight into the influence of gas properties on arc temperature, arc voltage and interrupting ability. We have seen the practicality of modeling the arc as a circuit element. We

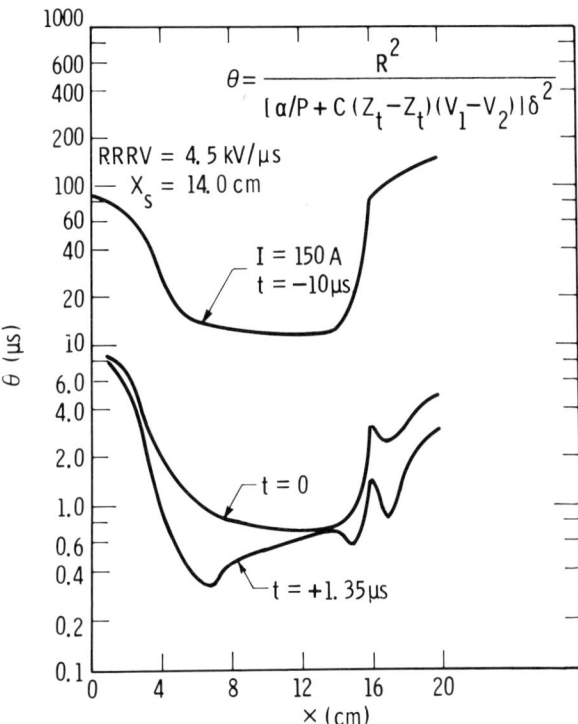

Fig. 30. Effect of compression shock on arc cooling time constant.

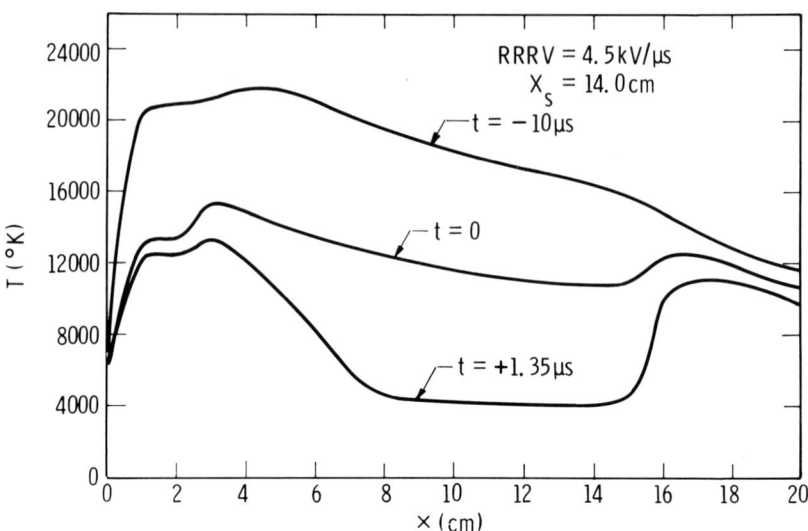

Fig. 31. Effect of compression shock on nozzle arc temperature profile.

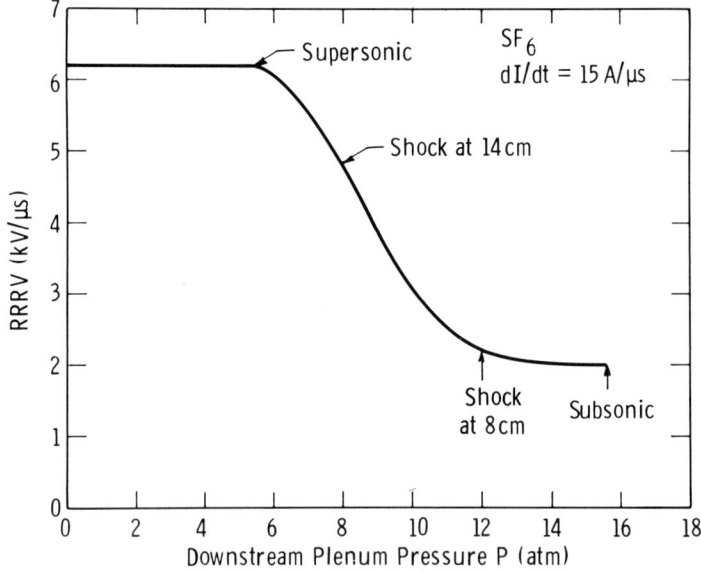

Fig. 32. Nozzle interrupting ability versus downstream pressure.

have seen the potential of these models in design application. From these remarks, the future holds promise for the continuing development of advanced nozzle arc models, and their application in circuit breaker design.

ACKNOWLEDGMENTS

The author gratefully acknowledges the support of the Westinghouse Power Circuit Breaker Division and their permission to publish this paper.

REFERENCES

1. A. M. Cassie, Report No. 102, CIGRE, Paris, France (1939)
2. O. Mayr, Archiv für Elektrotechnik 37 (1943) 588
3. T. E. Browne, Jr., AIEE Trans., 67 (1948) 141
4. T. E. Browne, Jr., Trans. AIEE, Power App. Syst., No. 40, pp 1508-1517
5. G. Frind, J. A. Rich, IEEE Trans. PAS (1974) 1675
6. T. E. Browne, Jr., Proc. IEEE Summer Power Meeting Paper (1977) No. F 77-626-5
7. L. S. Frost, Proc. IEEE Summer Power Meeting (1977) No. F 77-627-3
8. T. Cebeci and A. M. O. Smith, Analysis of Turbulent Boundary Layers, Academic Press (1974)
9. D. E. Abbott, J. D. A. Walker, R. E. York, 4th Int. Conf. Num. Meth. Fluid Dynamics, June (1974), Boulder, Colo.
10. L. Niemeyer and K. Ragaller, Z. Naturforsch. 28a (1973) 1281
11. S. Pai, Fluid Dynamics of Jets, Van Nostrand (1954)
12. H. Schlichting, Boundary Layer Theory, McGraw-Hill (1960)
13. W. Hermann, U. Kogelschatz, L. Niemeyer, K. Ragaller, and E. Schade, J. Phys. D: Appl. Phys. 7 (1974) 1703
14. B. W. Swanson and R. M. Roidt, Proc. IEEE 59 (1971) 493
15. D. T. Tuma, J. J. Lowke, J. Appl. Phys. 46 (1975) 3361
16. R. W. Liebermann and J. J. Lowke, J. Quant. Spectr. Rad. Transfer 16 (1976) 253
17. W. Hermann and K. Ragaller, IEEE Trans. PAS 96 (1977) 1546
18. W. Hermann, U. Kogelschatz, K. Ragaller and E. Schade, J. Phys. D: Appl. Phys. 7 (1974) 607
19. J. J. Lowke and H. C. Ludwig, J. Appl. Phys. 46 (1975) 3352
20. F. R. El-Akkari and D. T. Tuma (1977) Proc. IEEE Winter Power Meeting Paper No. F 77-126-6 21. B. W. Swanson, IEEE Trans. PAS 96 (1977) 1697
21. B. W. Swanson, IEEE Trans. PAS 96 (1977) 1967
22. R. E. Kinsinger and H. O. Noeske, EPRI Symposium on Fault Current Limiters, State University of New York at Buffalo, Sept. (1976)
23. L. C. Campbell, J. F. Perkins and J. L. Dallachy, Gas Discharges IEE Conference Publication No. 143 (1976) 44
24. A. H. Shapiro, The Dynamics and Thermodynamics of Compressible Fluid Flow, Vol. I, page 135, Ronald Press (1953)
25. H. T. Nagamatsu, R. E. Sheer, Jr. and E. C. Bigelow, Proc. IEEE Winter Power Meeting (1974) Paper No. C 74-184-8

DISCUSSION
(Chairman: D. M. Benenson, State University of New York at Buffalo)

M. D. Cowley (University of Cambridge)

It seems to me that an important difference between Swanson's model[1] and that of BBC is that Swanson uses only the energy equation whereas BBC solve simultaneously the energy and the continuity equation. As the continuity equation plays an important role in determining the area of the arc, this could explain the good performance of the BBC-model.

K. Ragaller (Brown Boveri)
This statement is not quite correct. Swanson starts with an integral form of the energy equation and derives equations (50) and (51) for the area and the temperature. These two equations contain the continuity equation in an implicit form, as can be shown by taking the difference between them. I think more important differences arise as a result of the different ways of radial discretisation, i.e. the assumption of a Bessel profile by Swanson; a 3-zone model by BBC (comment subsequent to meeting).

J. J. Lowke (University of Sydney)
I would like to comment on the general development of theoretical models of the arc in a circuit breaker. The first models were those of Cassie and Mayr in the 1940's which set the pattern for a whole series of models, for example, by Browne, ter Horst, Rieder and Urbanek, Möller, etc., each involving empirical constants which had to be determined from circuit-breaker oscillograms. Now the BBC group have started a new type of theoretical arc modelling based upon the basic conservation equations of mass, energy and momentum. The particular contribution of this group has been to point out the importance of turbulence in arc interruption, and to formulate a model for its representation. It is probable that other workers in the field will now continue the BBC work with different mathematical models of turbulence. Fluid dynamicists have worked for decades on the problem of representing turbulence and have achieved some success in being able to provide turbulence parameters as functions of the Reynolds number for a wide variety of physical situations, as is described in Schlichting's book[2] for example. At Sydney we have explained the decay in axial velocity of a free burning arc using a turbulence parameter significantly different from that of BBC but in agreement with previous determinations of such parameters for free turbulent jets. The model of Tuma[3] differs from that of BBC in that averages are made involving the density of the plasma and the cold gas surrounding the arc. It remains to be seen whether future workers can obtain more general and simpler parameters to represent turbulence for different circuit breakers or whether they are just 'muddying the water'.

K. Ragaller
I agree that it will be an important task in the future to find a good and reliable model to represent the turbulent heat exchange of an arc near current zero. It must be pointed out, however, that progress will require the use of very basic and general differential equations instead of the simple empirical differential equations, such as the following one, which have been used until now

$$\frac{1}{g}\frac{dg}{dt} = \frac{1}{\tau}\left(\frac{ui}{P} - 1\right).$$

The improvement, which results from the use of more accurate differential equations, is independent of the details of turbulence modelling. The error that is made in oversimplifying the differential equation is much greater than the error which results from inexact modelling of one term (W_t) in the correct differential equations. I would like to provide two examples which show the practical potential of modern arc models. The first (Fig. 1) shows the increase of the current limit in the slf by use of a capacitor parallel to the line (or the circuit breaker) for an SF_6 breaker.

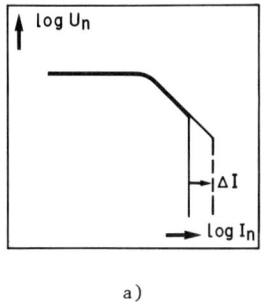
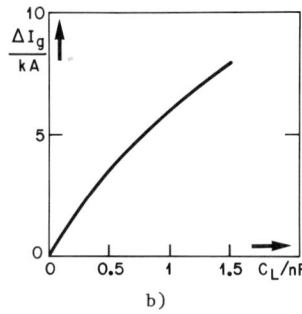

Fig. 1. Shift of the current limit in the short-line fault by a parallel capacitor.
(a) Representation in the log U_n/log I-plane.
(b) Calculated current increase in dependence on capacitance for an SF_6 circuit breaker.

The second example (Fig. 2) shows a calculation of the interruption of small inductive currents again using an SF_6 circuit breaker. The figure shows that the well-known phenomenon of current chopping is quite well predicted by theory.

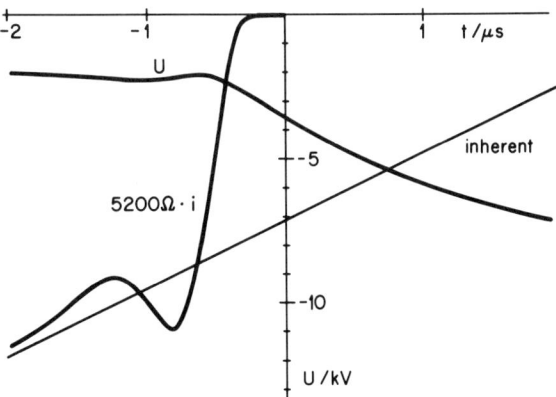

Fig. 2. Calculated result for the 'chopping' of a small inductive current with an SF_6 circuit breaker at 145 kV/1 kA.

In both examples, a check with experimental results is possible. The practical use of such theoretical models for a variety of different switching conditions, and the comparison with experimental data is another important task for the future.

B. W. Swanson

In support of Ragaller's first statement, I would like to add that it is a well-known principle in mathematical physics that a partial differential equation contains much more information than an ordinary differential equation.

W. Rieder (Technical University, Vienna)

I think if one compares different models and tries to answer the question whether one is better than another, it is also important to consider what the models are used for. For the purpose of studying the behavior of a given breaker in different circuits, the models of Cassie and Mayr have proven their usefulness. If the relationship between the electrical behavior and the design of the breaker is to be modeled, then obviously new and more complicated models must be used.

B. W. Swanson

I agree: for certain breakers, after careful fitting of constants, the simple Cassie-Mayr models can be used, whereas for more general problems, such as a variation of the breaker, more refined models are needed.

W. Hertz (Siemens)

You showed a figure with a post-arc current in SF_6 of 3 - 4 A for a time interval of about 1 µs. Was this a measured value and what was the current and the pressure in this case?

B. W. Swanson

This was a theoretical result, but the values are in agreement with those measured.

T. E. Browne (Westinghouse)

I can confirm that we have measurements of post-arc currents in SF_6 with the order of magnitude just mentioned.

G. Frind (General Electric)

I should like to raise the question whether we really have enough experimental evidence at the present time to confirm that turbulence is the dominating influence.

K. Ragaller

This question can best be answered with the aid of the measured variation of the electrical field strength along the nozzle axis for the arc near current zero (e.g. Fig. 15 in the paper by Ragaller and Reichert). The maximum electric field occurs at a position where all other effects, such as straining (see also discussion of Jone's paper) or radiation, have practically no importance. The electrical field is a very direct measure of the heat exchange between the arc and cold gas. On the other hand it is well proven that there exists extremely intense turbulence in this region of the arc. Also, there is no doubt that turbulence increases the classical heat exchange by orders of magnitude. If one looks into the details of the interaction of the arc with the turbulent flow field (Fig. 3) the arc column is seen to be extremely thin and embedded in the turbulent flow field. The transverse shift of 1 mm in 1 µs gives 1000 m/s as the order of magnitude of the large-scale intensity of the turbulence. It is inconceivable that the kind of interaction shown in Fig. 3 would not influence the current interruption.

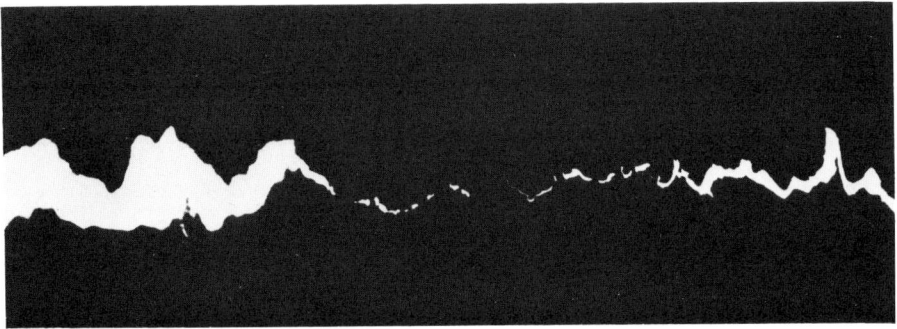

Fig. 3. Streak picture of a decaying arc in the turbulent flow section.

G. Frind

There remain some doubts concerning the method of measurement of the electric field near current zero. How were the arc diameter and the temperature determined? Also, other groups which have measured the electrical field strength, at the University of Liverpool,[4] Siemens,[5] and General Electric,[6] find the peak of the electric field further upstream near the nozzle throat or even close to the nozzle front edge.

K. Ragaller

As far as the measurement technique is concerned we used a double-scan technique an example of which is shown in Fig. 4. Two scanning slits separated by a distance of 0.4 mm are moved across the arc with a speed of 900 m/s. This gives a first radial profile of the arc within a time of 1 - 2 µs and a second profile 0.45 µs later. This type of measurement gives not only a very sensitive measure of the temperature (and therefore the conductance) profile, but also shows the change in the profile with time. The application of this technique in the upstream part of the arc, at the throat and in the turbulent section clearly confirms the dominating influence of turbulence.

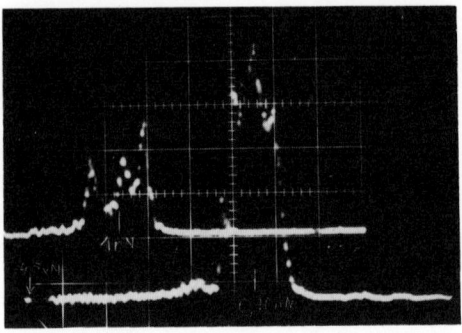

Fig. 4. Radial intensity profiles in the turbulent arc section taken with double-scan technique.
Right-hand profile: 0.7 µs after current zero
Left-hand profile: 1   µs after current zero.

Regarding the measurements of other groups I agree that the geometry of the experiment can influence the geometrical position of the dominating arc section because the point of fully devloped turbulence or of the largest straining effect, for example, are geometry-dependent.

L. H. A. King (GEC)

I would like to make a suggestion that may help to clear up the argument about the position of the high voltage-gradient region and its position along the arc column in the nozzle. In our work, we find the region of maximum gradient to be in the nozzle just in front of the vena-contracta as the current approaches zero. But because the arc temperature even at current zero is still very high, this region of the arc, with its associated high voltage gradient, moves along the nozzle with a very high velocity during the recovery period and thus enters the highly turbulent zone. Because of this, measurements made during the recovery period show the high gradient downstream of the throat. Thus we have a situation where both parties may be right and the real process contains the best of both arguments.

REFERENCES

1. B. W. Swanson, this volume
2. H. Schlichting, Boundary Layer Theory, McGraw Hill, New York (1960)
3. D. T. Tuma, J. Appl. Phys. 46 (1975) 3361
4. G. R. Jones, this volume
5. D. W. Branston and J. Mentel, 13th Int. Conf. Phen. Ionized Gases, Berlin 1977 6. EPRI Report EL-284 (1977) 2

INITIAL TRANSIENT RECOVERY VOLTAGE

C. DUBANTON
Electricité de France, Clamart, France

SUMMARY

The transient recovery voltage (trv) which appears at the opening of a circuit breaker is dependent on the lumped and distributed impedances of the elements of the network and busbars. It is important to determine whether the low impedances of the bars, or the presence of small capacitances connected to the bars, influences the breaking. It is intended to show how the different elements and arrangements of a station may be taken into account and how they may influence the trv. Due to the short lengths of these elements, the time under consideration is very short, less than 1 µs; consequently, special attention will be given on the representation of the impedances together with a description of possible methods to study this phenomenon.

A brief survey of the results obtained by a working group of CIGRE gives the main results and a simple method for evaluation. Simple estimates of the influence of the resistance of an arc permits a discussion of the true importance of the initial transient recovery voltage (itrv) in the network or in a test-station.

1. INTRODUCTION

The previous contributions to this Symposium have clearly shown the importance of the interaction between the network and the arc of a circuit breaker during the clearing of a fault. They have also pointed out that various effects may arise during breaking, according to the type of network, specially the nature of the impedances.

This paper focuses on these effects which are related to the impedances of the elements located close to the circuit breaker: busbars, isolators, measurement transformers, and so on. As the distances between the elements and the circuit breaker are very short, and the propagation speed for electrical phenomena high (nearly the speed of light), the times under consideration are very short, of order 1 microsecond.

It is intended to analyse the efects of the elements on the shape of the trv during its initital phase in various configurations of the network, independently of the interaction between arc and network conditions, which will be explained later. Thus, it is possible to determine the stresses imposed by the network, which must be simulated in a testing procedure in order to represent the ambient electrical conditions for the circuit breaker.

After a summary of extensive studies of CIGRE the true importance of the itrv and its representation is discussed briefly.

2. ANALYSIS OF THE CURRENT AND VOLTAGE DISTRIBUTION IN A SUBSTATION DURING A SHORT CIRCUIT

To determine the inherent conditions of the network, we suppose that the circuit breaker is ideal, which means that it has no arc impedance and an infinite dielectric breakdown voltage just after current zero. This consideration permits determination of the stress imposed by the network, independently of the type of circuit breaker and its design.

Let us consider now the single-line arrangement shown in Fig. 1, from which it is clear that very different currents flow through each element.

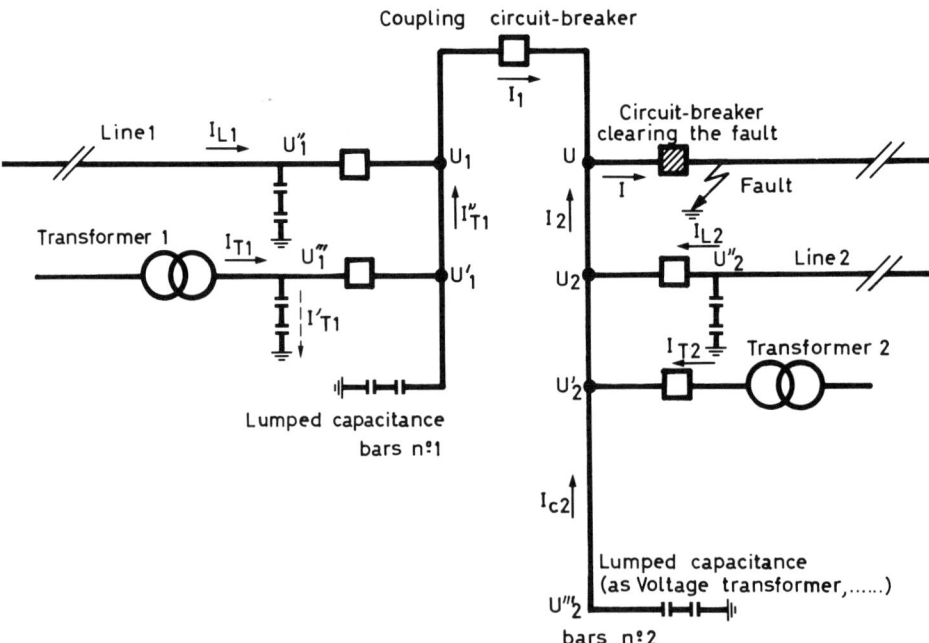

Fig. 1. Single-line diagram of a substation with two busbars.

In this example the total short-circuit current I is fed through either busbars no. 1 and the elements connected to it, or busbars no. 2 and the elements connected to it. According to Ohm's law, the voltages across the elements vary from one point to another, in relation to the fundamental frequency impedances. These low voltages U_1, U_2, ... may reach several hundred volts for high short-circuit currents, and the transient voltages are superimposed on the fundamental-frequency voltage at the moment of breaking.

The discontinuity in the current derivative due to breaking by the ideal circuit breaker, is propagated through the circuit with a speed just lower than that of light and some transient oscillations of voltage and current are produced in the high-frequency impedances of the station.

As far as voltage oscillations due to propagation in the bars are concerned, we are considering phenomena the duration of which is in the range from one tenths to a few microseconds, depending on the lengths of bars between lumped elements of significative surge impedance (the speed of light being 300 m/μs).

Such remarks show that it is impossible not to consider the exact layout of the substation, and the surge impedances of the elements, if we are interested in events which occur in this range of time and voltage.

Thus, in order to understand the effects of the different elements and to justify some simplifying assumptions, we propose to look at some special layouts. It will be then possible to determine the itrv in more complicated station configurations.

3. THE ITRV IN ANTENNA LAYOUT

Simplifications of a number of actual elements are represented in Fig. 2. For example, the voltage transformer is modeled by only a capacitance to ground with a value equal to the high-voltage capacitance of the transformer: the influence of the low-voltage part is totally neglected and also the effect of the very short connections to neighbouring elements, and so on. By comparing the values of the different impedances of the actual equipment, the validity of this kind of representation is easily demonstrated.

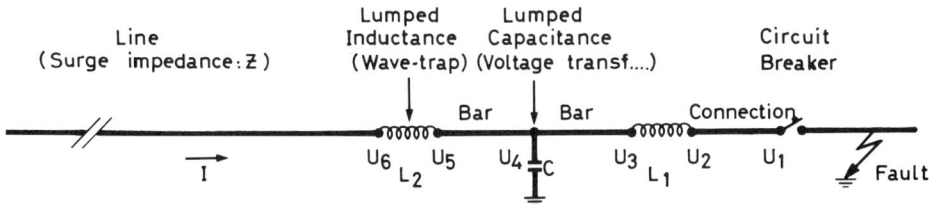

Fig. 2. Simplest diagram of a substation for the purpose of itrv calculation.

The simplification is less obvious when we consider a wave trap, as shown in Fig. 3, due to the capacitance to ground, and the tuning of the high-voltage inductance with a rather complicated system.

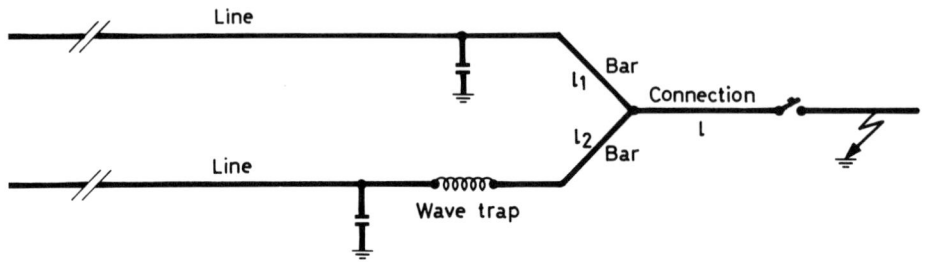

Fig. 3. Simplified diagram of a substation with tree layout.

This point will be discussed later and the choice of an inductance in this scheme is only a rough approximation included more to illustrate the influence of an inductance than that of an actual wave trap. If we consider the opening of a circuit breaker with a fault at one terminal, we can record the voltage at different points of this circuit. The first circuit considers L_1 and L_2 to be zero and the values of the impedances will be discussed later.

Figure 4 shows the shape of the itrv in the case of single-phase operation. The oscillation is due to the propagation of the voltage from the circuit breaker to the capacitance and to a reflection of the wave at this point, the capacitance for this very-high-frequency signal representing nearly a short circuit. The period of the first oscillation is about twice the travel time of the signal from the circuit breaker to the capacitance, and the slope of the first part is proportional to the surge impedance of the bars and to the slope of the current. If we neglect the initial oscillations, we can see that the trv tends to a curve which can be described by a line with a given slope (proportional to the surge impedance of line and to the current) together with an exponential curve at the beginning: this part of the trv has the shape of the trv described in the standards of I.E.C. by 'initial slope' and 'time delay'. For example, curve A gives a time delay of 0.75 µs, corresponding to the product of the capacitance (2 nF) and the surge impedance of the line (389 Ω). The itrv is superimposed on this trv and represents the effects of the details of the station configuration.

Figure 4 also shows the influence of the capacitance of the elements. The calculation of the inherent itrv was performed using two values of the capacitance in the

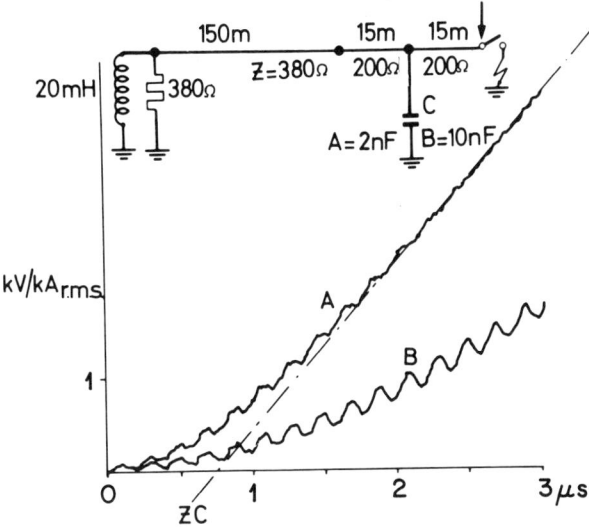

Fig. 4. Calculated itrv curves for a layout according to Fig. 2. Line length between breaker and capacitance is 15 m. The capacitance is 2 nF for curve A and 10 nF for curve B.

same circuit and we can observe the differences in the itrv for curves A and B. The initial part is identical for each case including the first peak of the itrv since the initial slope is independent of the values of the capacitance, and depends only on the surge impedance of the bars between the breaker and the distance of the capacitance to the circuit breaker. After this initial period, however, the influence of the capacitor is apparent: the larger capacitance delays the recovery voltage.

Curve A of Fig. 5 shows the effect on the itrv of a very low capacitance of only 100 pF. If the very beginning has exactly the same shape for the first tenth of a microsecond, the trv is practically not influenced by the capacitance, and the initial time delay is more a result of the surge impedance change between bars and line than an effect of the small capacitor.

Curve B of Fig. 5 gives another indication of the effects and is to be compared with curve A of Fig. 4. The trv and the time delay are not different, although the frequency of oscillation is lower due to the increased length of the bars between the circuit breaker and the capacitor, which is now 45 m instead of 15 m. In this case, the first peak of the itrv appears after 0.3 μs, that is 3 times the time to the first peak of curve A in Fig. 4. The slope is the same in both cases, and proportional to the surge impedance of the bars.

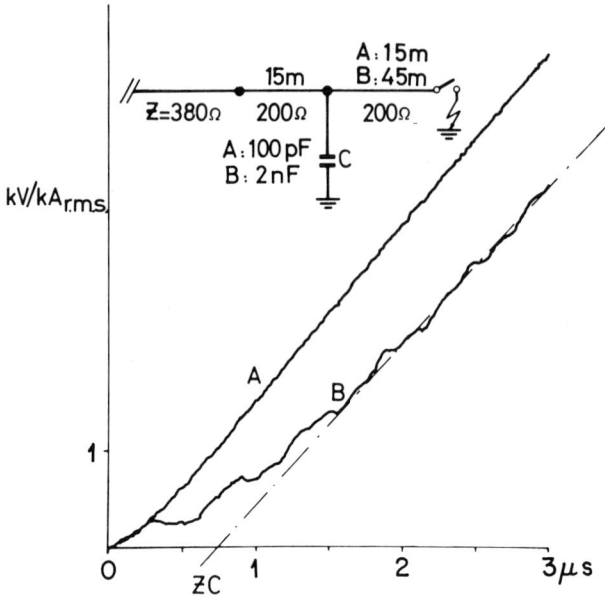

Fig. 5. Influence on itrv curves of distance to the capcitance and its size.
   Curve A: Distance 15 m
            Capacitance 100 pF
   Curve B: Distance 45 m
            Capacitance 2 nF

It is now possible to state some preliminary conclusions from this first approach:

- the initial slope depends only on the surge impedance of the bars connected to the circuit breaker,
- a capacitance of high value - more than 1 nF, for example - produces a reflection of the waves and an oscillation in the voltage,
- the first peak of the itrv appears at twice the travelling time of the waves between the circuit breaker and the capacitance.

We now look at the influence of a series inductance, which is especially important when we consider the effect of wave traps. Figure 6 shows the same network system as used in Figure 2, but the capacitances are absent and there is a small series inductance of 200 µH. In this case we observe several interesting influences.

During the first tenth of a microsecond, the wave shape has the same slope as that in Fig. 4 but the reflection of voltage is now positive and there is an increase in the voltage.

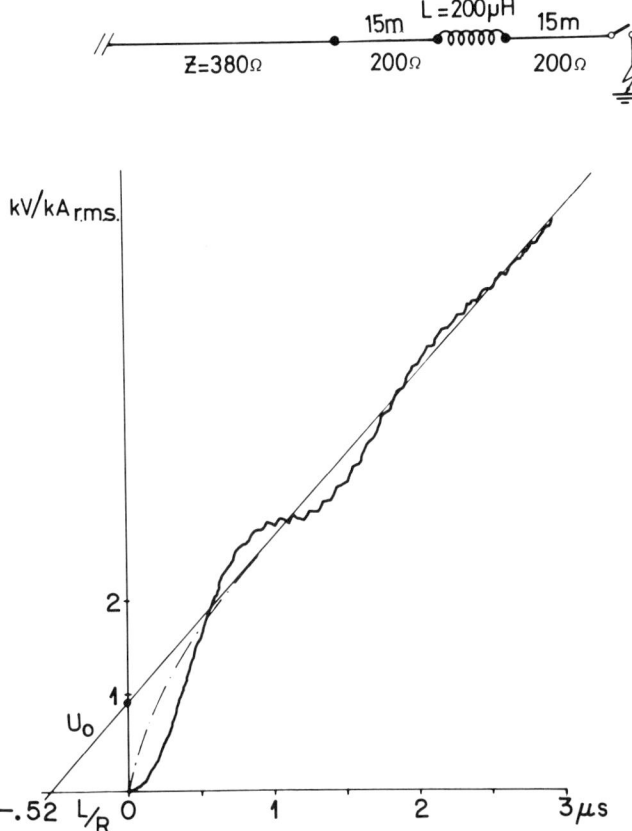

Fig. 6. Influence of a series inductance (wave trap) on itrv. Layout according to Fig. 2.

There is also an oscillation of the voltage around a line which has a slope proportional to the impedance of the semi-infinite line $Z = 380\ \Omega$ which defines the initial part of the recovery voltage when the layout of the station is neglected. This line intersects the time axis at a point which can be interpreted as a 'negative time delay' and the voltage axis it a point U_o which can be interpreted as a voltage offset.

In either case, it is only an interpretation of the fact that the series inductance gives theoretically a vertical tangent at the beginning of the wave in an ideal scheme with a semi-infinite line and a series inductance.

Finally we observe that the oscillation of the voltage around the exponential line, obtained by considering the ideal scheme with only an inductance in series with the line of impedance Z, is determined solely by this inductance and the capacitances of the bars. In the layout under consideration, the total capacitance of 15 m long bars is about 250 pF. Thus, if we assume that there is a voltage oscillation between the inductance and the capacitances of the two bars, which are then in series, we obtain a period for the voltage oscillation of

$$T \simeq 2\pi \sqrt{200 \cdot 10^{-6} \cdot 125 \cdot 10^{-12}} \simeq 1 \ \mu s.$$

This oscillation is damped by the presence of the surge impedance Z of the line, and the period is slightly shorter due to the damping of one of the capacitances of the preceding circuit.

From this analysis, the influence of a series inductance can be summarized as follows:

- a 'negative time delay' of the trv
- a very high itrv after the first peak
- an oscillation with the capacitances of the circuit.

4. THE ITRV IN A TREE LAYOUT

The configuration of the substation in the previous analysis was very simple and we shall now try to look at some more realistic network schemes.

We consider the layout represented in Fig. 3, which is the first step towards a complete representation of the substation.

Figure 7 gives the initial part of the trv on the same basis as used in the study above at three different points of the substation.

Curve A is the itrv at the terminal of the circuit breaker, and we see the same beginning as was observed in the other schemes: the part of the circuit breaker and the connection point to the two other bars. The slope is proportional to the surge impedance of this bar, and we reach what can be called the first peak, although the slope of the voltage is reduced after this point. In fact, the second part of the itrv for the time 0.1 µs and 0.4 µs has approximately the half of the initial slope, obviously due to the division of the voltage on the two bars. After this time there appears the combined effect of the series inductance on one branch and of the capacitances. The two curves B and C give an idea of the contribution of the two branches to the itrv and we can observe for these two 'components' the effect described in the previous analysis.

From this simple example we can derive some basic conclusions for this type of configuration:

- A connection with two bars gives a reduction of the surge impedance important enough to reduce the slope of the itrv and to give a noticeable reflection of the voltage wave.
- The elements situated on the two branches are still important and have a strong effect on the shape of the itrv.

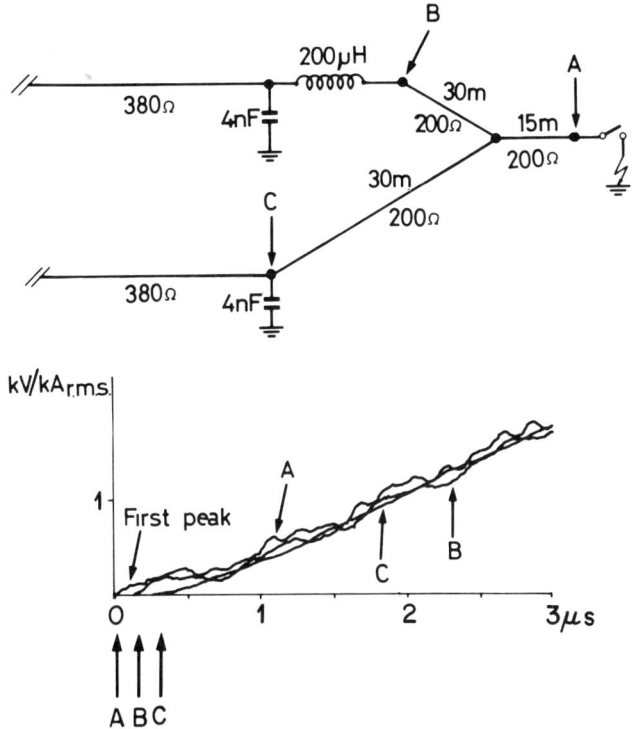

Fig. 7. Calculated itrv curves for a substation with tree layout according to Fig. 3. Curves A, B and C show the contribution at different locations.

All these calculations were performed with a single-phase representation of the impedances. It is of interest to compare the results with a three-phase representation.

5. COMPARISON BETWEEN SINGLE-PHASE AND THREE-PHASE SIMULATION

For the sake of example, the previous calculations were repeated on a three-phase basis, and the results shown in Fig. 8 should be compared with the case in Fig. 7.

This three-phase model includes all the couplings between the phases on the bars and on the line.

For the faulted phase, it is quite impossible to find any difference from a single-phase representation provided the right choice is made for the equivalent single-phase surge impedances. To a certain extent, this problem is simpler than for low-frequency phenomena.

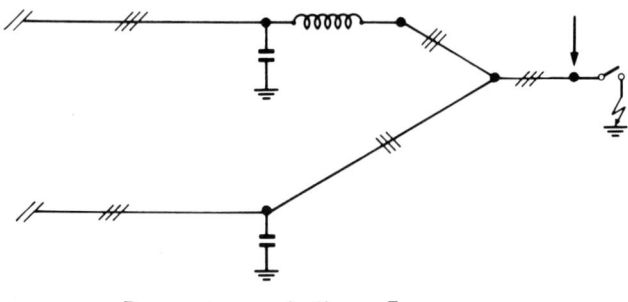

Parameters of figure 7

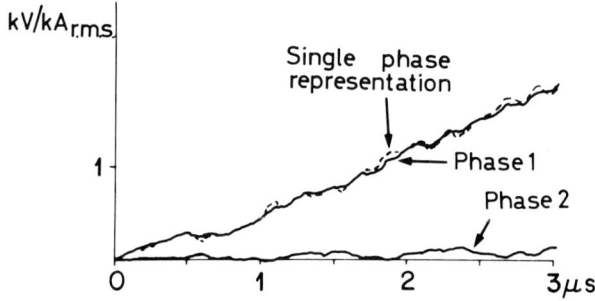

Fig. 8. Comparison of 3-phase itrv calculation with a single-phase calculation. Coupling effect between two phases.

The inductances were, in fact, determined for the high-frequency range, taking into account the skin effect: the variation of the inductance with frequency is negligible.

In the case chosen by way of example, we simulate the pole opening the fault, the two other poles of the circuit breaker being already open. In this case, it is easy to show that the equivalent surge impedance for the single-phase calculation is

$$Z = (Z_o + 2Z_d)/3$$

Z_d and Z_o being the direct and zero-sequence surge impedances, or, equivalently, Z is the surge impedance of the phase considered alone above ground.

All these definitions and comparisons can be found elsewhere in the literature, for example, in refs.[1,2].

In conclusion it may be said that provided the choice of the equivalent single-phase impedance is made correctly, the single-phase representation may be considered valid in most cases.

The induction effect on the other phases is low, and does not influence their trv greatly of the other phases.

6. THE PARTICULAR EFFECTS OF LINE WAVE TRAPS

For telecommunication purposes, one phase or more of a line may be equipped with a wave trap, tuned to a rather high frequency, usually in the range 80 kHz to 400 kHz. This lumped element has inductances, capacitances and resistors, plus some stray capacitances, which have to be evaluated.

For a basic understanding of what can occur, two particular circuits are considered in Figs. 9 and 10. The circuit of Fig. 9 shows a very simplified wave trap,

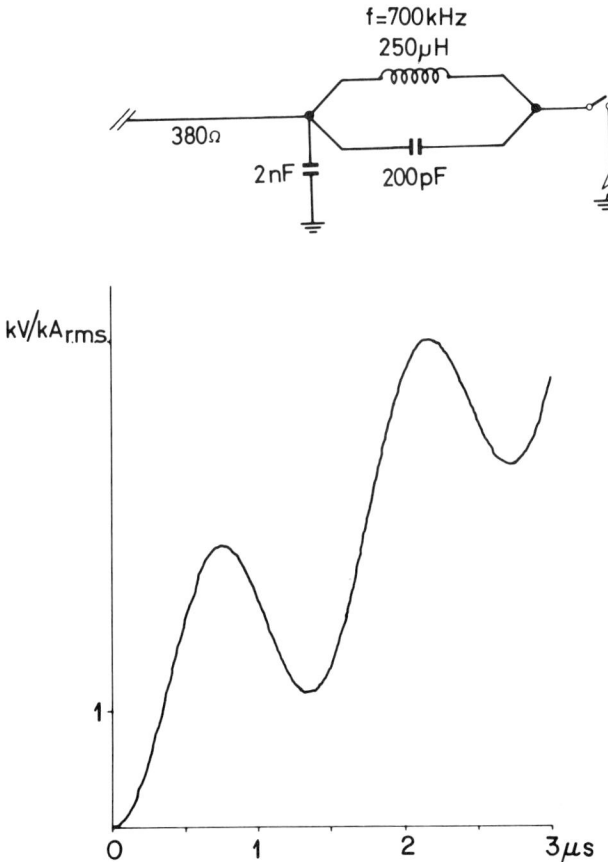

Fig. 9. Model circuit for a wave trap tuned to 700 kHz (top) and corresponding itrv curve (bottom).

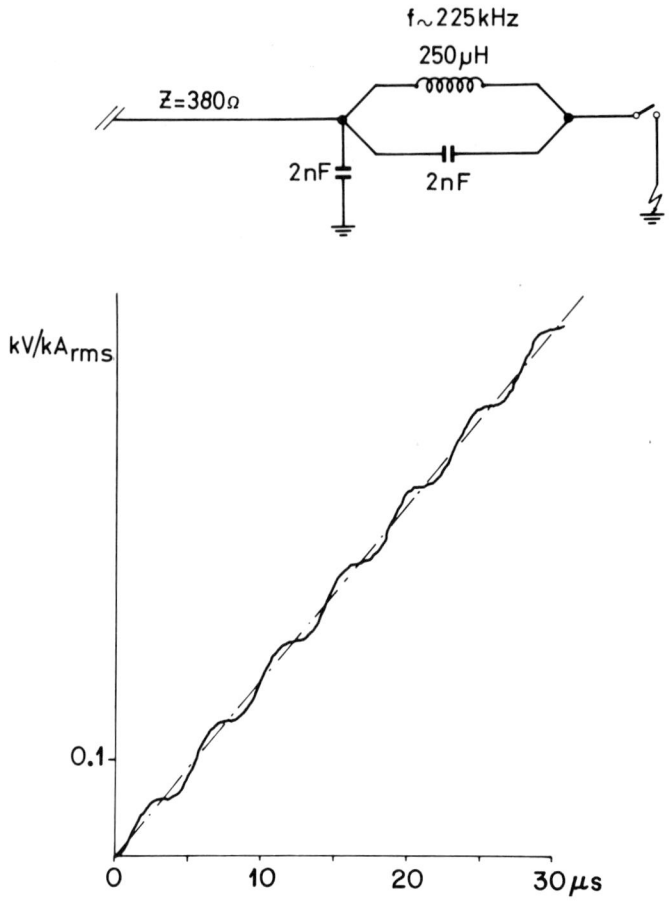

Fig. 10. Model circuit for a wave trap tuned to 225 kHz (increased parallel capacitance compared with Fig. 9) and corresponding itrv curve.

made up of only an inductance shunted by a capacitance to give a tuning frequency of 225 kHz. The capacitance to ground is the voltage divider to transfer the information to the ground. For this simple diagram, we observe a trv which presents an oscillation at the tuning frequency along a rather straight line with almost no time delay at the beginning; the influence of the capacitance to ground is negligible in this case, as expected.

In Fig. 10, the ordinate of which is expanded relative to that of Fig. 9, we have changed the tuning frequency by reducing the shunt capacitance. In this case, the oscillation of voltage has again the tuning frequency (about 700 kHz) and the time delay seems to be negative (the scale is not convenient for its evaluation). The amplitude of the oscillation is very significant and due only to the inductance.

In fact, such a representation is not absolutely correct as it neglects the different surge impedances of the connection and is an oversimplification of a wave trap.

Extensive studies of different types of wave trap included in different circuits have been performed by CIGRE Working Group 13.01 and the results are presented in reference[1]. Some of the conclusions which may be drawn regarding wave traps are as follows:

- A wave trap has a significant effect if it is connected directly to the circuit breaker.
- There is a significant influence if the current comes from the line and the wave trap.
- The main voltage oscillation is generally of the same order of frequency as the tuning frequency of the wave trap.
- The delay due to the capacitance to ground may be suppressed by a wave trap.

Considering in addition the results obtained in sections 2, 3 and 4, we arrive at the following conclusions:

- For a very high short-circuit current fed through the bars by many lines, in the case of a terminal fault the effect of wave traps is reduced.
- The effect of a wave trap on the short-line-fault (slf) condition is important. In fact, Figs. 9 and 10 represent the itrv contribution of the line side when a fault appears on the line a short distance away.
- The exact effect depends greatly on the exact constitution of the wave trap.

At this point we come to the end of the basic analysis of the influence of the different elements of a station and feel that we have obtained as many shapes of the itrv as there exist different layouts and construction of substations.

7. THE ITRV IN ACTUAL STATION LAYOUTS

The variety of stations, due not only to the voltage level, but also to the exploitation techniques and available devices, is so large than it seems impossible to define a general shape of the itrv.

The CIGRE Working Group 13.01, in an extensive study, considered nearly all basic configurations for each voltage level in different countries and established a kind of catalog of the different possible layouts which were used for the evaluation of itrv.

The main results, given in[1] can be summarized as follows:

- The initial part of the itrv depends only on the surge impedance of the connection, and on the current.
- The surge impedance to consider to obtain the initial slope is generally less than 260 Ω (this value is a good estimate for practical situations).
- The time to the first main modification of the wave is twice the travelling time of the wave to the 'main discontinuity'.
- In practice, the main discontinuity is either a division in two branches of the bars or the presence of a capacitance of more than 1.000 pF.
- After this initial part of the itrv, the wave may have very different shapes. As the presence of inductances is very rare on the bars, the itrv generally oscillates with a low amplitude to reach the recovery voltage which would be defined if we considered the station as a lumped element.
- When we assume that the station is a lumped element with a certain number of outgoing lines, the equivalent capacitance is always less than the sum of the capacitances of the individual elements, due to the effect of the inductance of the bars.

Finally, some figures can be assumed as representative of the majority of the networks. The justification for this table is given in[1].

TABLE 1

Network Voltage (kV)	145	245	300	362	420	525
Time to first peak t_1 (μs)	0.4	0.6	0.7	0.8	0.8	1

The rate of rise of the initial part is proportional to the surge impedance (260 Ω) multiplied by the current slope, $\sqrt{2}$, and a coefficient depending only on the type of fault and the opening sequence of the poles of the breaker.

In spite of these main results, the evaluation of the exact shape of the itrv might be necessary in cases where it is proved that its effect on the breaking is important. In such cases, a testing station, for example, a complete computation may be necessary and, as a consequence, it is of great interest to know how to determine the itrv.

8. SOME DETERMINATIONS OF ITRV

One of the most interesting methods of determination is obviously measurement in the network. However, it is one of the most difficult methods and perhaps the least accurate.

It is obvious that to determine this inherent itrv, you must use an ideal circuit breaker which - as is indicated by the name - does not exist. There is always an interaction between the arc and the network. Thus, it becomes necessary to use the low-voltage, or low-current, injection method. That means that all phenomena are linear, - which can easily be accepted in this case - and that the connections of low-voltage devices to the h.v. network are appropriately designed. Such tests have been performed.

For all tests, whether at high or low voltage, one of the greatest difficulties is the measurement of the current and voltage in this high-frequency range, taking into account the dimension of the station and the possible external perturbations.

Nevertheless, some results are available[1,3,4] which permit the evaluation and checking of other methods.

In fact, the most practical method is computation with the aid of a transient-phenomenon program. Such techniques are now available and give very accurate results if appropriate assumptions are made for the representation of the circuit and if the values of the parameters are selected with care. All the examples which were described in the first part of this paper were calculated with a computer program especially written for this purpose.

For the choice of the parameters, special measurements and computation methods are necessary due to the high-frequency range of the voltage and current oscillations. For example, the computation of surge impedances is made taking into account the skin effect on the inductances and resistances,[5,6] and checked with field measurements.

If some parameters are used from field measurements, it is necessary to take care that the measurement technique employed gives accurate values. For example, values may be derived by considering the current and the voltage during the first microsecond of an impulse test. The surge impedance calculated from this first part of the recording may differ largely from the values usually expected.

9. IS THE ITRV IMPORTANT?

If we assume that the foregoing considerations have given a more precise idea of the itrv phenomenon and its origin, it is very important to know if this consequence of the internal layout of the substation greatly influences the breaking of the current.

We can make two remarks:

- First, certain types of circuit breaker have an arc during the interruption which distorts the current so that the current just before current zero has a reduced slope. In such a case, it is obvious that the itrv (proportional to the slope of the current just before the interruption) is reduced and that the voltage reached during the first microsecond is very low and does not hamper the breaking.

- Secondly, many differences were observed between the results of testing in different testing stations with apparently the same wave shape of the trv. By considering the exact layout of the testing stations, differences may appear in some cases which can be explained in terms of the itrv.

These two contradictory remarks lead to the conclusion that some breakers might be more sensitive to the itrv than others, and that some attention must be paid to the exact itrv when a failure of breaking occurs in a testing station with a very high short-circuit current.

It can also be said parenthetically that if no breaking failure is attributed to the itrv in the network for the time being, it is due to the fact that for a circuit breaker which could have been sensitive to the itrv, without special care, the practical short circuit in the network is much lower than the rating and so the itrv (which is proportional to the short-circuit current) is very low.

In fact, the previous analysis gives a partial answer to the question of the importance of the itrv. If we notice that the amplitude of the itrv during the first microseconds is proportional to the short-circuit current, the problem must be taken into account only for high ratings. Furthermore, the amplitude of the voltage being in the range of few kilovolts for these high short-circuit currents, the only circuit breakers which might be sensitive are those for which the arc does not reduce the slope of the current before the clearing or those where the recovery withstand voltage is not sufficient during the first microsecond.

These considerations limit the problem to only some cases of circuit breakers and, in many cases, the itrv is not of great importance. Nevertheless, especially for new technology and for high ratings, this phenomenon has to be taken into account. From a design point of view, it is generally sufficient to consider the initial part and to consider that the circuit breaker has to clear the short-line fault. With these two considerations, the transient recovery withstand voltage of the circuit breaker is approximately defined.

The three conclusions resulting from these considerations are:

- For the design of stations, care has to be taken to avoid the presence of inductances very close to the circuit breaker.
- In many cases, the itrv has no great influence on the design of a circuit breaker.
- Some attention has to be paid to the itrv in cases where high ratings are specified and with some techniques which do not disturb the current before zero.

10. CONCLUSIONS

The analysis of the influence of the different elements which constitute the busbars of a substation shows the relative importance of these elements and consequences for the itrv.

Capacitances delay the recovery voltage and allow some oscillation, and reduce the stress.

Inductances increase the stress and provoke some oscillations with capacitances which can be of a relatively high amplitude.

The connections give some voltage ramps with a constant slope.

Wave traps generally increase the stresses on the circuit breaker.

There exist as many itrv wave shapes as station layouts, although some simplifications may be assumed for the initial part.

Accurate estimation of the itrv in any network is possible, especially with the aid of computer programs. Some care has to be taken over the estimates of the values for the parameters.

However, the consideration of itrv is only important in a high-voltage network for high short-circuit currents. Some breakers might be sensitive to this stress, as has already been demonstrated.[1,4]

11. ACKNOWLEDGEMENTS

The examples were chosen and calculated by the staff of the T.N.A. (Transient Network Analysis) of EdF and thanks are expressed here to every member, particularly G. Le Roy, M. Desforges, C. Cagnet and F. Piche.

Most of the results were arrived at by those members of Working Group 13.01 of CIGRE, who have worked extensively on this subject: M. Braun, Dwek, Harner, Hinterthür, Humphries, Lageman, Mazzoleni, Panek, Völcker reference[1] in the report of this Working Group.

REFERENCES

1. G. Catenacci, Electra 46 (1976) 39
2. G. Köppl and P. Geng, Brown Boveri Review 53 (1966) 311
3. E. Bolton, M. J. Battisson, B. J. Bickford, M. G. Dweck, R. L. Jackson and M. Scott, Proc. IEEE 117 (1970) 771
4. B. Calvino, G. Mazza, B. Mazzoleni and V. Villa, Some Aspects of the Stresses Supported by HV Circuit-Breakers Clearing a Short-Circuit, Cigré Report 13-08 (1974)
5. D. E. Hedman, IEEE Trans PAS 84 (1965) 200
6. C. Dubanton, Calcul approché des parametres d'une ligne de transport, Valeurs homopolaires, Bulletin de la Direction des Etudes et Recherches d'E.d.F., Série no. 1 (1969) 53

DISCUSSION
(Chairman: E. Ruoss, Brown Boveri)

G. Catenacci (CESI)

Can you tell us the difference in itrv for different station layouts, such as single-breaker or breaker-and-a-half arrangements?

M. Dubanton
I restricted my lecture to the situation where the fault occurs at the terminals of a circuit breaker for a single-breaker scheme. The itrv is then determined by the source-side busbars and elements alone. If, in the case of a breaker-and-a-half or ring-bus scheme, a fault occurs at the terminals of one breaker the situation is again the same. For the adjacent breaker, however, there is not only on the line side but also a short bus connection, of 50-100 m length, between breaker and fault which produces a contribution to the itrv. As the slope for this side is also proportional to 260 Ω, the total slope from both sides is proportional to twice this value, namely 520 Ω. A similar situation occurs in single-breaker schemes if a fault occurs between the breaker and an outgoing line.

J. Urbanek (General Electric)
Isn't it a general rule that the short line fault stresses a circuit breaker as far as the thermal mode is concerned more than the terminal fault with itrv, so that this requirement could be omitted if the short line fault is tested?

M. Dubanton
In most cases the short line fault covers the different itrv-situations. The situation is not so easy if we consider the possibility of oscillations with wave traps. Also, in the aforementioned case of an itrv on both sides of the breaker, at least the inherent trv in the first microsecond is above the short line fault trv (see also paper by Hermann and Ragaller).

D. M. Benenson (State University of New York at Buffalo)
Could you comment on the accuracy of the modelling together with possible results from experiment?

M. Dubanton
What I showed in my presentation was all calculated. In the CIGRE working group we did some comparisons with field tests. It must be said, however, that field tests in the itrv-regime are very difficult. We have no ideal circuit breaker for this regime, and measurements at lower voltages are disturbed by radio noise.

E. Ruoss
The present test standards for short-line fault include neither a time delay nor an itrv. In reality we have both influences. However, if one considers itrv and time delay together, the resulting trv is still below the present short line fault testing standards. Is that correct?

M. Dubanton

I feel that you are right in most cases, but not necessarily in every case.

M. B. Humphries (CEGB)

Could I just comment on that point. Where you have a capacitor voltage transformer connected to the line termination, then the practical combination of the itrv and short line fault components is generally contained within the theoretical IEC 450 Ω short line fault sawtooth waveform. If you have only a gas-filled current transformer, on the line terminals, then the combined stress will be greater than the IEC duty.

INTERACTION BETWEEN ARC AND NETWORK IN THE ITRV-REGIME*

W. HERMANN and K. RAGALLER
Brown Boveri & Company Ltd., Baden, Switzerland

SUMMARY

A theoretical study of the intensive interaction between a SF_6-circuit breaker and the high voltage network during the itrv-regime is presented. The influence on the recovery voltage of the time variation of the arc voltage and the arc current before and after current zero are analyzed using as an example the normal short line fault. The tools for a theoretical study of the interaction are described: a physical arc model which is applicable to the thermal interruption mode and a simplified circuit representation of the network situation in which an itrv appears.

The variations of current and voltage across the circuit breaker in the close neighbourhood of current zero are presented for the tf and the slf with different types of itrv. It is shown that the inherent voltage oscillation of the itrv is transformed to a steadily increasing variation on a strongly reduced voltage level.

Different network configurations yielding an itrv are calculated to explore the consequences of the superimposed itrv wave on the current limit of a circuit breaker. Tests are simulated using existing test standards for a short line fault as well as proposed values for the itrv to determine the influence of the itrv on the range of application of a circuit breaker. It turns out that a SF_6 circuit breaker which masters the slf according to present test standards, has no problem in mastering tests including itrv according to an IEC-proposal. In some cases the breaking limit is even raised compared to present slf test specifications.

The influence of the quantitaties which mainly determine the effect of the itrv on a circuit breaker, the time to the first peak and the value of the reflecting capacitance is analyzed.

---

*) Presented at the symposium by W. Hermann.

1. INTRODUCTION

In the preceding paper by Dubanton the appearance of the itrv has been discussed in detail. It has been shown that for several quite different layouts of an outdoor substation a rather similar voltage oscillation appears within the first µs after current zero. This can be specified by only a few characteristic parameters for a large variety of cases as has been done in an IEC-publication.[1]

Here the influence of the itrv on the breaking limit of a circuit breaker shall be investigated. These limits are given in the graphic representation of the range of application of a breaker as it has been discussed in the paper by K. Ragaller and K. Reichert.[2] In a double logarithmic drawing of the nominal voltage and the rated fault current there appear two principal failure modes which limit the applicability of a circuit breaker. The dielectric failure mode usually shows up in the tf (terminal fault) or out-of-phase switching and in most cases limits the breaker towards higher voltages. The thermal failure mode usually shows up in the slf (short line fault) and yields the vertical current limit. If the itrv appears in addition to the usual voltage stress, the question arises whether and in which way it changes these limits.

The itrv is a physical phenomenon which is very similar to the slf. It is a voltage oscillation which appears on the very short line in the vicinity of the circuit breaker, i.e. in the busbar of the substation. Compared to the slf the first voltage peak is rather low, but the time to the first peak is extremely short. Thus the main stress of the itrv is a high rate of rise of the recovery voltage within the first µs after current zero. One expects therefore an influence on the thermal mode of interruption. But it is not obvious whether and in which direction it shifts the thermal limit. Furthermore it is an open question which fault location gives the highest stress regarding the thermal interruption mode if the itrv is taken in consideration; is it still the slf or does there show up a thermal limit in the tf?

Close to current zero, where the decaying arc conductance passes values of the order of magnitude of the other circuit elements, there is a strong interaction between the arc and the network. Since the critical time period of the itrv is in the very first µs after current zero, the effect of the itrv on the breaker limit is strongly determined by this interaction. The basic feature of this interaction will be discussed first with the help of a simple circuit with one line.

2. INFLUENCE OF THE ARC ON THE TRV IN A CIRCUIT WITH A LINE

For a general discussion of the interaction the simplified one phase equivalent circuit, as shown in Fig. 1 is used. On the left hand side of the circuit breaker is the voltage source and the lumped representation of the feeding line inductance and the corresponding capacitance. After current interruption this side yields the trv as given in the tf. Due to the shorted line on the right-hand side of the circuit breaker a triangular voltage variation is superimposed on the recovery voltage which is caused by travelling wave effects on the short-circuited line.[2,3]

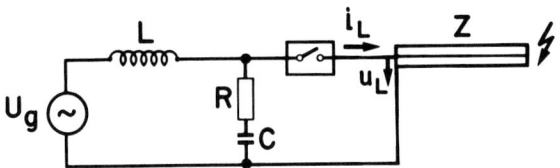

Fig. 1. Simple network to study the interaction of a circuit breaker with a line.

For the following calculations an ideal, lossless line is used where the waves are travelling without damping or distortion. In this case the momentary values of voltage and current along the line can be represented by the corresponding values at the entrance of the line at an earlier instant. Instead of following the momentary u-, i-distribution along the complete line it is thus possible to store and use the values at the line entrance for a time period equal to the travelling time $t_L = 2\ell/c$, where ℓ is the length of the line and c is the velocity of wave propagation. In this way, the following relation between the momentary voltage u_L and current i_L at the line entrance is obtained[4]

$$u_L(t) = Z i_L(t) - (u_L(t-t_L) + Z i_L(t-t_L)) \tag{1}$$

This formula is very useful for the investigation of the influence of the arc on the trv.

The voltage variation without any interaction, the so-called inherent stress, is obtained by opening the circuit with an ideal circuit breaker. This breaker has zero resistance before current zero and infinite resistance afterwards.

The current and voltage variation across an ideal circuit breaker around current zero are shown in Fig. 2a. Here as in some of the following figures the arc current is multiplied by the surge impedance to yield a voltage. With the ideal breaker one obtains a current decay to zero with constant slope and no current afterwards. Before current zero the voltage across the breaker is zero, while the quasisteady voltage across the line entrance u_{L0} and the current i_L are given by

$$u_{L0} = -\frac{t_L}{2} Z \left(\frac{di}{dt}\right)_0$$

$$i_L = \left(\frac{di}{dt}\right)_0 t \tag{2}$$

where t = 0 at current zero. In the observed short period around current zero the source side voltage is practically constant due to the normally large source side ca-

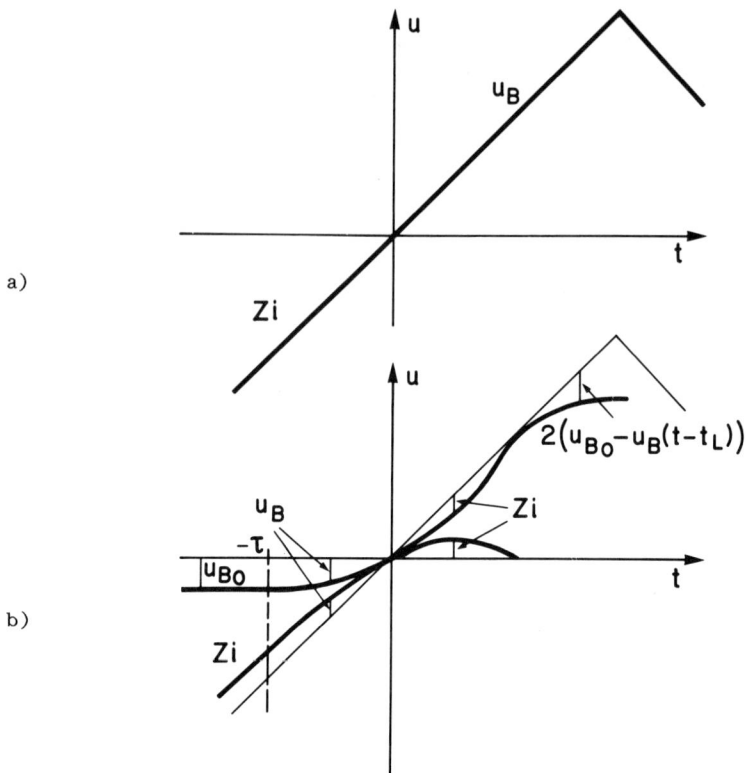

Fig. 2a,b. Interaction of the arc with the line.
a) Inherent current and voltage
b) Change of inherent values.

pacitance. Therefore after current zero the triangular voltage variation at the line entrance appears across the breaker: $u_B = u_{L0} - u_L$. The triangular variation of u_L can easily be constructed using formula (1).

Combining u_L and Zi_L and shifting this curve in time by the value t_L one obtains $(u_L + Zi_L)_{t-t_L}$ of equ. (1). A prescription of the current variation then directly yields the line voltage variation. The voltage variation across the breaker, as shown in Fig. 2a, is then given by $u_B = u_{L0} - u_L$.

Contrary to the ideal breaker a real circuit breaker has a decaying but limited conductance around current zero which is the conductance of the plasma column between the separated contacts. This causes in contrast to the ideal breaker an arc voltage before current zero and a post-arc current afterwards. Both influence the trv.

With the current decaying to zero the arc voltage has to drop to zero too. The form of this voltage drop and the deformation of the current before zero caused by this effect can easily be discussed for the following assumptions: Up to a time $t = -\tau$ the arc voltage stays constant, i.e. $u_B = u_{B0}$ for $t < -\tau$; the time interval τ within which the arc voltage decays to zero is smaller than the wave travelling time t_L back and forth on the line: $\tau < t_L$. Equ. (1) gives then after some algebraic manipulation a current deformation compared to the inherent case before current zero

$$\Delta i = \frac{u_B(t)}{Z}. \tag{3}$$

This is shown in Fig. 2b. The result is a shift of current zero compared to the current ramp of approximately

$$\Delta t = \frac{u_{B0}}{Z \cdot di/dt}. \tag{4}$$

This effect clearly reduces the stress on the breaker in the thermal regime by producing an interval of reduced current shortly before current zero.

This current deformation and the corresponding voltage change produce travelling waves on the line which are felt by the breaker again after the time t_L.

Using equ. (1) again it can be shown that this effect reduces the peak of the inherent line oscillation by Δu, where

$$\Delta u(t) = 2(u_{B0} - u_B(t-t_L)) \tag{5}$$

for $0 < t < t_L$ and $\tau < t_L$.

Close to the first voltage peak, where $t \simeq t_L$, and thus $u_B(t-t_L) = u_B(0) = 0$, the reduction is $\Delta u = 2u_{B0}$.

Figure 2b shows the graphic representation of equ. (5). The peak itself is reduced by twice the arc voltage u_{B0}.

In addition to these effects the post-arc current influences the trv. This influence can be shown to be

$$\Delta u(t) = Z \cdot i(t), \tag{6}$$

if the post-arc current interval is smaller than the travelling time t_L.

This change of voltage also shows up in the reflected wave after the first peak of the line side trv, where there is an increase in the trv of twice the value of equ. (6).

In this discussion it has been possible to separate the different contributions to the trv deformation by assuming a large enough value for t_L. In a normal short-line fault this condition would usually be fulfilled. In the case of the itrv, however, both interaction effects are effective at the same time due to the short travelling time $t_L \approx 1$ µs.

The first voltage peak is reduced due to the arc voltage and at the same time due to the post-arc current. The contribution of the post-arc current in the reflected wave leads to an increase of the voltage stress in the valley after the first peak. Thus arc voltage and post-arc current have the tendency to transform the inherent voltage oscillation to a voltage variation which is delayed, slowly increasing and approaching a quasisteady value far below the inherent peak value.

In SF_6 typical values for the arc voltage and the post-arc current times surge impedance are each about 1-2 kV per chamber. This is a relatively small voltage reduction compared to the value of the first peak in the normal slf but a rather large reduction compared to the first peak voltage of the itrv. Therefore a rather strong influence of arc voltage and post-arc current is expected in the case of the itrv.

3. REPRESENTATION OF THE CIRCUIT BREAKER

The coupling link in the interaction between arc and network is the time variation of the arc conductance. This conductance determines the arc voltage and the post-arc current and thus influences the voltage stress. But on the other hand, the voltage stress applied by the network produces a current through the arc and thus changes the arc conductance by ohmic heating.

In a numerical simulation of such an interaction a realistic representation of the arc in the circuit breaker is of special importance.

A main feature of the interaction is that the time variation of the arc conductance is changed during the observed time period. The way in which it is changed depends on the strength of the interaction and thus partially on the network. Here it is necessary to apply a physical arc model where the main processes determining the physical behavior of the plasma column are taken into account so that the variations of the characteristic quantities adjust themselves according to the process of interaction.

To build up such a physical arc model it is necessary to understand the interaction of the different physical processes occurring within the arc column, like ohmic heating, expansion of the quenching medium, radiative exchange of energy, turbulent exchange of energy, convection and so on. The processes determining the time variation of the conductance around current zero have then to be represented in a simplified but still realistic form. Basic studies[5] did show that the radial exchange by turbulent mixing of energy and momentum across the arc boundary is of determining importance for the period around current zero. A practicable representation of this process is a main point of the physical model used for the study of the interaction.

This model has been described in detail in[6] and is discussed in Swanson's contribution of this volume.

The model is able to describe limiting curves in the thermal mode dependent on fault current, pressure and quenching medium. A comparison of theoretical results using this model with experimental data is given in Frind's contribution in this volume.

The model can be used for the representation of the circuit breaker in a circuit as given for instance in Fig. 1, to study the interaction between network and circuit breaker. But it can also be applied to study the characteristics of a circuit breaker separately. As an example,[6] Fig. 3a shows the variation of the post-arc current in a SF_6 breaker for the case where a current ramp with constant slope is applied before current zero and a voltage rise with constant slope is applied after zero. The current slope corresponds to 50 kA at 50 Hz and the voltage slope is varied between 2.6 and 4.2 kV/μs.

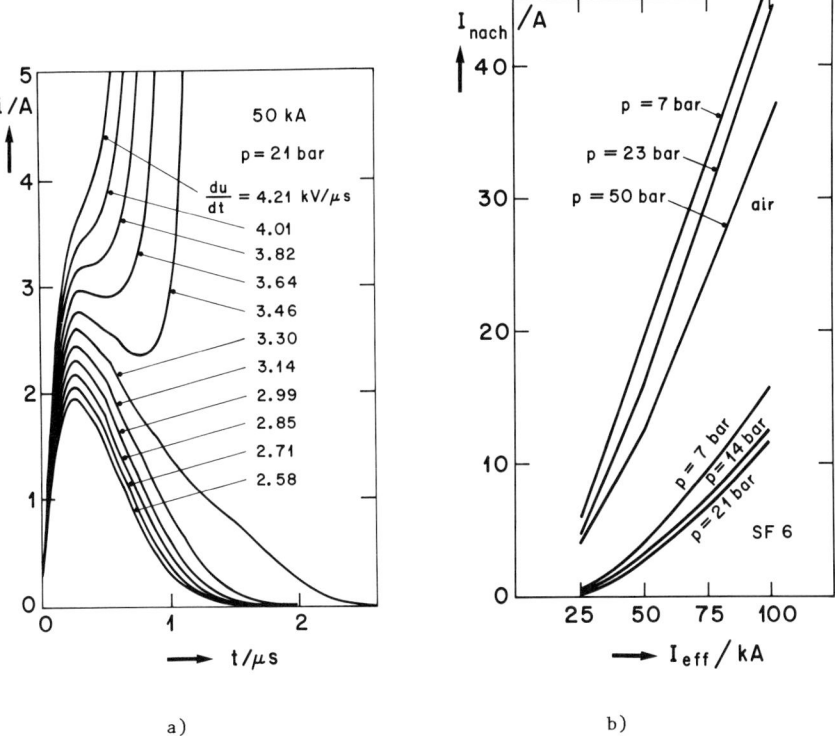

a) b)

Fig. 3a,b. a) Post-arc current for SF_6, 21 bar and I_n = 50 kA for different du/dt values.
b) Maximal post-arc current versus dependence on switching current for air and SF_6 at different pressures.

The post-arc current curves are formed by the decaying conductance which tends to reduce the current and the rising voltage which tends to increase the current. The typical form of the lower curves is determined by the dominance of the voltage rise immediately after current zero and the dominance of the conductance decay later on. There is a deviation from this form if the conductance variation is influenced by ohmic heating. As long as the conductance decay is not influenced by ohmic heating, the curves run similar to each other, increased in proportion to the increased voltage rise. But when the ohmic heating becomes effective the arc current increases at a greater rate than the applied voltage for a fixed value of time. A further increase of the rrrv leads to a stronger reduction of the conductance decay, a corresponding rise of the post-arc current and thus to thermal failure of the breaker.

In Fig. 3b the peak value of the post-arc current is plotted against the fault current for the case of interruption close at the thermal limit.[6] The curves are given for air and SF_6 for three typical values of upstream pressure for each. For SF_6 the peak value is nearly one order of magnitude smaller than in air; the post-arc currents in SF_6 are so small that it is very difficult to measure them. Since the reduction of the momentary voltage stress due to the post-arc current is given by current times surge impedance, even for the low SF_6 post-arc currents a strong influence in the itrv-regime is expected since there the voltage peak values are rather small. The influence of the SF_6 post-arc current in the case of a slf, on the other hand, is rather small. Contrary to SF_6 in an air breaker there is a strong contribution of the post-arc current to the slf while an itrv oscillation is completely damped out.

In all the calculations in the following an SF_6 double-nozzle breaker with 14 bar stagnation pressure was assumed giving a current limit in the 90 % short-line fault of 48 kA for one chamber.

4. REPRESENTATION OF THE DIFFERENT ITRV-CASES IN A SIMPLIFIED ONE-PHASE CIRCUIT

To investigate the interaction between arc and network in the itrv-regime the process of interruption is simulated numerically for several network configurations in which itrv can appear. For these simulations a simplified one-phase equivalent circuit as shown in Fig. 4 is used. On the left-hand side is the source side with the voltage source, feeding line inductance and capacitance. On the right-hand side is the short-circuited line. Between both there is the circuit breaker.

On both sides of the circuit breaker the possibility of an itrv oscillation is considered. For the source side this is justified in detail in Dubanton's paper. Corresponding information for the line side is mentioned in the discussion of this paper and in[1]. The itrv on both sides is modelled by an ideal line with a surge impedance of 260 Ω. The length of the line is adjusted to give a time to peak according to Table 1 of Dubanton's paper.

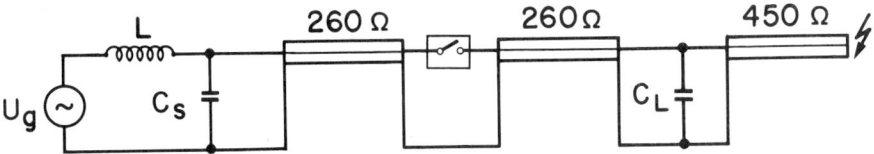

Fig. 4. General model for modelling different networks and fault conditions including itrv.

Between the outgoing line and the itrv-line a capacitor C_L is inserted in the model network. C_L represents the 'first major capacitor' in the vicinity of the breaker which is able to reflect the itrv wave.

The size of this capacitor is chosen according to[1] where a time delay t_d for the line oscillation is specified, namely 0.2 µs for 52-245 kV and 0.5 µs for voltages above 245 kV. From $C_L = t_d/450\ \Omega$, this results in a capacitance of $C_L \sim 0.4$ nF and $C_L \sim 1$ nF respectively.

The source-side capacitance C_s was chosen to be $C_s = 1.8$ µF in all cases.

By choosing different locations of the short circuit in the scheme of Fig. 4 the following types of fault can be modelled:

   Terminal fault with itrv from one side,
   Terminal fault with itrv from both sides,
   Short line fault with itrv from both sides.

In the latter case the possibility of eliminating one of the two itrv lines was also provided in the program. Thus in addition the following cases were studied:

   Short line fault with itrv from the source side only,
   Short line fault with itrv from the line side only.

The results for these different cases will be discussed in the following.

5. TERMINAL FAULT WITH ITRV

The interaction arc-network in the itrv regime can be illustrated especially clearly in the case of a terminal fault with itrv since, in addition to the itrv, there is no further voltage rise in the critical time period. There is no line voltage and the rise of the source voltage is delayed beyond the period of interaction.

In the case of the terminal fault only the configuration with itrv from both sides is treated, since for the itrv from one side, the inherent voltage variation is already below the slf stress. For reason of simplicity the two line sections with a surge impedance of 260 Ω are substituted by a single one with a surge impedance of 520 Ω which yields an equivalent stress up to the first voltage peak and a slightly increased stress later on. In the following figures the arc current is multiplied by the effective surge impedance of 520 Ω.

In Fig. 5 the voltage and current variation around current zero for a terminal fault with itrv from both sides are drawn for a 145 kV one-chamber breaker. The fault current is 63 kA which is close to the thermal limit, the time to the first peak τ_1 is chosen according to the IEC-publication[1] τ_1 = 0.4 μs.

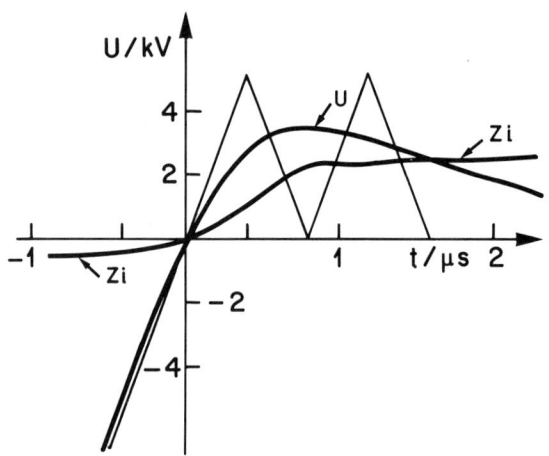

Fig. 5. Voltage and current for tf with itrv from both sides. Thin line: inherent values. U_n = 145 kV, I_n = 63 kA, τ_1 = 0.4 μs. The current is multiplied by 520 Ω.

The inherent voltage variation is depicted by a thin line. Compared to this the real voltage wave has a strongly distorted form. The arc voltage of the real breaker produces a current deformation with a delay of current zero. Furthermore, it reduces the inherent voltage stress in the region of the first peak by about 0.8 kV which ist 16 % of the inherent peak. The main lowering of voltage is caused by a post-arc current which rises to about 7 A. In the region of the inherent peak where it reaches about 5 A it reduces the voltage by about 2.7 kV which is 53 % of the inherent stress. Thus only a small fraction (about 30 %) of the inherent stress shows up across the circuit breaker in reality. In the valley between two subsequent voltage peaks the voltage variation is increased compared to the inherent stress. Thus the triangular wave pattern is replaced by a steady rise at a strongly reduced level.

From the voltage and current variation in Fig. 5 it can be derived that the decay of the conductance is slowed down by the ohmic heating due to the post-arc current. Between current zero (t = 0) and 2 μs later (t = 2 μs) the arc resistance increases only by a factor of 4 (150 Ω to 700 Ω) while without heating the resistance would increase by a factor of 20 within the same time. Thus the period of interaction is strongly extended and a situation close to equilibrium between heating and cooling is obtained.

An example of the voltage and current variation around current zero for a multi-chamber breaker is shown in Fig. 6 for a 420 kV four-chamber breaker and for the same type of fault. For a terminal fault with itrv from both sides the thermal limit is reached at approximately 94 kA. The voltage and current pattern after current zero are quite similar to Fig. 5. Here a quasistationary situation is obtained for an even longer time.

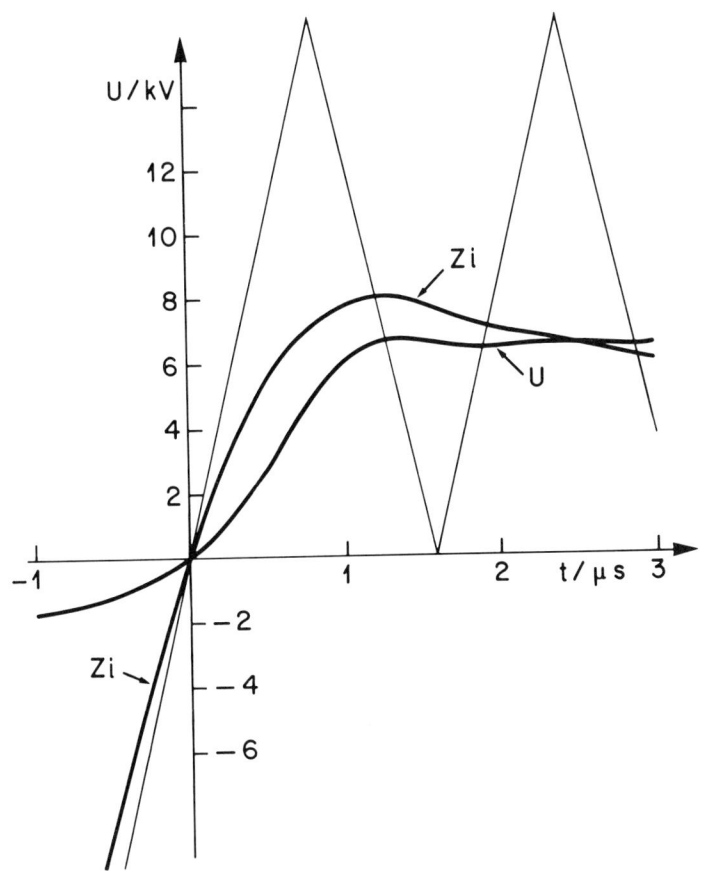

Fig. 6. As in Fig. 5, but for U_n = 420 kV, I_n = 93 kA, τ_1 = 0.8 µs.

Due to the essential reduction in the rate of rise of the recovery voltage the thermal limit of the breaker is strongly increased compared to the limit which would result from the inherent values.

Compared to this reduction the influence of the post-arc current in the slf is negligible. Therefore the thermal limit in the case of a terminal fault with itrv lies far above the current where this limit is reached in the slf. This situation is represented in Fig. 7 where the area of application for a 420 kV four-chamber breaker is shown. The fault current is normalized by the current at which the thermal limit is reached in the slf without itrv (in this example approximately 69 kA). The thermal limit for the tf with itrv from both sides lies about 35 % higher in current. For a 145 kV one-chamber breaker the corresponding relation is 63 kA for tf with itrv

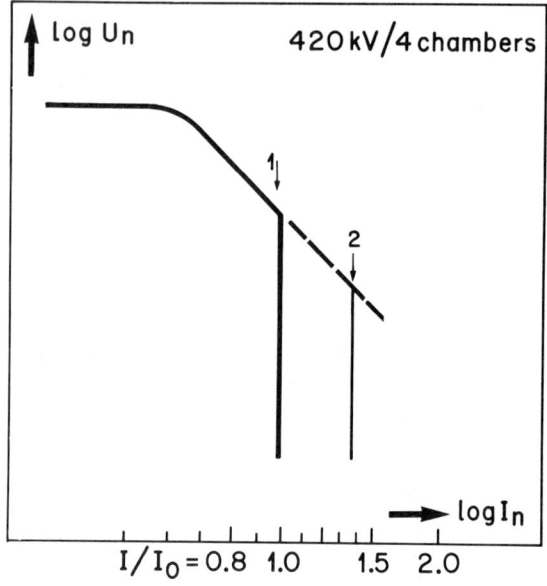

Fig. 7. Comparison of current limit in the log U_n/log I_n-plane for slf (1) and tf with itrv from both sides (2).

from both sides to 48 kA for the slf without itrv. Thus the tf with itrv from both sides is covered by the present slf test specifications and no additional tests are necessary to verify this fault behavior.

6. SHORT LINE FAULT WITH ITRV

6.1. Comparison of Inherent Voltage Variations

In the case of a slf with line side itrv the contributions of the line and the itrv line section modify each other even in the inherent case. But for a slf with source side itrv both contributions just add and therefore can be sketched directly.

As an example the inherent voltage variation for a 420 kV breaker in the case of a 90 % slf with itrv from the source side for a fault current of 70 kA at 50 Hz is shown in Fig. 8a and 8b. In Fig. 8a according to the IEC-proposal the contribution of the short-circuited line is given by a voltage rise after a time delay of $t_d = 0.5$ µs with a constant slope of du/dt = 450 Ω·di/dt and the itrv is superimposed from t = 0 with a constant slope of du/dt = 260 Ω·di/dt up to $\tau_1 = 0.8$ µs. The combined stress is always below the inherent stress as given by the present slf test specifications. If one takes into account that the time delay is caused by a capacitance, then one obtains a contribution of the short-circuited line which begins at t = 0 and which has a slope continuously approaching the value without capacitance (Fig. 8b). In this case the combined inherent stress slightly exceeds the stress according to present slf specifications during a limited time period.

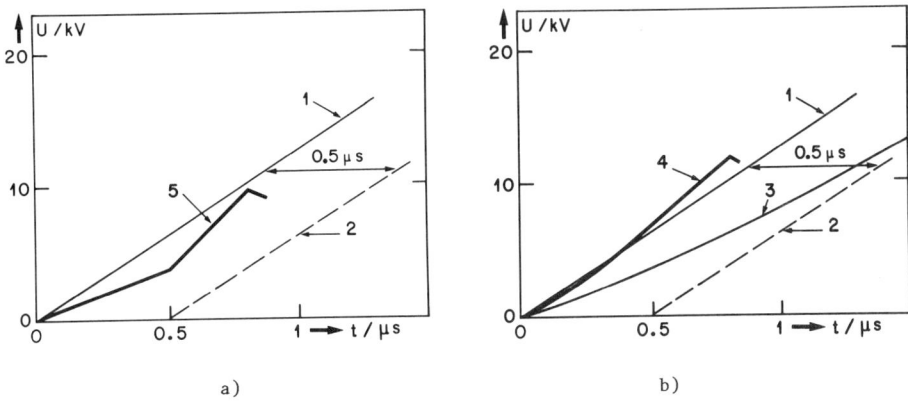

Fig. 8a,b. Comparison of inherent voltages for a 420 kV network for 70 kA.
1 90 % slf
2 slf with time delay of 0.5 µs according to IEC
3 slf with time delay of 0.5 µs caused by a capacitor at the line entrance
4 source side itrv (Z = 260 Ω, $\tau_1 = 0.8$ µs) superimposed on curve 3
5 source side itrv (Z = 260 Ω, $\tau_1 = 0.8$ µs) superimposed on curve 2.

6.2. Voltage and Current Variation around Current Zero

To illustrate the modification of the inherent stress by post-arc current and arc voltage a comparison between inherent and real voltage and current variations is shown for the different itrv-cases in a slf.

In Fig. 9 an example for a slf with itrv from the source side ($\tau_1 = 0.5$ µs, $C_L = 0.4$ nF, Z = 260 Ω) is shown for a 145 kV breaker. The fault current is 90 % of 48.1 kA which is close to the thermal limit. The inherent voltage variation is represented by the thin line in Fig. 9. The arc voltage and the post-arc current times

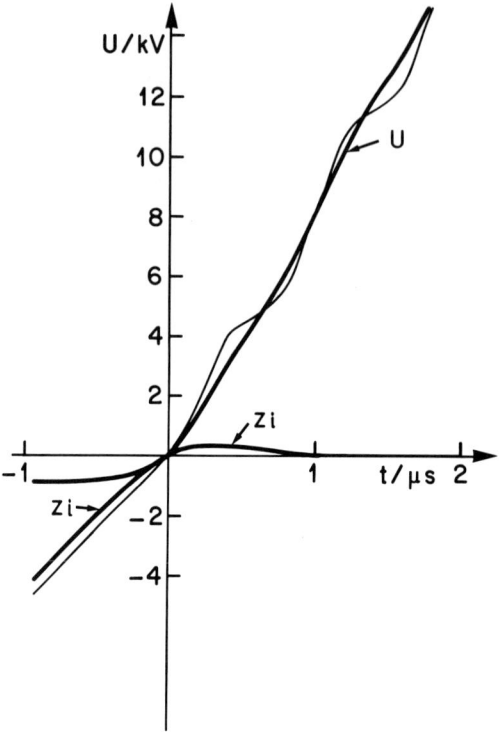

Fig. 9. Voltage and current traces for short-line fault with source side itrv. U_n = 145 kV, I_n = 48 kA, τ_1 = 0.5 µs, C_L = 0.4 nF. The current is multiplied by 260 Ω.

surge impedance have about the same values as the amplitude of the itrv oscillation. Therefore they completely damp out the additional itrv stress and enforce a voltage rise with nearly constant slope.

The most critical parameter in the description of the itrv stress is the time to peak τ_1. Increasing this time corresponds to raising the voltage of the first peak. For equal parameters as in Fig. 9 an increase of τ_1 from τ_1 = 0.5 µs to τ_1 = 0.8 µs leads to a failure in interruption as shown in Fig. 10. The recovery voltage is reduced compared to the inherent stress but after a time of about 0.6 µs the heating of the arc plasma by the current leads to a failure of the breaker. The conductance increases strongly and the current rises while the voltage drops.

For the same values of U_n, I and C_L as in Fig. 9, but with τ_1 = 0.4 µs the variation of voltage and current for a slf but with a line-side itrv is shown in Fig. 11. Here the wave pattern is already distorted in the inherent case (thin line). The de-

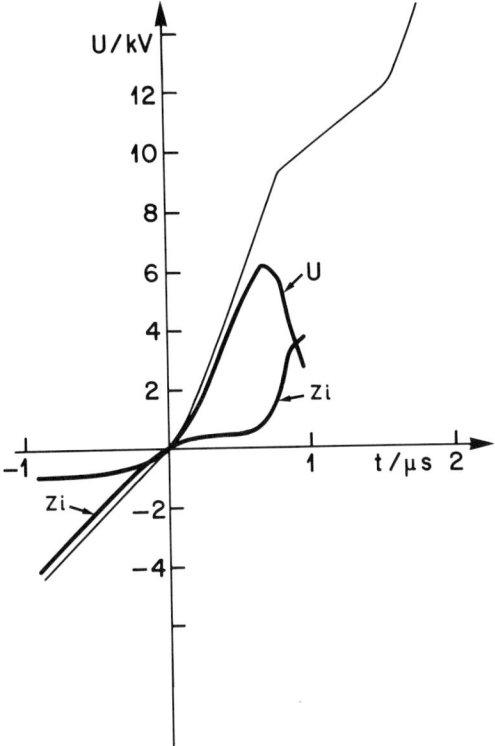

Fig. 10. Same as in Fig. 9 with τ_1 increased to $\tau_1 = 0.8$ μs.

lay in the rise of the main line voltage due to the short line section (0.2 μs) and due to the reflecting capacitance (0.2 μs) are recognizable. The current deformation and the corresponding time delay before current zero are larger in the case of the line side than in the case of the source side itrv due to the influence of the short-line section. This implies a further aid in the process of interruption. The real stress is below the inherent one before the first peak of the itrv but above later on. Nevertheless, the inherent as well as the real stress are below the corresponding curves resulting from short line fault tests according to the present specifications.

A comparison between slf with source side and slf with line side itrv for a 420 kV breaker is shown in Fig. 12 and Fig. 13 ($\tau_1 = 0.8$ μs, $C_L = 1$ nF). The fault current is 90 % of 69 kA which is close to the thermal limit. For the line-side itrv the time delay due to the line section (0.4 μs) and due to the reflecting capacitance (0.5 μs) is very pronounced. The current deformation is larger than for the source-side itrv and yields a longer time delay before current zero. In the case of the source-side itrv the faster increase of the line voltage in the first μs after zero

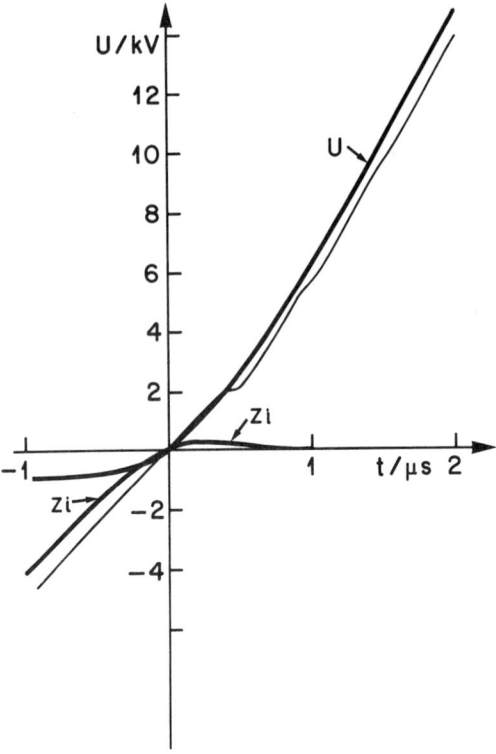

Fig. 11. Same as in Fig. 9, but for the case of short-line fault with line-side itrv.

causes an especially high post-arc current which leads to a great reduction of the real voltage stress compared to the inherent one.

6.3. Consequence of the itrv Stress on the Breaking Limit

By numerical simulation the shift of the thermal limit of a circuit breaker due to the different cases of itrv has been determined. For this purpose the fault current has been varied until a fault current and a current yielding interruption had been found close to each other. The results for the cases of a 90 % slf for a 145 kV one-chamber breaker and a 420 kV four-chamber breaker are given in Tables 1 and 2.

The results according to Table 2 are shown in the double logarithmic graphic representation of the breaker limits, Fig. 14. The limit in the dielectric mode and in the thermal mode as obtained by tests according to present short-line fault test spec-

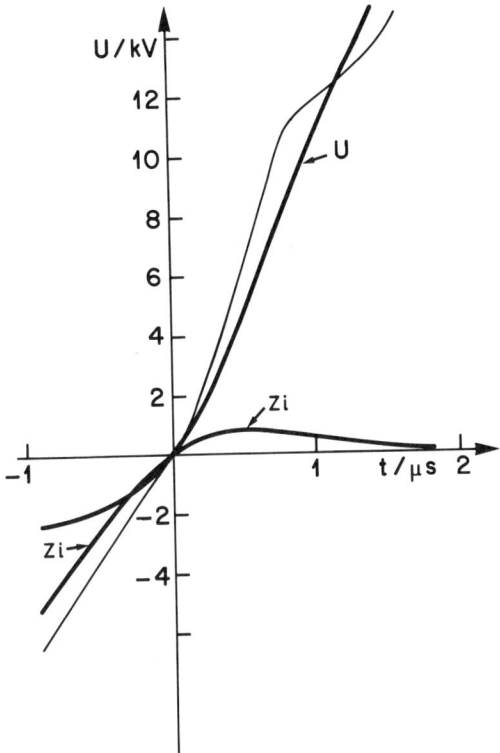

Fig. 12. Voltage and current for short-line fault with source side itrv. U_n = 420 kV, I_n = 69 kA, τ_1 = 0.8 μs, C_L = 1 nF. The current is multiplied by 260 Ω.

TABLE 1

U_n = 145 kV   one-chamber breaker
τ_1^n = 0.4 μs   90 % slf

	slf with source side itrv	slf with line side itrv	slf with itrv from both sides	slf present test	slf with source side itrv
C_L/nF	0.4	0.4	0.4	0	0
thermal limit I_n/kA	48.5	53.5	47.5	48	45.5

TABLE 2

U_n = 420 kV four-chamber breaker
τ_1^n = 0.8 µs 90 % slf

	slf with source side itrv	slf with line side itrv	slf with itrv from both sides	slf present test	slf with source side itrv
C_L/nF	1.0	1.0	1.0	0	0
thermal limit I_n/kA	69.5	79.5	69	69	63

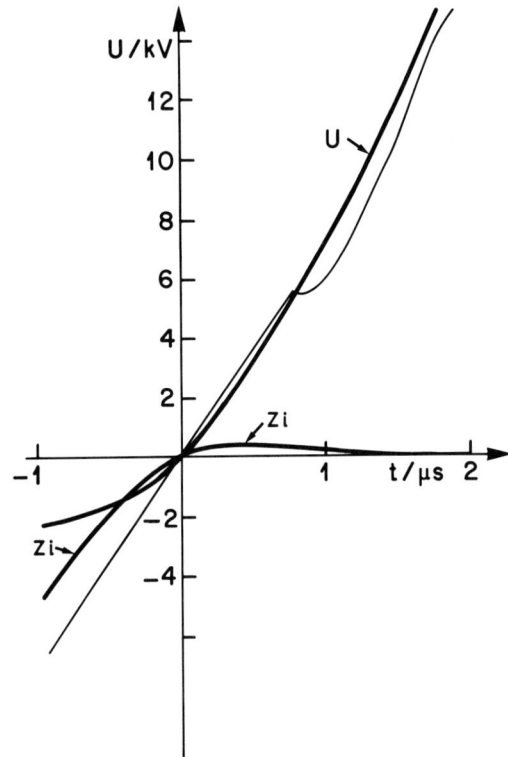

Fig. 13. Same as in Fig. 12, but for the case of short line fault with line-side itrv.

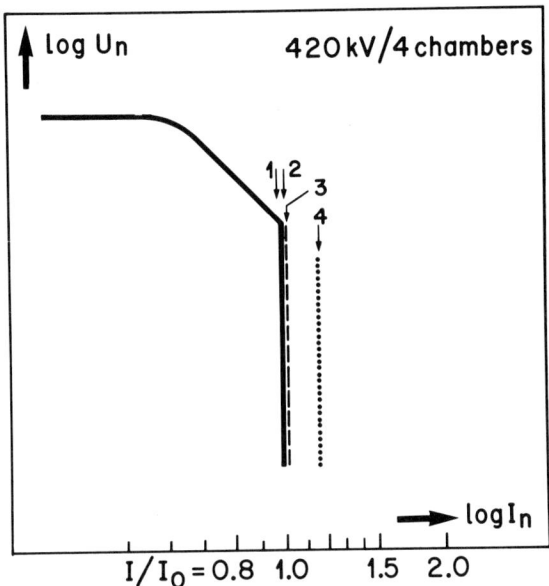

Fig. 14. Comparison of the current limit in the log U_n/log I_n-plane for short line faults with different itrv contributions.
1 = slf limit
2 = slf and itrv both sides
3 = slf and itrv source side
4 = slf and itrv line side.

ifications are drawn as solid lines. The fault current is normalized by the limiting current in the present short line fault test.

The line-side itrv is not critical at all. The voltage rise due to the main line is delayed by the reflecting capacitance as well as by the itrv line section. Therefore the thermal limit of the breaker lies well above the slf limit according to present test specifications.

In the case of the slf with source-side itrv only the itrv yields an additional voltage stress, but the line-side time delay brings a reduction of the voltage stress compared to present test requirements. Both effects approximately compensate each other so that compared to present slf rules the current limit is raised only negligibly.

In the case of the slf with itrv from both sides the additional stress due to the source-side itrv and the reduction of the line-side stress due to the time delay caused by C_L and τ_1 compensate each other. Thus the same limit approximately results as in the present short-line fault test. Due to the enhanced rate of rise corresponding to twice 260 Ω, i.e. Z = 520 Ω the inherent voltage variation runs above the present slf specification from zero up to the first peak of the itrv. But due to the strong interaction between arc and network this increased initial stress is compensated.

6.4. Main Parameters Describing the itrv Stress

The main quantities which determine the influence of the itrv on the breaker limit are the time to the first peak τ_1, which corresponds to the length of the itrv line section, and C_L, the value of the reflecting capacitance. Both quantities are independent of each other. For substations of different voltage levels they have characteristic values. The influence of these two main parameters on the breaking limit of SF_6 breakers is illustrated in Fig. 15 and Fig. 16.

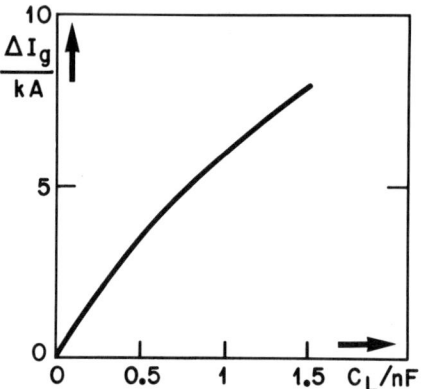

Fig. 15. Increase of the current limit with increasing capacity C_L for the case of short line fault with source-side itrv.

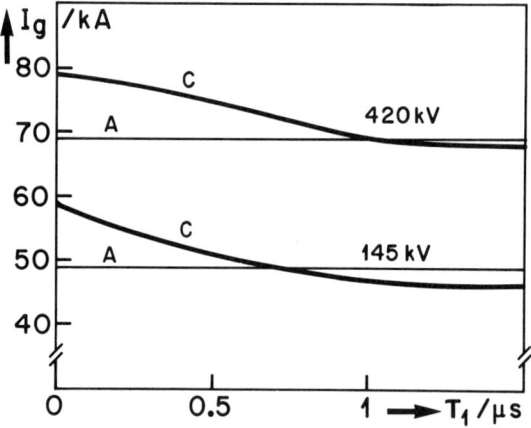

Fig. 16. Dependence of the limiting current I_g on the time to the first itrv peak τ_1 for the case of a short line fault with itrv from both sides (Curve C). Line A shows the current limit in the common short line fault.

In Fig. 15 for a slf with itrv from the source side the increase in the current limit in the thermal mode with increasing capacitance C_L is shown. This curve is identical for a 145 kV and a 420 kV breaker. When $C_L = 0$, the reduction of the current limit due to the superposition of a source-side itrv would be 2.5 kA for the 145 kV breaker and 6 kA for the 420 kV breaker (see Tables 1 and 2). The capacitance C_L which corresponds in both cases to the time delay t_d proposed by IEC for the two voltage levels approximately compensates the influence of the itrv.

In Fig. 16 the variation of the breaking limit with τ_1 the time to the first itrv peak is shown for a 145 kV and for a 420 kV breaker. The breaking limit in the slf according to existing specifications is drawn as a straight line A. The curve C gives the limiting current for the slf with itrv from both sides. The inductance of the complete line is chosen in such a way that the fault current is reduced to 90 % of the tf current. For $\tau_1 = 0$, i.e. without itrv, the breaking limit as given by curve C is raised above the slf limit due to the parallel capacitance C_L which in this example is chosen as $C_L = 1.5$ nF for both voltage levels. An increase of τ_1, i.e. of the two short-line sections results in a decrease of the breaking limit. This is mainly caused by the source-sinde line section. If there is only an itrv contribution from the line side, the limit is increasing with the length of the line, i.e. with τ_1.

7. CONCLUSION

A physical circuit breaker model has been applied to study the interaction between circuit breaker and network in the neighbourhood close to current zero. Especially for the case of an interaction extending over a prolonged time period, only a model of this type ensures realistic results.

It has been shown that for a SF_6 breaker a strong interaction between breaker and network is effective in the time period up to 1 μs after current zero. The variation of the arc voltage and the corresponding current deformation before zero as well as the residual current after zero reduce the voltage peaks in the itrv regime very effectively. The superimposed wave pattern is damped away to a quasicontinuous voltage stress.

The existence of a lumped capacitance within a short distance of the breaker causes a time delay of the line-side voltage stress and thus aids the breaker in the process of interruption. An additional time delay is caused by the short section on the line side.

Thus the existence of the itrv causes no increased stress for an SF_6 breaker compared to present short-line fault specifications. In the case of the line-side itrv the breaking limit in the thermal mode is strongly increased. Thus the existence of itrv oscillations as described in the IEC paper[1] yields no additional problem to an SF_6 breaker and does not require additional test procedures once the control of the slf according to the existing requirements has been verified.

ACKNOWLEDGEMENT

The help of W. Egli in performing the numerical calculations is gratefully acknowledged.

REFERENCES

1. IEC paper 17 A 150
2. K. Ragaller, K. Reichert, this volume
3. M. B. Humphries, this volume
4. L. Ferschl, H. Kopplin, H. H. Schramm, E. Slamecka, J. D. Welly, Cigré report 13-07 1974
5. W. Hermann, U. Kogelschatz, L. Niemeyer, K. Ragaller and E. Schade, IEEE Trans. PAS 95 (1976) 1165
6. W. Hermann, K. Ragaller, IEEE Trans. PAS 96 (1977) 1546

DISCUSSION
(Chairman: E. Ruoss, Brown Boveri & Cie. Ltd., Baden)

J. Urbanek (General Electric)

I have three questions. First, did you include the proper amplitude factors in your model network? For the same trv rates the higher the damping the smaller the effect of the arc on the trv. Second, your failure-mode graph, showing dielectric and thermal failure modes, would imply that a case with high current would be more likely to be thermal than one with a lower current. With practical circuit breakers, however, though the terminal fault has a higher current than the short-line fault, the former represents a dielectric limit, and the latter a thermal one. And finally, could you give some information on the axial position of the arc at which the interruption occurs in these cases?

W. Hermann

The answer to your first question is that we did not include any damping in our models. The accuracy to which itrv is known does not justify that. We cannot agree with your statement about circuit-breaker-network interaction in the general case. There are situations where higher damping causes a stronger interaction at the same rate of rise.

As far as your second question is concerned, this is a misunderstanding. We plot in our diagram the rated short circuit current. This nominal value is the same for terminal fault and short line fault. When evaluating the limiting curves the relation between rated current and test current has to be taken into account.

The answer to the third question can be given only by the results of arc models which include axial variation (models of Swanson or Tuma and Lowke).[1]

D. T. Tuma (Carnegie-Mellon University)

I can show results of calculations we did with our arc model and which give an answer to this question (Figs. 1 and 2). Figures 1 and 2 are axial profiles of the electric field in the Brown Boveri nozzle in the post current zero regime calculated using the model of El-Akkari and Tuma.[2] The results are for an arc in air starting at a current of 2000 A and linearly ramped to zero at the rate of 24 A/μs. After current zero a linear voltage ramp with a characteristic rrrv is applied across the decaying arc to explore the possibility of thermal reignition.

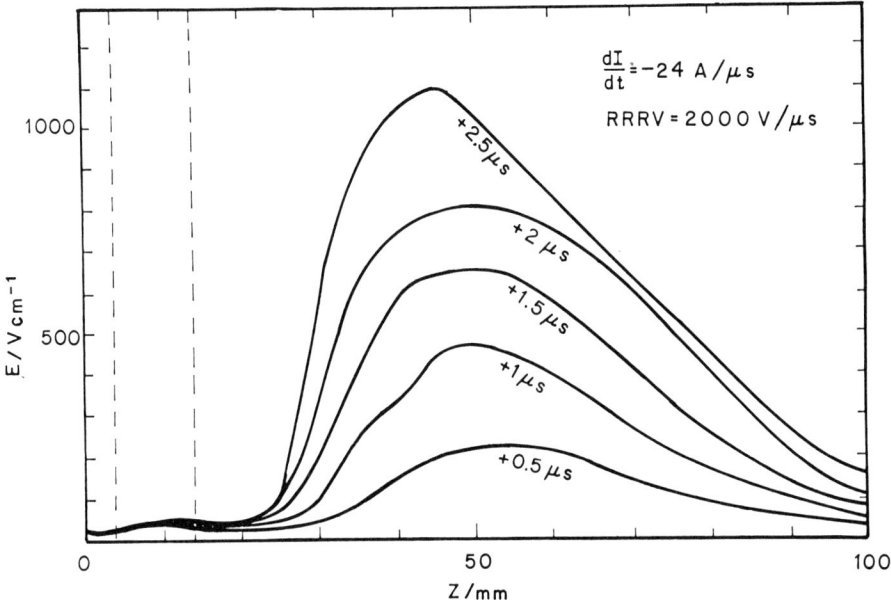

Fig. 1. Axial profiles of the axial electric field at 0.5 μs intervals after current zero with rrrv = 2000 V/μs. The peak electric field is seen to move towards the nozzle throat (shown as dashed vertical lines) as time evolves. These results correspond to case 1 of Fig. 7 of[2] which shows successful interruption of the arc.

G. Frind (General Electric)

Did you make measurements on either one of the mentioned post-arc currents? A lot of people have trouble measuring post-arc currents in SF_6 which seem to be very small.

E. v. Bonin (AEG)

My remark concerns the same problem: your Fig. 5 with the high post-arc current of 7 A and strong damping of the trv looks very atypical. I have never seen a measured oscillogram with that appearance.

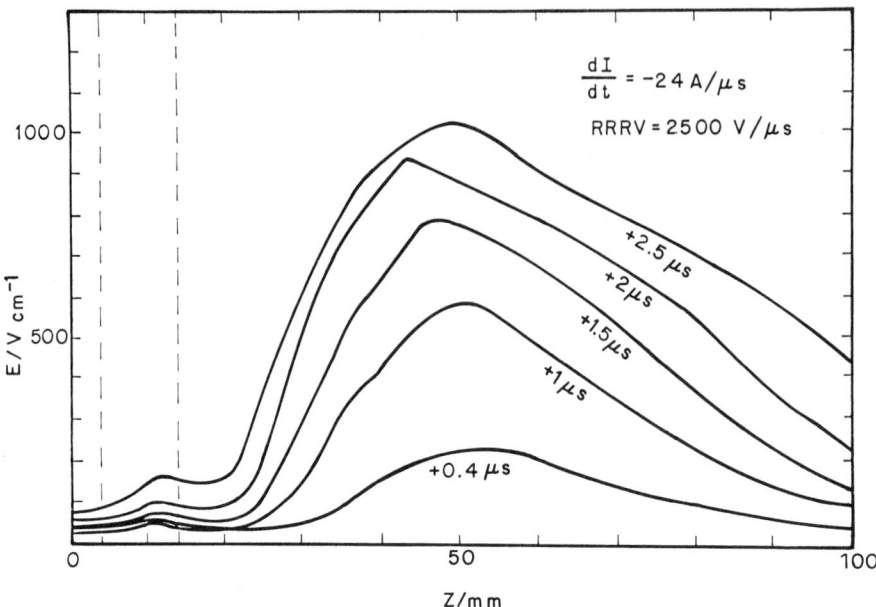

Fig. 2. Same conditions as Fig. 1 except that an rrrv = 2500 V/μs is applied. The peak electric field first moves towards the nozzle throat, but suddenly shifts toward downstream at t = 2.5 μs indicating the beginning of thermal reignition. These results correspond to case 2 of Fig. 7 of[2] which shows a reignition of the arc.

K. Ragaller (Brown Boveri)

Figure 3 shows measured oscillograms[3] for an SF_6 breaker in a slf test. The scales are: current 40 A/div, voltage 5.4 kV/div, time 2 μs/div. Except that in this case the source side voltage is not negligible during the interaction period, the qualitative appearance is very similar to Fig. 5. (Comment subsequent to meeting).

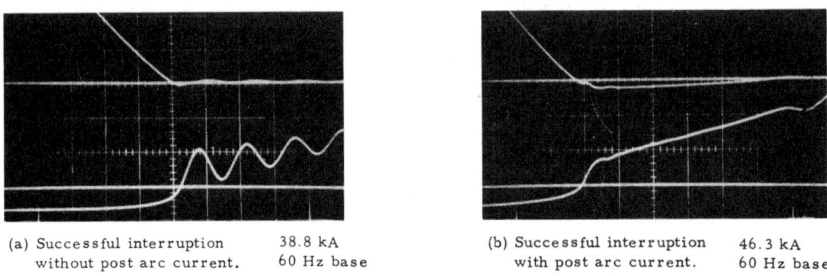

(a) Successful interruption without post arc current. 38.8 kA 60 Hz base

(b) Successful interruption with post arc current. 46.3 kA 60 Hz base

Fig. 3. Current zero oscillograms from ref.[3].

REFERENCES

1 B. W. Swanson, this volume
2 F. R. El-Akkari and D. T. Tuma, IEEE Paper No. F 77 126-6 (1977)
3 M. Murano, H. Nishikawa, A. Kobayashi, T. Okazaki, S. Yamashita, IEEE Trans. PAS 94 (1975) 1890

DETERMINATION OF THE PEAK TRANSIENT RECOVERY VOLTAGE*

J. L. DIESENDORF and S. K. LOWE
Electricity Commission of New South Wales, Australia

L. SAUNDERS
State Electricity Commission of Victoria, Australia

SUMMARY

The inherent trv of a system at a given location may be predominantly oscillatory or exponential. Interaction effects of the circuit breaker, other than switching resistors, and high frequency transients due to short line faults or local system elements are of little significance in the peak regime. Exponential trv's are frequently disguised by a major superposed oscillation caused by parts of the faulted sub-system which lie outside the primary fault circuit.

The system trv should be assessed beyond the highest peak. With an oscillatory trv several 100 μs will suffice but about 2 ms is required for exponential trv's. This suggests a system model extending to a radius of 300 km! Fortunately this is not necessary in complex interconnected systems. The model has to cover the whole of the system directly connected to the fault point at the one voltage, however the models of lines going out from remote busbars can usually be markedly simplified. Care must be taken in modelling local transformers although in general an 'equivalent line' model suffices for the system behind a transformer. Local loads can have a very significant effect after a few 100 μs despite several stages of transformation to low voltage.

Some simple rules for the calculation of inherent trv using a single phase travelling wave computer program are elucidated.

1. INTRODUCTION

1.1. TRV and Inherent trv

The transient recovery voltage (trv) obtained on the interruption of fault current in a high voltage AC network or system depends on:

---

*) Presented at the symposium by J. L. Diesendorf

i. the characteristics of the system
ii. the type and characteristics of the fault, and
iii. the instant of initiation, pole opening sequence, and interaction characteristics of the circuit breaker.

The duty imposed on a circuit breaker by a system has to be characterised for specification and test purposes in a manner which depends on the circuit, that is on the parameters of the system and of the fault, but which is independent of the circuit breaker. For this purpose the inherent trv of a system for a particular fault current interruption is defined as that trv which would occur for an ideal interruption, that is an instantaneous interruption of the fault current at the natural current zero such that there is no distortion of the fault current due to the action of the circuit breaker prior to the interruption and no post arc current flows subsequent to the current zero.

1.2. Interaction and the Total trv

A real circuit breaker interacts with the system. The voltage of the intensely cooled arc between the circuit breaker contacts tends to force the current to a premature zero while any current flowing after the zero can be considered to create a superposed component trv as it flows into the network. Both the these interactions can be represented in the current injection method of trv calculation.

Interaction effects to not explicitely appear in the specification of circuit breaker duty but must be understood by the breaker designer or manufacturer when he selects a version of his breaker to offer, by the utility engineer when he examines test certificates during tender analysis and by all parties to type testing. The economical and flexible building block approach to circuit breaker manufacture allows different versions of a breaker to be assembled to meet the system voltage, duty cycle, rated current, fault level and trv requirements of the purchaser. At the tendering stage, available test certificates will not necessarily cover the precise version offered let alone demonstrate the detailed requirements of the customer. To compare test performance with system requirements an appreciation of the nature and significance of interaction effects can be crucial.

The change of trv magnitude in the peak regime due to interaction effects is somewhat less than the change in magnitude in the initial trv and short line fault regimes. Expressed as a percentage the interaction is very much smaller in the peak regime and can usually be neglected. Two exceptions come to mind:

i. the deliberate modification of the trv by the use of damping resistors or capacitors permanently shunting the interrupters, or of resistors switched in two stages;
ii. the case of field testing to confirm calculation techniques.

The simplified methods of trv calculation presented by the authors have been checked in field tests for cases of faults remote from sources of generation, and where interaction was seen to be significant the trv waveshapes were accurately reproduced when current distortion was taken into account.

1.3. Circuit Breaker Fault Duty

Circuit breaker specifications impose a wide variety of type test duties on a circuit breaker. These cover from 10 % to 100 % of the rated short-circuit breaking current each with its corresponding trv characteristics, and some special duties. The result is a standard breaker for the rating which is capable of meeting the vast majority of possible system applications. The same breaker may be applied to switch transformers or feeders or it may be moved from one application to another. In a single application a variety of fault types and locations are possible. Mal-operations or changes of protection may result in different fault duties. With the development of a power system, multiplication of lines, increasing fault level, and all of the variables mentioned above, the fault duty on an interrupter may be quite different at each fault current interruption over its life.

Increased rates of rise of recovery voltage (rrrv) are specified for the test duties where the fault current is less than the rated value. These figures are a compromise between the capabilities of the circuit breaker and the range of system requirements, as it would be uneconomical to standardise on breakers that are capable of meeting rare extreme fault switching duties. When special duties arise they are met by special circuit breakers with additional interrupters and/or switching resistors. In some cases the system trv may be modified, for example, by addition of capacitance to reduce a resonant frequency or to introduce a time delay in the first microseconds of the system trv.

1.4. Characteristics of System Transient Recovery Voltages

Faults approaching the rated capacity of a breaker may occur on the main transmission system in a few typical locations.

At stations where the fault current is primarily contributed over transmission lines (switching stations) the trv is basically exponential with moderate rate of rise due to the several lines required to make up the fault current. As lines trip out on a busbar fault the circuit breakers will successively break smaller fault currents but will be stressed with steeper initial slopes of trv. This situation is characterised by eq. (1):

$$\frac{du}{dt} = \frac{k_p \sqrt{2} \; 2\pi f \; I \; Z}{n} \tag{1}$$

where   du/dt   = initial slope of trv
        I   = current interrupted rms
        f   = system frequency
        Z   = average effective line surge impedance
        n   = number of lines
        k_p = phase factor for type of fault (first pole to clear factor).

At stations where fault current is supplied mainly through local transformers (generation stations and transformation stations) the trv is oscillatory when no transmission lines are connected. The presence of shunt lines overdamps the response to give an exponential waveform, but the superposition of sawtooth oscillations from these lines may contribute a significant oscillatory component. Series transmission lines cannot normally damp out the oscillatory response of the source. The ramp response of the line is added to the oscillation.

If the line is on one terminal of the breaker and the source on the other, the responses can only add, in the manner discussed in relation to short line faults.[35] If the series lines are on the source side of step-down transformers, their surge impedance referred to the lower voltage is usually below the critical damping value.

Where generator circuit breakers are installed at large modern power stations they may be subjected to very severe oscillatory trv's.[28] This also applies to circuit breakers adjacent to series reactors which may be installed to limit the fault current.[34]

Faults cleared on the h.v. side of substations fed over lines from a switching or main transformation station have trv shapes dominated by the sawtooth response of the line and superposed on the source station response. Of course the fault level is reduced due to the line impedance. In addition, damping due to local loads becomes apparent after some tenths of a millisecond.

Faults on the low voltage side of substations and zone substations supplying local loads may impose rated (although usually lower) fault current on the breaker but in a radial system this is not reduced as lines are switched out. A very severe oscillatory response can result.

Elementary circuits illustrating the various possibilities are presented in section 4.1.

The presence of cables or capacitor banks will result in local oscillations which can change the form of the trv. The slope of the trv is always reduced when the effect of capacitance is first seen, but increased oscillation occurs and can result in a higher peak voltage.

1.5. TRV Specification and Calculation

Oscillatory trv's are primarily characterised by the co-ordinates of the first peak. The time to peak follows from the oscillation frequency and usually a high amplitude factor (peak factor) applies. The frequency follows from the effective inductance and capacitance for the fault case in question.

The complex trv's composed of an exponential waveshape with superposed oscillations are characterised by an envelope fixed by two points in addition to the origin. Frequently a first major peak occurs at about one-third the time of the highest peak and has a magnitude of 50 to 90 % of the highest crest. This first major peak will often be the critical one for the circuit breaker dielectric due to the high temperature remaining in the former arc channel. Calculation of the first peak of the trv is relatively straightforward as the models used can be markedly simplified. This is fortunate as the extent and complexity of the system needing to be modelled is still very great.

For all these calculations it is necessary to develop a simplified representation of the system to reduce the task of system modelling to manageable proportions. This art is founded on an understanding of the influence of the different parts of the system and of particular system elements on the trv. It is hoped that the simple models developed in the present paper will assist others to learn this art.

2. THE CALCULATION TECHNIQUE

2.1. The Basic Procedure

TRV's may be determined using a physical model of the power system or a mathematical one. The transient network analyser (TNA), a physical model, is well suited to the determination of statistical distributions of switching surge magnitudes for system design and is normally used for this purpose. Digital computers are normally used for trv calculations. The system may be represented for the calculation by lumped or distributed constant elements or by a combination of both.

The action of the circuit breaker is normally represented using the principle of superposition. This principle applies without approximation to linear networks for which the output response is directly proportional to the input stimulus. The current interruption is represented by the sum of the undistorted prospective fault current and a cancellation current. The trv is then obtained as the sum of the prospective fault voltage and the response to the cancellation current.

For a solid fault at the circuit breaker terminals the prospective fault voltage is the arc voltage. This is zero for an ideal circuit breaker assumed for calculation of inherent trv. In this case the inherent trv is simply the response to the cancellation current. Where localised current distortion occurs in the last 100 μs before the current zero, this may be represented in the cancellation current and the corresponding transient arc voltage will be obtained.

Where the fault is remote from the breaker and comparison with measured voltages to earth at the breaker terminals is required, the prospective fault voltage may be a significant proportion of the system voltage determined by the ratio of voltage between fault and breaker and between breaker and source. Such a case is illustrated in Fig. 1 reproduced from a previous paper.[1] In this case the line side transient is the dominant component of the trv.

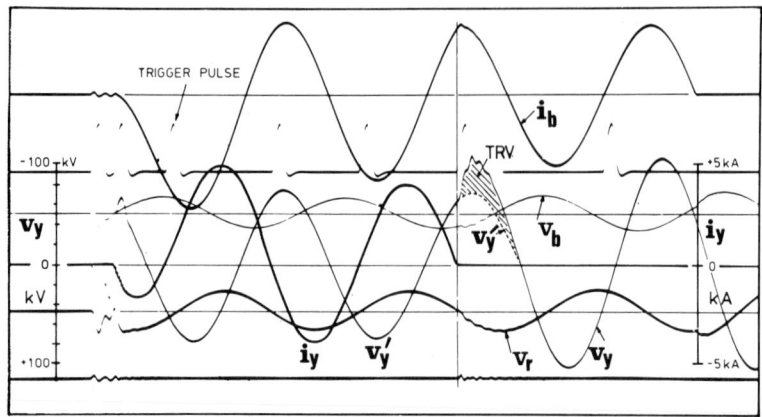

Fig. 1. Typical test recordings of trv for three-phase ungrounded remote fault. Galvanometer oscillograph of fault current (i_y), and source side recovery voltage for first phase to clear (v_y) and other phase quantities. Shown in dashed outline is the reduced trv under remote fault conditions. It is measured as the difference between the recovery voltage (v_y) and the projected fault voltage (v'_y).

2.2. The Injection Current

The cancellation current is given by:

$$i_c = -u_{-1}(t)\, i_f(t) \qquad (2)$$

where $u_{-1}(t)$ is the unit step function of $t = 0$
       $i_f(t)$ is the fault current.

The fault current is interrupted at the instant of injection of $i_c(t)$, that is at $t = 0$. For calculation of inherent trv the injected current used is given by:

$$i_c = \sqrt{2}\,\omega\, I\, t \qquad (3)$$

Is this the right current?

As can be seen in Fig. 1, a real fault current contains a d.c. component, and the alternating component has a decreasing amplitude. The former is associated with the phase at the instant of fault initiation and the latter with the transition of generator reactance from its initial sub-transient value towards the transient reactance, a matter taken into account in deciding the fault current for the calculation.

The slope of the ramp current given by differentiation of eq. (3) corresponds to the steepest slope of the symmetrical sinoidal current of rms value I. The offset caused by the d.c. component may allow the current zero to occur at a phase far from 0 or π at which this slope applies. This may lead to an over-estimate of the trv. On the other hand in the worst case the slope of the decaying d.c. component may add to the steepest slope of the a.c. component to give a current slope steeper than the ramp current. In practice for zero phase angle to lie in the important trv period the d.c. component has almost disappeared and the ramp approximation turns out to be almost exact. For example for current zero 0.3 ms premature the d.c. offset is less than 10 % and its slope is less than 0.04 % with X/R as low as 14. The deviation of the ramp injection current from the sinoidal cancellation current is less than 0.12 % over the first 600 µs.

Thus the ramp current injection accurately represents the worst cancellation current.

From eq. (2) if the symmetrical fault current is expressed in kA_{rms} the slope of the injection current becomes

$$\frac{di}{dt} = 0.444 \, I \qquad kA/ms \text{ (or } A/\mu s\text{)} \tag{4}$$

for a 50 Hz system, and

$$\frac{di}{dt} = 0.533 \, I \qquad kA/ms \tag{4a}$$

for a 60 Hz system.

In a favourable case the reduction of trv due to d.c. offset can be substantial. In matching the field test result for the case illustrated in section 4.4 the true cancellation current was used. Use of a ramp injection current would have resulted in an increase in trv magnitude at the first peak (300 - 400 µs) of 2 - 3 % and at the highest peak (1200 µs) of 8 - 9 %. Clearly for such actual trv calculations current asymmetry must be represented.

2.3. Wave Propagation on Transmission Lines

Digital computer programs of varying complexity have been developed for trv calculations or for electromagnetic transient problems in general.[16] Representation of transmission lines so as to take account realistically of attenuation and distortion is a matter which has received considerable attention. The propagation of surges on multi-conductor systems may be considered to take place in three modes.[2,4,11,13,18,20] Boundary conditions determine the significance of each mode of propagation. Since Adams[13] revived interest in unbalanced (loss free) wave propagation many writers have taken up the matter.

For each mode the effective surge impedance and the modal velocity are different. The system impedances are also frequency dependent. Carson[12] provided for this with his earth impedance correction terms and others have followed.[19,33] Conductor skin effects impinge on all modes of propagation. The frequency dependence of a mode with earth return is also a result of the changing permittivity and permeability of the ground. These properties account for the attenuation and distortion of travelling waves on transmission lines. More simplistically one may think in terms of energy coupled from one phase to the others and to ground and of the losses in each of these circuits.

The frequency dependence of transmission line parameters may be handled using the convolution theorem of Fourier transformation theory.[14] Weighting functions are obtained which characterise the frequency dependent response of the line to impulse excitation. Appropriate integration[15] provides a transfer function for the particular transmission line which can be built into an electromagnetic transients program.[17,31]

Wave propagation theory has also been generalised and extended to cover non-homogeneous systems such as multi-layer earth, line transpositions, discrete bonding of earth wires, and cross-bonded cable networks.[21,22,23]

2.4. Singe Phase Equivalent Models and the Lattice Diagram

Wedepohl[20] has shown that for the special case of a fully symmetrical three phase system, symmetrical components[5] (designated 1, 2, 0) conform to the general solution for travelling wave phenomena in polyphase systems. An asymmetrical three phase system (e.g. horizontal configuration three phase transmission line) is better represented using α, β, 0 components.[9] These two sets of components are identical under balanced conditions:

For a balanced ground mode in phase current i_a, i_b, i_c are equal.
It is readily demonstrated that

$$i_a = i_b = i_c = i_0 \quad \text{and}$$
$$i_\alpha = i_\beta = i_1 = i_2 = 0$$

For a balanced phase mode $i_a = a^2 i_b = a i_c$ (where $a = -1/2 + j\sqrt{3}/2$ applies a phase rotation of 120°). It is readily shown that

$$i_1 = i_\alpha = i_a,$$
$$i_2 = i_\beta = 0 \text{ and}$$
$$i_0 = 0 \text{ for both component systems.}$$

Application of the component transformation effectively decouples the phases allowing the three phase problem to be treated by a single phase equivalent representation. This is true at surge frequencies as it is at 50 or 60 Hz.[3-9] Symmetrical components are widely used for trv calculation simply because their power frequency applications have made them so well known. Just as the absence of transpositions may be overlooked in the application of symmetrical components to fault calculations, symmetry may be assumed for trv calculation. Furthermore, very worthwhile results can be obtained utilising only power frequency impedances which are often tabulated by utilities for other purposes. Positive and negative sequence modes are combined to give a single 'line', 'phase', or 'conductor' mode of propagation for which the dependence of line surge impedance on frequency is minimal.[32]

The simplest means of handling travelling wave propagation is by means of the lattice diagram.[10] A basic time interval τ that is small compared with system element response times is adopted for the calculation. Response to an arbitrary waveshape is synthesised from the model response to an elementary step function. The lattice diagram model does not allow modelling of frequency dependent parameters although effective surge impedances and resistive damping can be evaluated for the dominant trv frequency and used in the model. This is only considered worthwhile for the zero sequence model.

A recent paper[32] demonstrates that the rrrv and the overall trv up to the first major peak can be calculated accurately using conventional Laplace transform calculations which take no account of frequency dependent parameters. An interesting observation was that errors due to bundle conductor clashing and ground mode frequency dependence appeared to compensate. The so-called conventional approach has been found to be accurate enough for most purposes.

In recent Australian trv studies[1] frequency independent positive and zero sequence network models were used for line and ground modes of propagation respectively. Very good agreement with full scale test data was achieved for the terminal fault cases investigated. The method is capable of application to remote faults and some cases of this kind have been studied.

The boundary conditions governing the connection of the distributed sequence network for various types of fault have been outlined in a recent Electra paper[30] in the context of determining the appropriate effective surge impedances for trv calculations. In application to total trv calculation a difficulty is encountered in relation to remote faults. The boundary conditions at the circuit breaker are not the same as at the fault. Until the first reflection from the fault point the trv is independent of the type of fault. At longer times the fault condition becomes very significant. Single phase equivalent models cannot represent these changes however fault side and source side component trv's may be calculated using the larger of the possible impedances and the faster mode velocity to obtain a conservative result. Fortunately these cases are not the ones of greatest significance.

In the cases which follow a number of terminal fault trv calculations are discussed. The calculations were carried out using the procedures discussed above, and for the greater part using tabulated power frequency impedances of the ECNSW system.

Calculations of this type have been carried out for many years[2,10] and recent reports[25,26] indicate their continuing acceptability in the industry.

Terminal faults. Sequence network connections for a symmetrical three phase terminal fault and for a single phase to earth fault are shown in Fig. 2. These are the same connections as are used for calculating the fault currents. The fault is assumed to be adjacent to the circuit breaker. Any continuation of the line beyond the fault can only marginally alleviate the trv stress and is usually neglected. An approximate representation of a continuing line is indicated in the case examined in section 4.4.

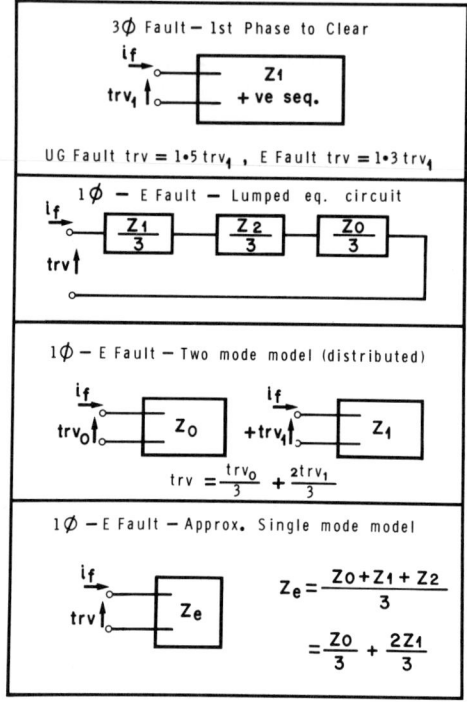

Fig. 2. Sequence network connections for trv calculations.

To obtain the trv across the breaker contacts on the first phase to clear on a three phase ungrounded fault the familiar phase factor k_p of 1.5 must be applied. It is simpler when different fault cases are to be studied to apply the factor k_p to the calculated trv or to the injected current rather than to scale up the model impedances. For a three phase grounded fault a factor $k_p = 1.3$ is applied to obtain the trv specified for a single phase test.

The series interconnection of sequence networks for the single phase to earth fault cannot be applied to distributed models as it is in lumped parameter calculations. This circuit is a representation of the boundary conditions requiring the injection current to be identical for each sequence and indicating that the trv is given by the sum of the responses of the three sequence networks. Assuming the equivalence of positive and negative sequence networks the trv can be obtained as the scaled sum of the responses of the positive and zero sequence networks:

$$trv = 2/3 \ trv_1 + 1/3 \ trv_0$$

where trv_1 and trv_0 are the component trv's indicated in Fig. 2.

It should be noted that the replacement of the two sequence component models by an approximate single mode model as indicated at the bottom of Fig. 2 cannot be made without additional approximations. These are the neglect of the difference of mode propagation velocities and the neglect of differences in the sequence network particularly where transformers are involved. The latter may result in gross errors in final recovery voltage. Nevertheless a conservative result can be obtained that is generally of adequate accuracy through the first 300 -400 µs of the terminal fault trv in cases where line oscillations dominate the response.

2.5. System Modelling for the Lattice Diagram

Travelling wave programs using lattice diagram techniques generally model system elements in three ways:

 i. finite lines
 ii. infinite lines
 iii. transfer elements

These model elements are illustrated in Fig. 3 a, b, and c.

A finite line between nodes k and n is a distributed contant lossless transmission line having a fixed surge impedance Z_{kn} (= Z_{nk}) and travel time T_{kn} (= $n_{kn}\tau$) where τ is the basic time interval adopted for the calculation. Reflection and transmission coefficients are defined for the branch nodes in the usual way.[2,10]

An infinite line emanating from a node is a distributed constant reflection free transmission line characterised solely by a surge impedance. Shunt resistors and loads are represented by this element with the surge impedance Z_c set equal to the lumped resistance R.

A transfer element is a pair of nodes i and j for which any surge arriving at the one node is instantaneously transferred to the other node scaled up or down by a multiplying factor or transfer coefficient K_{ij}. The nodes are provided with matching resistors R_i and R_j equal to the input impedance of the other node referred to the first node. A transfer element is an ideal transformer and can be used to link system models developed for different voltages.

The representation of system elements by model elements may be carried out internally in the computer program or may be done manually in data preparation.

Transmission lines and lengths of cable are represented directly by finite lines of appropriate surge impedance and travel time.

Shunt capacitance (lumped) and short cables are represented as shown in Fig. 3d by very short open circuit finite lines of surge impedance Z_C and travel time T_C given by:

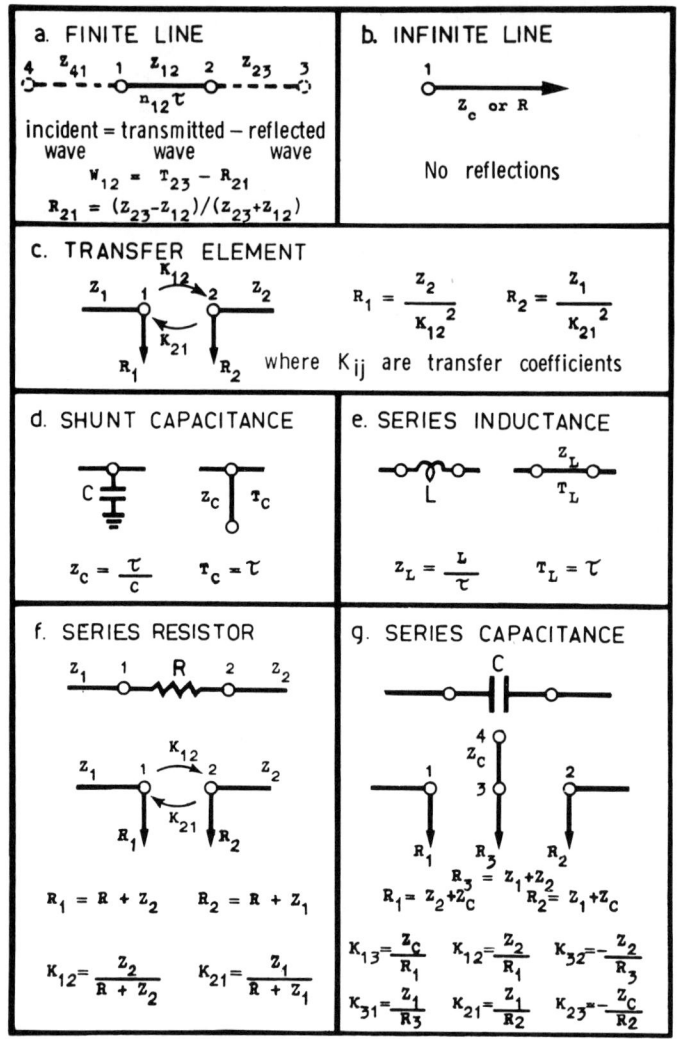

Fig. 3. Representation of system elements for a travelling wave program based on the lattice diagram.
a-c the basic elements,
d-g the system elements.

$$Z_C = \tau/C \quad \text{and} \quad T_C = \tau$$

Use of a distributed model brings in some inductance which may be calculated from $Z = \sqrt{L/C}$. For example if $C = 0.1$ µF and $\tau = 2$ µs, $L = 0.067$ µH/m.

Series inductance (lumped) such as transformer leakage reactance is represented as shown in Fig. 3e by a very short finite line with surge impedance Z_L and travel time T_L given by:

$$Z_L = L/\tau \quad \text{and} \quad T_L = \tau$$

Z_L incorporates some capacitance: 0.67 pf/m for 10 mH, $\tau = 2$ μs.

Fig. 3f and g show representations of series resistors and series capacitors utilising single and double transfer elements.

2.6. The Extent of the Models

The extent of the model is determined in the first instance by the extent of the interconnected system and by the time required to ensure that the highest peak of the trv is included. Calculation of the trv beyond 2 ms after interruption can only be of academic interest as the roll off of the sine wave precludes a highest peak after this time. Given propagation at the velocity of light a model extending to 300 km is indicated. Fortunately the whole of the system within this range does not normally have to be modelled.

Two techniques that can assist in assessing the extent or detail of modelling required will now be outlined. Both will be illustrated from the ECNSW system[36] as it was in 1974. The main load centre of the system is Sydney which is ringed by 330/132 kV stations. The 330 kV main transmission system is meshed. The 132 kV subsystems are radial. The main sources of generation are 100 to 150 km north of Sydney and also 500 km to the south at peak times. Faults at three typical points in the Sydney West 132 kV sub-transmission system may be considered:

    i.   at Hawkesbury, at the extremity of a radial feeder
    ii. at Carlingford, a typical 132/66 kV substation in a residential area
    iii. at Sydney West, the major station supplying the sub-system.

Fig. 4 shows the network sketched as for fault current calculation. From some simple sums with the fault levels and/or impedances the voltage drops at significant points in this network for solid faults at these three locations are obtained. The injection of the cancellation current will ultimately restore all these voltages back to 100 %.

Looking at the system voltages for the remote fault near Hawkesbury, it can be seen that 70 % of the voltage drop occurs in the final radial line from Carlingford. The drop at Sydney West 132 kV busbar is 19 % and at the 330 kV busbar is only 7 %. Clearly the modelling of Carlingford is very important, Sydney West less important, the 330 kV system unimportant.

The voltages for the fault at Carlingford show the local modelling to be critical and Sydney West 132 kV (voltage drop 62 %) very important.

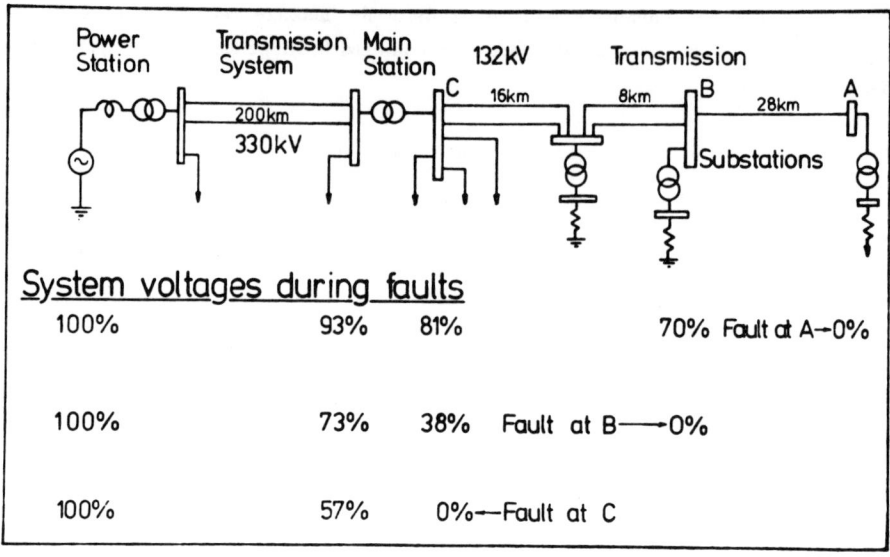

Fig. 4. Faulted system single line diagram and system voltages.

For the fault at Sydney West representation of the 330 kV transformers is of considerable importance.

To this point we are thinking primarily of power frequency considerations. The model must provide sufficiently accurate representation of these impedances in the long term in order to get the final recovery voltage right.

The second step is to think of your model network and ask "Where will the injected current go?" One observes immediately that it must charge up all parts of the network whose voltages are depressed by the fault and in doing so will restore full current to all loads. It immediately becomes clear that subsystems elesewhere, off the 330 kV system, may be neglected and the 330 kV system itself represented in an elementary way rather than in detail, although the 330/132 kV transformers are very important as the fault current is supplied through them. The transient oscillation associated with the charging up of the remainder of the 132 kV network is of fundamental importance so care must be taken in its modelling to obtain the correct times to peak. The damping effect of loads and losses has to be considered also.

These considerations are quantified in the examples and sensitivity analyses which follow.

3. THE SYSTEM MODEL

3.1. Transmission Lines and Cables

Transmission lines and cables are modelled directly as 'finite lines' characterised by surge impedance and travel time.

a) From power frequency parameters

The sequence surge impedances Z_i, velocities v_i and travel times T_i are given by eq. (4)-(6).

$$Z_i = Z_b \sqrt{X_i/B_i} \qquad (5)$$

$$v_i = \omega/\sqrt{X_i B_i} \qquad (6)$$

$$T_i = L/v_i \qquad (7)$$

where Z_b is the base impedance in ohms
X_i is the i-sequence reactance of the whole line in percent on 100 MVA
B_i is the i-sequence susceptance of the whole line in percent on 100 MVA
ω is the system frequency in radians per second
L is the line or cable length in metres.

Typical sequence mode velocities obtained are:

$$v_1 = 95\ \%\ \text{to}\ 98\ \%\ \text{of}\ c,$$

$$v_0 = 60\ \%\ \text{to}\ 80\ \%\ \text{of}\ c\ \text{at}\ 50\ \text{Hz}.$$

For 50 kHz one would obtain:

$$v_1 \simeq 100\ \%\ c,\ \text{and}$$

$$v_0 = 90\ \%\ c$$

where c = velocity of light (300 m/µs).

It is important to note that double circuit transmission lines have very large zero sequence mutual impedances and should be calculated from the parallel impedance terms given by:

$$Z_D = (Z_0 + Z_{m_0})/2 \qquad (8)$$

which follow from the equivalent circuit of Fig. 5.

When account is taken of the mutuals, one confirms that $Z_{D1} \simeq Z_1/2$ and $Z_{D0} > Z_0/2$.

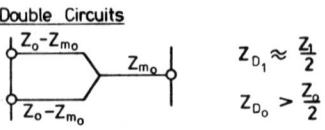

Fig. 5

b) From conductor geometry

For conductors and earth wires i, j etc., with image conductors i', j' etc., and heights and separations defined as in the sketch, one may write:

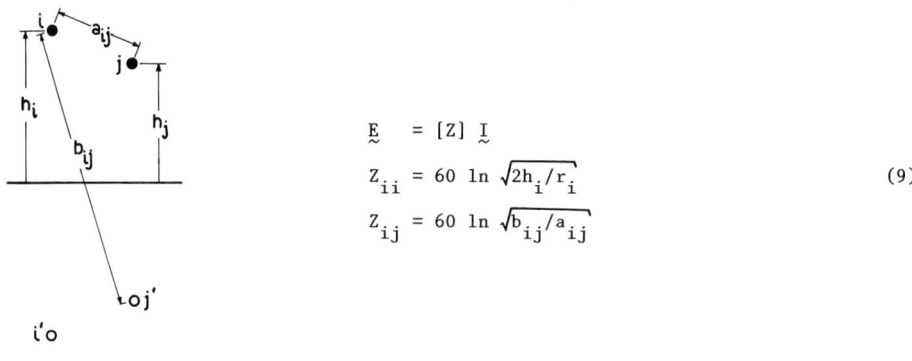

$$\underset{\sim}{E} = [Z] \underset{\sim}{I}$$
$$Z_{ii} = 60 \ln \sqrt{2h_i/r_i} \tag{9}$$
$$Z_{ij} = 60 \ln \sqrt{b_{ij}/a_{ij}}$$

Fig. 6

and the dimension of the vectors is equal to the number of conductors and earth wires involved. For double circuit lines the dimension of the Z matrix can be reduced from 8 to 4 by representing it as an equivalent single circuit. The earthwire terms are then eliminated using the transformation

$$Z_{ij'} = Z_{ij} - Z_{ig} Z_{gj}/Z_{gg} \tag{10}$$

where   i, j are conductor subscripts and
          g   is the earthwire subscript.

The symmetrical component transformation then gives the surge frequency sequence surge impedances on the diagonal of the matrix and the size of the other terms, if non-zero, is a measure of the asymmetry of the case.

c) For trv frequencies

The surge impedances can be determined for the fundamental frequency of the trv by replacing the image conductors i', j' of b) above with idealised return conductors at some greater depth D_e determined from Carson's formulae. The corresponding impedance matrix is built up using equations (9) but substituting D_e for the separation terms. Carson's simplified formulae also provide the frequency dependent earth resistance terms which dominate the ground mode damping.

3.2. Transformers

The transfer inductance of transformers may be caculated directly from the power frequency transformer leakage reactance. This value is applicable to cases of over damped response and to low frequency oscillations. The transformer modelling may be complicated by the presence of additional windings. In the positive sequence model, tertiary windings supplying station auxiliaries can usually be neglected. In the zero sequence, delta windings provide a path for circulating currents and behave as if short circuited. The primary to tertiary leakage inductance is used. A delta primary winding as typically found at zone substations in New South Wales behaves as an open circuit.

Major substations usually have a spare transformer energised from one side only. If such a transformer has a delta secondary or tertiary winding this forms part of the zero sequence network and should not be overlooked.

Where the trv is dominated by the source transformer oscillation, as may occur where there is no transmission line (remaining) connected on the transformer side of the breaker, the effect of the distributed capacitances through the transformer is to reduce the effective inductance.[7,27] The resonant frequency of the transformer may be 5 to 25 kHz and the amplitude factor is high (1.6 to 1.9). Fortunately in these cases the modelling of the remainder of the system is not critical and usually only a relatively low fault current is involved. Field studies on such systems[27,29] have been carried out and a standardised procedure for estimating the frequency exists.[24] The effective inductance of a phase is typically 0.75 times its leakage and the equivalent parallel capacitance is generally taken as 0.4 C where C is the winding capacitance to ground.

3.3. Distribution Systems and Loads

The modelling of distribution systems becomes important when calculating the trv in or near a load centre. Where the breaker is in such a station several stages of transformation may need to be modelled. Generally parallel transmission lines or distribution lines on the remote side of transformers can be combined to form a single equivalent line supplying the total load. Distribution transformers are modelled assuming typical percentage impedances and with appropriate allowance for diversity in assessing the total rating. The sub-transmission system and distribution lines may be neglected except where there are sufficient cables to make a significant capacitance when referred to the circuit breaker voltage.

4. SOME SAMPLE TRV's

4.1. Elementary Circuit Responses

Figs. 7 and 8 show some elementary circuit responses to ramp current injection. The ramp is assumed to have a slope of 1 kA/ms for time t in ms.

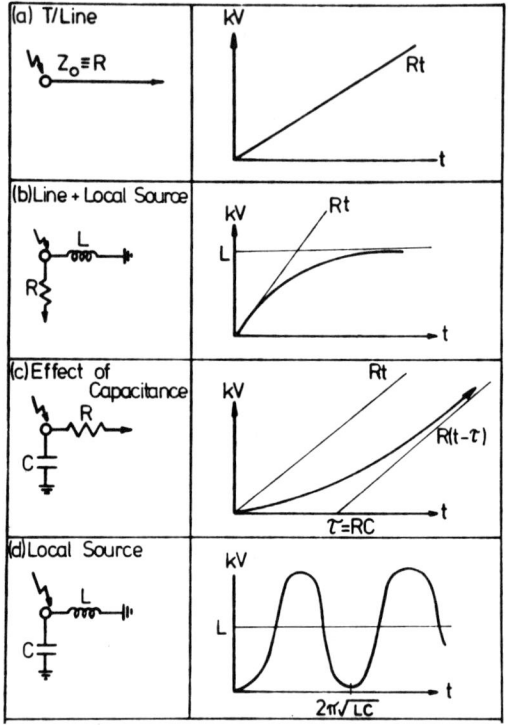

Fig. 7. Responses of elementary circuits to ramp currents - one and two element circuits.

The ramp response of a transmission line (Fig. 7(a)) becomes a sawtooth wave for a short-circuit termination and approaches the response of a capacitor over a time span long compared with the travel time when the termination is an open circuit. The response of a capacitor is given by Fig. 7(c) if R is made to approach infinity. A local source or a shunt reactor responds as an open circuit initially but it is penetrated after a time characterised by the L/R ratio (Fig. 7(b)). With no damping a local source has an oscillatory response (Fig. 7(d)).

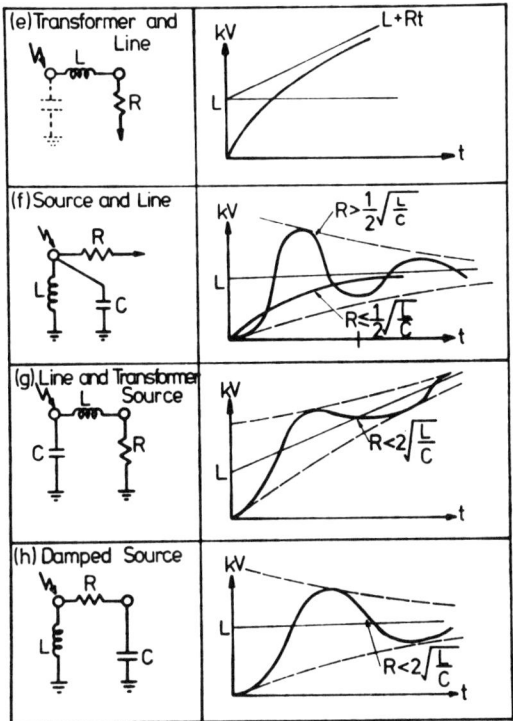

Fig. 8. Responses of elementary circuits to ramp currents - three element circuits.

Fig. 8 shows the effects of series and shunt damping on various source configurations. A low value of shunt resistor or a high value of series resistor is required to overdamp an oscillatory response. The response shown in (e) applies when the inductance is modelled by a finite line. The theoretical respone for an ideal lumped inductor is a triplet step to L followed by a ramp. However a real inductor has the response shown.

In times large compared with the travel time, any series transmission line behaves like an inductance.

The trv's encountered in any complex network can be understood in terms of these elementary circuit responses, although different circuits will apply in different time periods, and the total response may be the sum of more than one elementary response.

4.2. Line Fault Supplied from Switching Station

Figure 9 illustrates a case where a fault occurs at the end of a long double circuit line emanating from a switching station with several long double and single circuits connected.

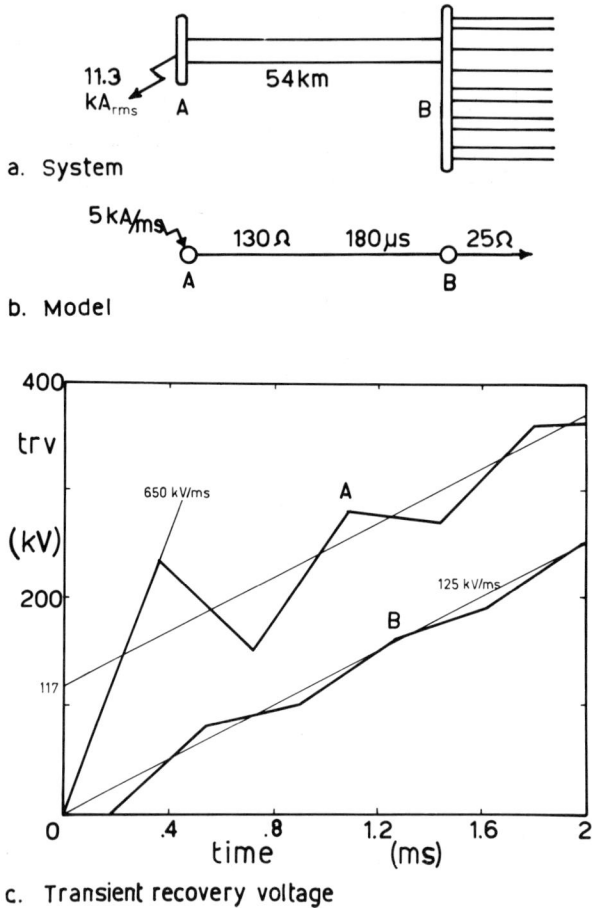

Fig. 9. Elementary trv for line fault supplied over lines from switching station.

The initial response is a ramp determined by the line surge impedance and the injection curent. The low termination impedance converts the response to a sawtooth. The final slope is determined by the surge impedance seen at the switching station. The long term response approaches the case of the elementary circuit of Fig. 8(e) and the intercept of the asymptote on the voltage axis is determined by the equivalent inductance of the line.

4.3. Fault Supplied Over Double Circuit Line from Transformer Station

In Fig. 10 busbar D is assumed to be the high voltage side of the station. Referred to the low voltage side the parallel surge impedance of the lines is very low. To the surge arriving at B the transformers look initially like an open circuit (doubling the trv slope) but this soon changes towards a short circuit as the 4 ohm surge impedance becomes the effective termination. The waveform is dominated by a sawtooth of quarter period 230 µs, being the line travel time plus 50 µs for penetration of the transformers.

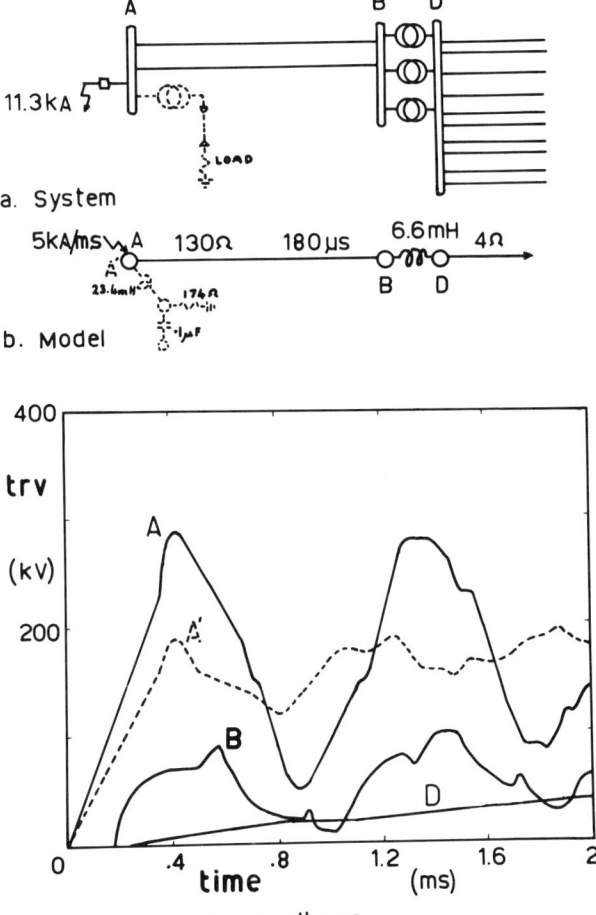

Fig. 10. Elementary trv for line fault supplied over lines from transformer station. Dashed curve indicates the effect of local distribution system and load.

252 J. DIESENDORF, S. K. LOWE, AND L. SAUNDERS

The waveform at B shows the classic response of transformers whilst D is the delayed ramp determined by the 4 ohm surge impedance of the h.v. side lines.

The dashed curve A' shows the response when a distribution system and load are connected at the faulted busbar. After a short delay nearly half of the injection current flows to the load and the sawtooth is heavily damped although in the long term the 4 ohm source lines dominate so that the final voltage is hardly changed.

4.4. A Real System

Figure 11 shows the single line diagram of the 132 kV Sydney West sub-system and its positive sequence travelling wave model. The figure is reproduced from an earlier paper.[1]

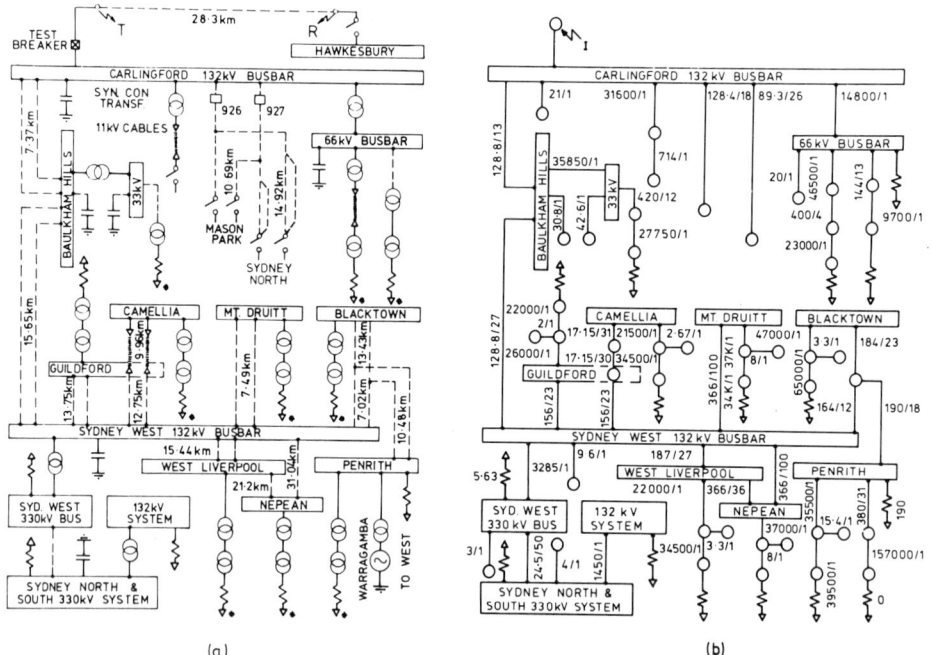

Fig. 11. System network configuration for trv tests at Carlingford 132 kV substation.
(a) Single line diagram of the 132 kV Sydney West Subsystem of the Electricity Commission of New South Wales.
   T = fault location for full terminal faults
   R = fault location for remote faults
   X = test breaker
   * = load
(b) Positive sequence (line mode) travelling wave model of test network for three-phase ungrounded trv calculations. The parameters for each branch are given by a number pair Z/τ where
   Z = branch surge impedance in ohms
   τ = travel time (multiples of 2 μs)
   —W— = load resistance
   O □ = nodes of network model

Figure 12 shows the measured transients for an 11.3 kA three phase ungrounded fault at Carlingford (Station B of the simplified sketch Fig. 10). On clearance of the first phase the other phases undergo a transient that is similar to the cleared phase but only half the magnitude and slightly lower in frequency. The difference of frequency is caused by the faulted three phase line that is connected to the fault side of the breaker. The line acts in the long term like its capacitance as it is open circuit at its remote end. The effect of the line on the total trv is small, and in the case of a short line would not be observable.

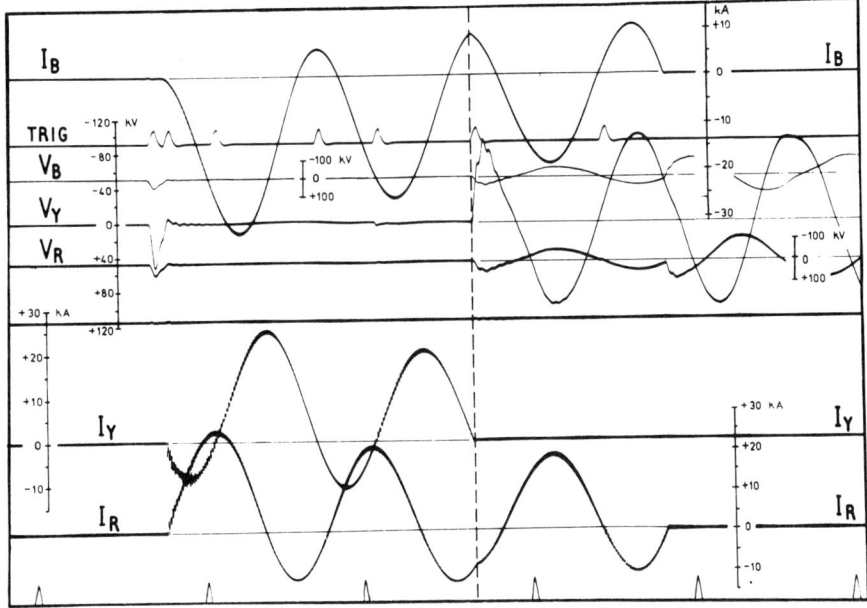

Fig. 12. Galvanometer oscillograph recordings for interruption of a three-phase ungrounded terminal fault, Y-phase first to clear. Current asymmetry moderate on phases Y and R but extreme on B-phase.

Figure 13 shows the measured trv to earth on the cleared phase and a computer plot of the calculated response obtained by ECNSW. For this calculation the true prospective fault current was injected. An independent calculation by the SECV (Fig. 14) shows an almost perfect agreement with the measurement. For this calculation a representation of the transient coupled from the faulted line was included.

The model which produced this remarkable agreement is shown in Fig. 15.

Except for the mutual from the line side the models are similar in form and detail. The SECV model differs in representation of the 330 kV system and in the distribution system. With the mutual removed its response is only marginally less accurate than the ECNSW model. Both models are based on power frequency parameters -

254 J. DIESENDORF, S. K. LOWE, AND L. SAUNDERS

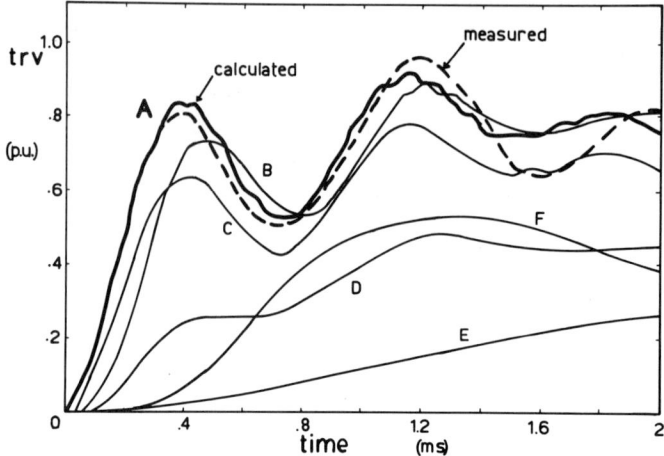

Fig. 13. Source side component of trv for first-phase-to-clear on a three phase ungrounded fault at Carlingford (Fig. 12). Comparison of measured and calculated curves. Curves B-F: voltages at other nodes - see text.

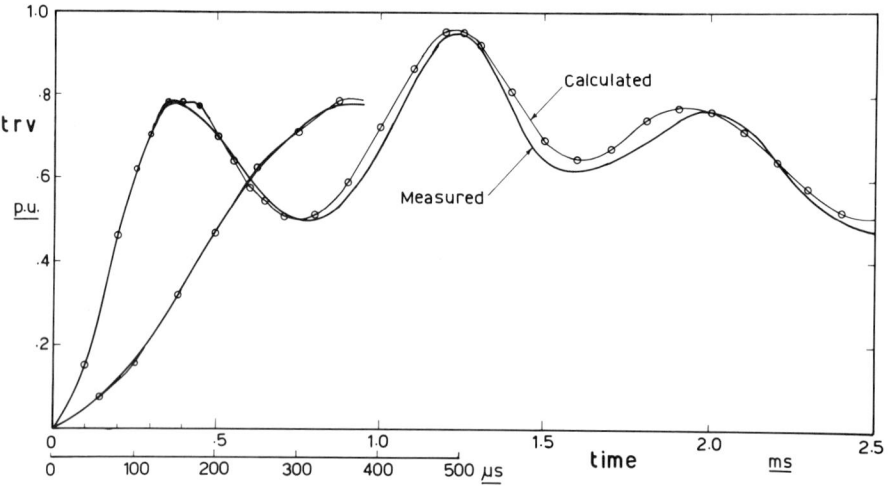

Fig. 14. Source side component of trv for first-phase-to-clear on a three phase ungrounded fault at Carlingford. Comparison of measured curve with calculation including correction for effect of the faulted line.

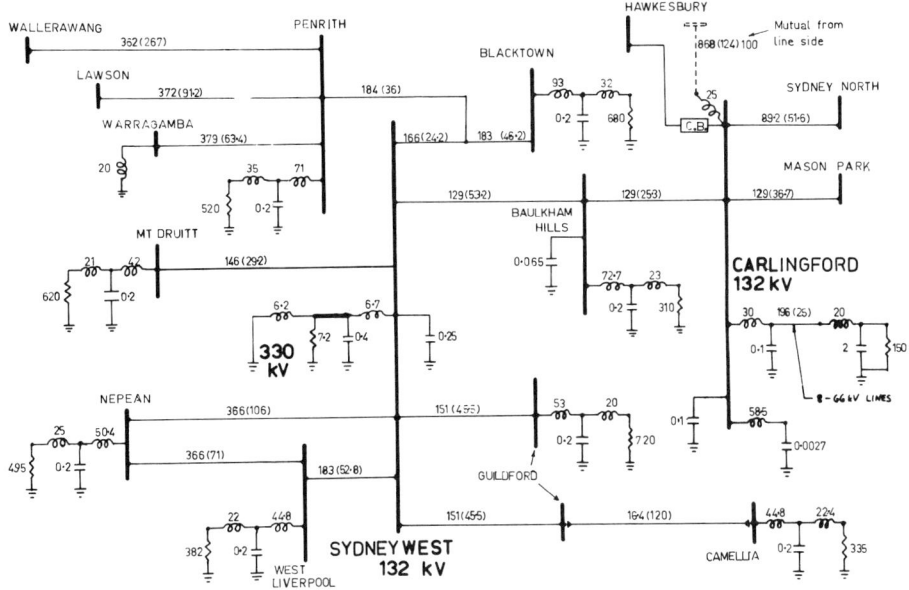

Fig. 15. Model of positive sequence network used for the trv calculation in Fig. 14. Lines are characterised by surge impedances in ohms and travel times in µs (in parentheses), and in addition the resistive damping of the Hawkesbury line. Inductances are in millihenries and capacitances in microfarads. Resistive loads are shown in ohms.

surge impedances and travel times calculated from power frequency reactance and susceptance, and transformers represented by their power frequency leakage inductances. The models include extra detail at the injection node and of the distribution system connected to it. All of the 132 kV lines involved are fully modelled, but 33 kV and lower voltage distribution systems are modelled by a series of two leakage inductances loaded with a resistor and with a capacitor representing 33 kV transmission lines and cables.

The unbalance caused by the 132 kV faulted line was represented in the SECV model by linking the line to the bus via an inductance which introduced an appropriate delay.

Referring again to Fig. 13 one can see what is happening at other nodes in the system. Curve E shows the 330 kV busbar in the ECNSW model. Dominated by the low resistor representing the parallel surge impedances of long lines (referred to 132 kV) it rises almost to the appropriate .27 p.u. corresponding to the power frequency voltage drop on the busbar. The important 132 kV busbar at Sydney West is shown as Curve D. The response of this busbar is tied to the slow response of the

132 kV cables not far away which are shown as Curve F. Curves B and C represent the Carlingford 66 kV busbar and the nearby Baulkham Hills 132 kV busbar.

Figure 16 shows the measured and calculated responses of the same system for the clearance of a single phase to earth fault. The trv is obtained as the scaled sum of the positive and zero sequence network responses shown. The SECV zero sequence model is shown in Fig. 17. Again it is entirely based on 50 Hz parameters except for the inclusion of loss resistors estimated for the trv frequency.

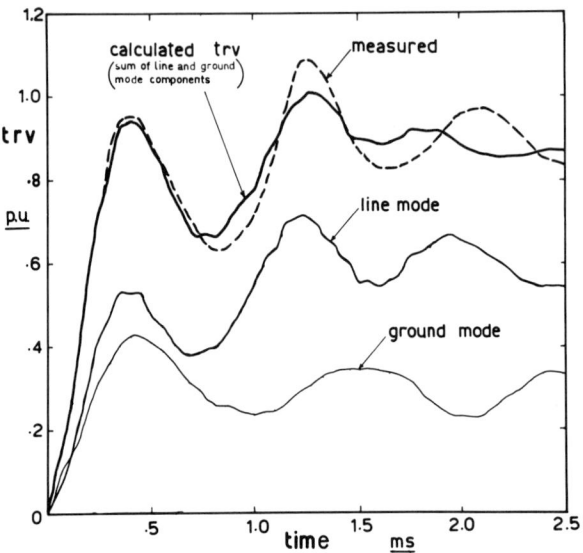

Fig. 16. Measured and calculated trv's for a single-phase-to-earth fault of Carlingford showing component trv's for phase and ground modes.

It will be noted that the networks and responses of the line and ground modes are generally similar to two of the elementary circuits of Fig. 8. The parallel LCR network (f) corresponds to the zero sequence model where transformer leakage inductances are short circuits due to the delta winding tertiaries. The series LCR connection (g) corresponds to the positive sequence model where transformers feed the loads.

5. SENSITIVITY ANALYSIS AND SIMPLIFICATIONS

5.1. Sensitivity to Load

The omission of system load from the models was found to have a very pronounced effect on the calculated trv. The effect of load is shown in Fig. 18 for the three-

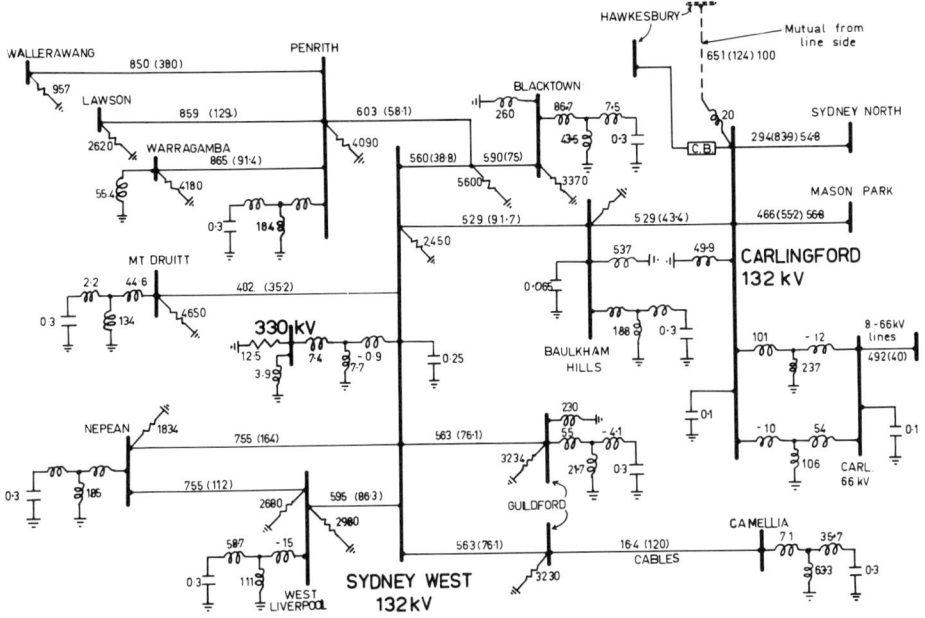

Fig. 17. Network model based on zero sequence 50 Hz parameters for calculation of ground mode component trv shown in Fig. 16.

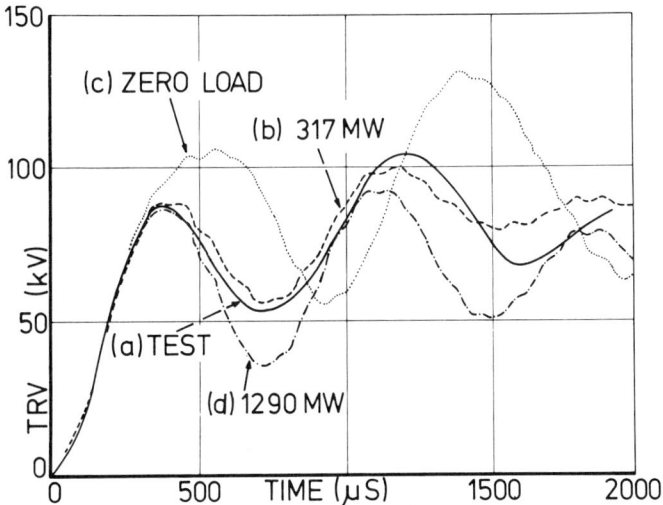

Fig. 18. Effect of the system load on the overall trv.
(a) Test measurement at near-minimum sub-system load of 317 MW.
(b) Calculation of Case (a)
(c) Calculation for zero system load.
(d) Calculation for maximum transformer loading of 1.290 MW.

phase ungrounded fault at Carlingford. The sub-system load at the time of the test measurement was near minimum at 317 MW. Omission of this load gave an increase in peak trv of 30 %, although the rate of rise to the first crest was unaffected. Increasing the load by a factor of four to the maximum transformer rating gave only slight additional crest damping.

Independent check calculations showed the trv to be insensitive to load until after the first peak. This was found to be due to the representation of cables in the Carlingford distribution system as a capacitor parallel to the load resistor.

With both models the omission of load was achieved simply by the omission of the load resistor from the models. Calculations showed also that the omission of remote 33 kV distribution systems and loads had little effect on the trv compared with changes in the Carlingford distribution system.

The sensitivity to load observed in this case may not occur at other locations. The fault in this case is in the midst of a load centre. Faults on the main transmission grid and certainly faults close to sources of generation will be much less sensitive or else insensitive to load. Nevertheless very many circuit breakers are installed in locations similar to the one modelled above.

5.2. Effect of Cables and Capacitor Banks

The effect on the trv at Carlingford of taking the remote 132 kV cables out of service is shown in Fig. 19 (solid curve A).

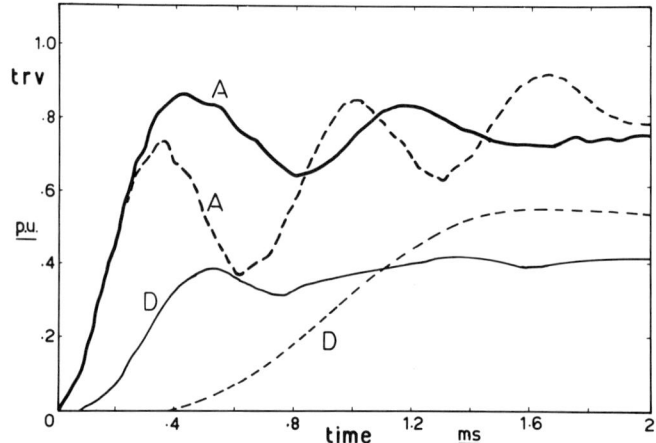

Fig. 19. Effect of cables and capacitor bank on source side trv for three-phase-ungrounded-fault at Carlingford - for comparison with curves A and D of Fig. 13.
Node A - Carlingford 132 kV busbar.
Node D - Sydney West 132 kV busbar.
Solid curves: Remote 132 kV cables (to Camellia) removed.
Dashed curves: Capacitor bank at Sydney West switched in. See later text.

The first peak is slightly increased due to the quicker response of the Sydney West busbar (Curve D) but the second peak is reduced due to the elimination of the slow charging oscillation of the cables.

For a fault at Sydney West (Fig. 20) the removal of the cables brings the response close to a simple exponential determined by the transformers and 330 kV system - a series LR circuit (compare Fig. 8(e)).

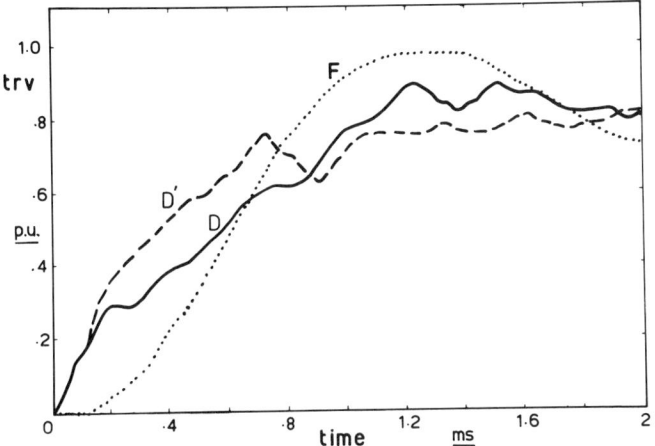

Fig. 20. Effect of Camellia cables on source side trv for a three-phase ungrounded line fault at Sydney West.
Curve D - Sydney West busbar with cables in service.
Curve F - Node voltage at cables.
Curve D' - Sydney West busbar with cables out-of-service.

The effect of the cables may be further illustrated by a simple case:

If the 132 kV system is treated as all open circuit, i.e. loading through transformers neglected, the system resembles a capacitor connected to the Sydney West busbar. The network becomes a series LCR circuit (Fig. 21) and the 330 kV side cannot overdamp it. The peak factor reaches 1.85. A shunt resistance of 18 ohms or less would be needed to damp out the response. Maximum load would fully damp the oscillation whilst representation of even light load (25 - 30 %) provides significant damping.

With capacitor banks connected to the busbar, the damping effect of the load is reduced. One may have a high peak factor but the rrrv is still low.

5.3. Model Simplifications

For parts of the system that are not close-up to the fault parallel similar circuits may be combined into a single line network. Parallel transmission lines are replaced by a line of average length, the surge impedance of the parallel combination, and total capacitance.

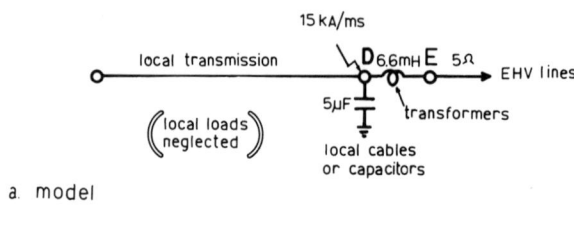

a. model

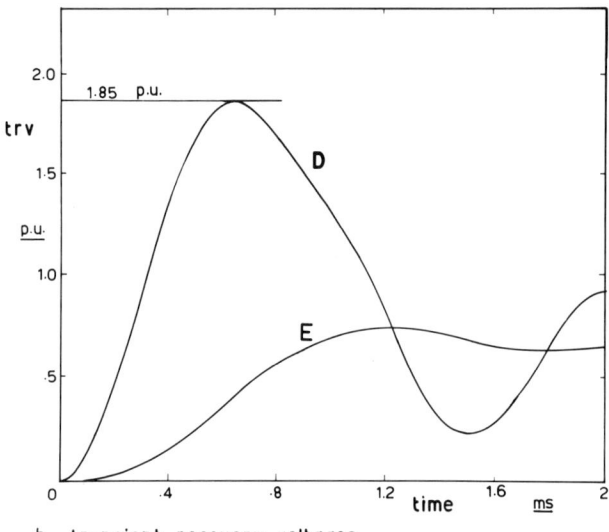

b. transient recovery voltages

Fig. 21. Effect of <u>local</u> capacitor bank on trv at transformer station.

Fig. 22 shows the effect on the calculated trv of replacing all 132 kV lines on the Sydney West busbar by a single line. The model simplification is shown in Fig. 23. The total power load was represented on the low voltage terminals of the combined distribution transformers. It can be seen that this great simplification only slightly increased the trv crest.

Simplifications can only be applied with great caution to parts of the system close to the fault and to the trunk supply route over which the fault current is supplied. Nevertheless, long remote transmission lines for which reflections can be neglected in the time of interest can always be modelled as infinite lines.

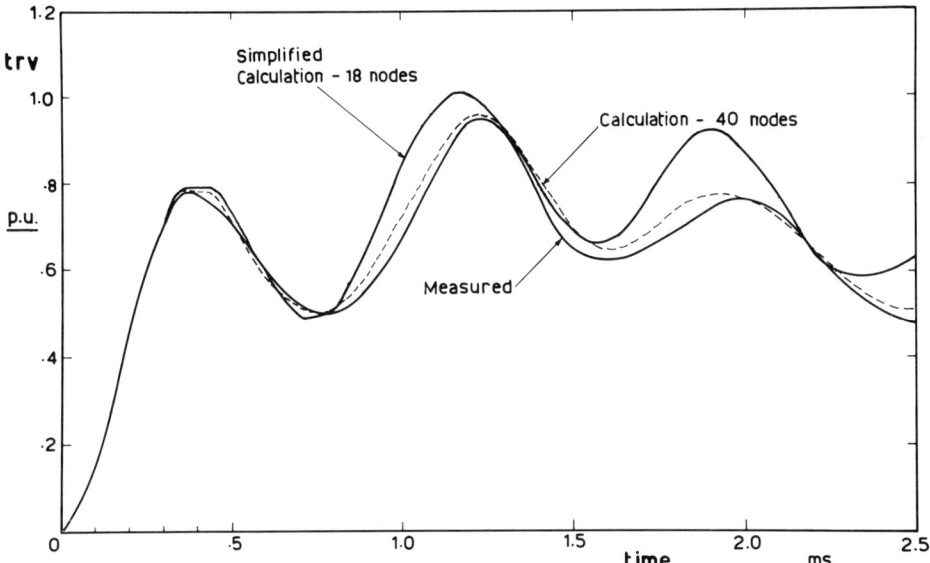

Fig. 22. Effect of model simplification on trv for three-phase-ungrounded-fault at Carlingford - 132 kV lines out of Sydney West represented by a single line (see Fig. 23).

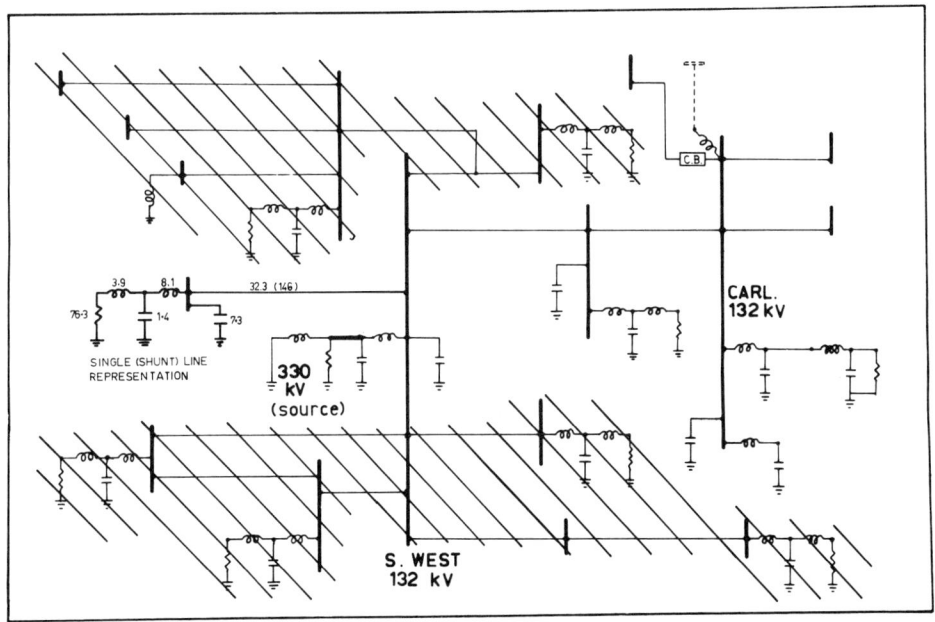

Fig. 23. Model simplification for calculation of trv in Fig. 22 - shunt lines at Sydney West represented by a single circuit.

6. SOME SIMPLE RULES FOR COMPLEX TRV ESTIMATION

The following rules are applicable only to the estimation of complex trv's comprising exponential and superposed oscillatory components. TRV's dominated by highly oscillatory transformer or machine responses can be estimated in other ways.[24]

1. Calculate the trv appearing at the supply side terminal of the circuit breaker on clearance of the first phase using only the phase mode network.
2. Estimate the total trv across the breaker terminals by applying a phase factor of 1.5 for ungrounded faults and 1.3 for three phase to earth faults where applicable.
3. Make the calculation using a travelling wave type of computer program based on the lattice diagram.
4. Use a ramp injection current given by $\sqrt{2}\ I\omega t$ where I is the design maximum symmetrical fault current for single or three phase faults.
5. Use the positive sequence power frequency parameters to model the phase mode network and neglect any circuitry connected to the faulted terminals.
6. Model the system in detail especially close up to the fault and in the main fault current path. Take account of distribution systems and loads by appropriate shunt networks unless all loads are remote.

7. CONCLUSION

The calculation of trv for very many practical cases can be made with acceptable accuracy using a simple travelling wave program based on the lattice diagram and phase and ground mode network models based on the symmetrical component positive and zero sequence networks with power frequency parameters.

8. ACKNOWLEDGEMENTS

The authors undertook trv calculations at the instigation of the Australian CIGRE Panel No. 13 as part of a larger co-operative investigation into system trv's which provided the rare opportunity of comparing calculations with full scale field test results obtained on the ECNSW system. The authors wish to thank the ECNSW and the SECV for their various contributions which made the investigation possible and for sponsorship of their own work. Particular thanks are due to Messrs. P. J. Medhurst and A. N. Bird for their experienced guidance and critical evaluations in the course of the work.

REFERENCES

1. R. O. Caldwell, J. L. Diesendorf, S. K. Lowe, P. J. Medhurst, J. R. Mortiss, E. Truupold, A. N. Bird, L. Saunders and A. D. Stokes, Australian studies of trv and the influence of breaker interaction by field tests and calculation, CIGRE Report 13-05 (1976)
2. L. V. Bewley, Travelling waves on transmission systems, Dover (1951)
3. A. N. Greenwood, Electrical transients in power systems, Wiley - Interscience (1971)
4. C. G. Wagner and G. D. McCann/C. F. Wagner, Wave propagation on transmission lines, Electrical T and D Reference Book, Westinghouse (1964)
5. C. F. Wagner and R. D. Evans, Symmetrical components, McGraw Hill (1933)
6. C. L. Fortescue, Discussion on E. W. Boehne, Trans. AIEE 55 (1936) 192, appearing in 56 (1937) 530
7. P. Hammarland, Transient recovery voltages, Acta Polytechnica, El. Eng. Series, 1, 1 (1947)
8. L. Gosland, ERA Report G/T 104 (1939)
9. W. P. Lewis, Proc. IEE 113 (1966) 2012
10. R. G. Colclaser and D. E. Buettner, Trans. IEEE, PAS 88 (1969) 1028
11. C. F. Wagner, Paper No. 63-1020, IEEE Summer General Meeting (1963)
12. J. R. Carson, Electric circuit theory and operational calculus, McGraw Hill (1926)
13. G. E. Adams, Trans. AIEE, 78 (1959) 639
14. A. Budner, IEEE Trans. PAS 89 (1970) 88
15. J. K. Snelson, IEEE Trans. PAS 91 (1972) 85
16. H. W. Dommel, IEEE Trans. PAS 88 (1969) 388
17. H. W. Dommel, IEEE Trans. PAS 90 (1971) 2561
18. D. E. Hedman, IEEE Trans. PAS 84 (1965) 200, discussion (1965) 489
19. D. E. Hedman, IEEE Trans. PAS 84 (1965) 205
20. L. M. Wedepohl, Proc. IEE, 110 (1963) 2200
21. L. M. Wedepohl and R. G. Wasley, Proc. IEE 112 (1965) 2113
22. L. M. Wedepohl, Proc. IEE, 113 (1966) 622
23. L. M. Wedepohl and C. S. Indulkar, Proc. IEE 121 (1974) 997
24. ANSI Standard C37.0721 - 1971
25. R. H. Harner, W. C. Potempa, T. J. Tobin, Investigation of total trv - study of system modelling on trv calculation and summary of trv parameters, CIGRE 13-74 (WG-01) IWD
26. R. H. Harner and E. W. Schmunk, Investigation of total trv - study of total trv parameters on the Allegheny power system 500 kV grid, CIGRE13-74 (WG-01) 23 IWD
27. R. H. Harner and J. Rodriguez, IEEE Trans. PAS 91 (1972) 1887
28. A. Braun, H. Huber and H. Suiter, Determination of the trv occurring across generator circuit breakers in large modern power stations, CIGRE Report 13-03 (1976)
29. R. G. Colclaser, J. E. Beehler and T. F. Garrity, IEEE Trans. PAS 92 (1975) 943
30. G. Catenacci, Contribution on the study of the initial part of the trv, Electra 46 (1976) 39
31. W. S. Meyer and H. W. Dommel, IEEE Trans. PAS 93 (1974) 1401
32. D. E. Hedman and S. R. Lambert, IEEE Trans. PAS 95 (1976) 197
33. D. K. Tran, Parts I & II, Proc. IEEE 124 (1977) 133
34. I. E. C. Publication 56, third edition, esp. 56-2 & 56-4, and current draft revision proposals
35. M. B. Humphries, This volume
36. A. J. Potter, Electra 34 (1974) 138

DISCUSSION
(Chairman: A. N. Bird, SECV)

J. Urbanek (General Electric)

I assume you tried to correlate the transient recovery voltages (trv) you calculated with the short-circuit currents. Is there any general statement you can make regarding the peak of the trv, and the time to peak, as a function of fault current? For the highest currents, for example, is the trv exponential, as rated in ANSI?

J. L. Diesendorf

Yes, it is our experience that you have a high short-circuit current at locations where you are supplying a lot of things. If it is close to the load center (Sydney-West or Sydney-South in our case) then we have very many lines going out. And, although the short-circuit currents are very high, there are so many lines that the trv is not too bad.

I think the more severe cases arise in special cases not treated here where, for example, you are close to generation and you have series reactors. This can be an incredible problem. We are just building a new power station as an extension to the existing one where the switchyard was designed for 50 kA and if we built the new station adjacent to it and connected it to the existing one we would exceed the rating of the breakers. Therefore we proposed to put series reactors between the two stations. Although this will limit the fault current just to the maximum rating of the breakers, the trv in this case is quite extreme.

Our general experience from surveying our system is that in the bulk of applications our trv's are well below what is specified in IEC. But, in some applications in the network at large, we approach the current proposals and there are a few applications of an unusual kind, such as where the series reactors are involved, where we have to think even of network modifications to reduce the trv.

E. Ruoss (Brown Boveri)

It was mentioned that the power frequency parameters were used for the trv calculation without considering the frequency dependence of some characteristic data. In overvoltage determination it is found that the frequency dependence of some parameters may exert a significant influence. This dependence is especially pronounced when the zero-sequence system is involved in the phenomena to be studied.

J. L. Diesendorf

We did some investigations with faults at the remote station Hawkesbury which were interrupted at Carlingford. In that case we had a very significant component introduced by the transmission line ('large short line fault'). We have a line-side transient and a source-side transient, and the voltage drop at Carlingford, for the fault at Hawkesbury 28 km away, was perhaps 30 %. This means that 70 % of the voltage drop is in the transmission line. Now in that circumstance, if you want to get close to the right answer, you have to take the frequency dependence into account. And the simplest way, which we have adopted, is simply to look at what the dominant frequency is and to evaluate the parameters for that frequency. This gave reasonable results. Of course, it would be very nice to do all our calculations taking account of the frequency dependence, but when you survey the network your data preparation becomes quite a substantial problem.

M. B. Humphries (CEGB)

First, I'd like to back up what you said about the interchangeability of the trv for the 3-phase earthed and unearthed fault. Fig. 1 shows some tests we did on a 275 kV system. We compared the trv for the 3-phase earthed and unearthed conditions for the same value of fault current. This is an actual system test record and you can see the fairly close agreement between the two trv wave shapes for this very simple system. There is just one line on the source side and one transformer. But in general we think you can interchange between the 3-phase earthed and unearthed conditions using phase factors or multiplying factors between the two.

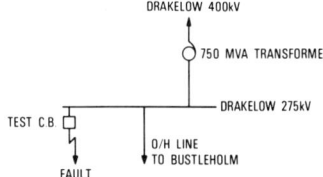

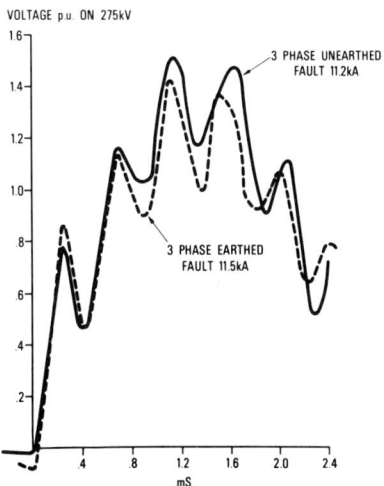

Fig. 1. Comparison of 3-phase earthed and unearthed fault source side trv for CEGB tests on a 275 kV system (Drakelow station).

Figure 2 for the same test series shows the 1.5 phase factor which should be used for the 3-phase unearthed fault condition. The top trace shows the source-side first-pole-to-clear oscillogram (which corresponds to the single phase fault), and the bottom trace shows the line side. If you add these two together you get values which exceed by more than 50 % the source-side oscillogram. I'm not so happy about your use of multiplying factors to go from 3-phase condition to the single phase. We don't feel you can do this and we've done some extensive studies to try to obtain multiplying factors and we can't for our 420 kV system.

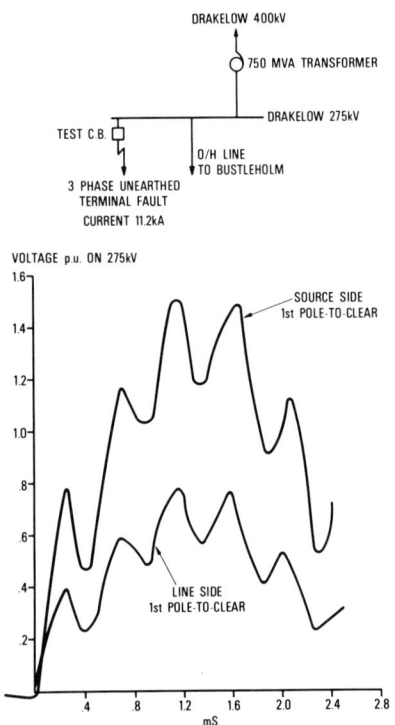

Fig. 2. Source and line side trv for 3-phase unearthed terminal fault.

In Fig. 3 I'd like to show you some trv wave forms which have been calculated for our 420 kV system in the current range 40 - 50 kA and illustrate the wide variety of wave shapes which you get in a large mesh system. In general they conform with your series LR, parallel C wave shape and therefore also conform to the IEC 4-parameter waveforms. Figure 4 shows the IEC equivalents of these waveforms. They are mostly 4-parameter shapes, although I believe there are two in there which are 2-parameter.

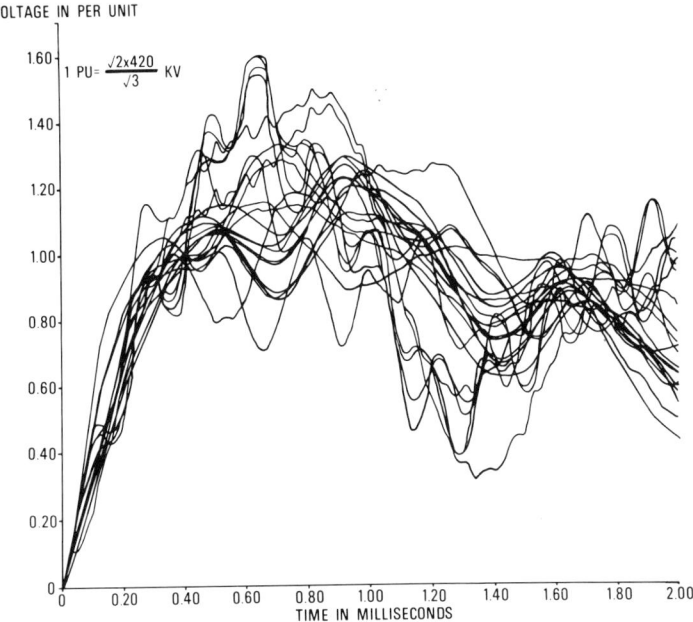

Fig. 3. 3-phase earthed fault trv waveforms in 40 - 50 kA current range in the CEGB 420 kV system.

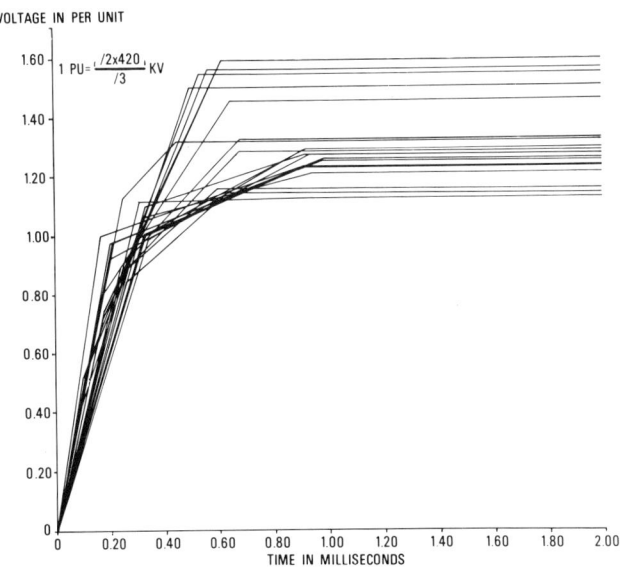

Fig. 4. 4-parameter envelopes of the trv waveforms of Fig. 3.

P. G. Parrott (CEGB)

I would like to confirm from our experience that a transformer right on the faulty busbar makes a very big difference indeed and I think none of us should neglect this particular case. On the other hand, we have found by studies in our high-voltage network that the representation of the ongoing lower-voltage network (in our case 275 kV) gives a reduction in the peak of something like 5 %. I feel this is the order of magnitude to which the load does influence the trv.

J. L. Diesendorf

I think I must simply agree with that. One should mention, however, that in the case of transformers it takes time to penetrate. And unless your transformers are right at the point where the fault occurs then their significance is very much reduced. In our own case I imagine that the effect of the lower voltage networks on the circuit breaker in the 330 kV network could be more significant if we are looking at the network from one of the stations which supplies many 132 kV lines. In this case the transformer rating is very high, its impedance therefore reduced, and the penetration time lowered.

I think you do have to actually examine your network to see whether, in any individual case, you need to take into account the lower-voltage networks. I'm very confident of the calculations we did at Carlingford mainly because the calculations with load agreed with the test result, and the calculation without load nobody disagrees with anyway. So I'm sure that the effect of load does have to be taken into account when you're close to it in such locations.

THE DIELECTRIC STRENGTH OF AN SF_6 GAP*

T. H. TEICH** and W. S. ZAENGL
High Voltage Laboratory, ETH Zurich, Switzerland

SUMMARY

This contribution is intended to give a brief introduction, for the non-specialist, into the early stages of gas breakdown at moderate temperatures. One of the techniques for the determination of discharge parameters will be described and some examples of data obtained by means of this technique at the University of Manchester, Institute of Science and Technology, are presented. Breakdown criteria which make use of such basic gas data will then be introduced and some classical illustrations will serve as evidence for the validity of models used. Measurements carried out in the High Voltage Laboratory of the Federal Institute of Technology at Zürich demonstrate the validity of the models and criteria previously mentioned over a wide range of conditions at elevated pressures in SF_6.

1. FUNDAMENTALS OF CONDUCTION AND DISCHARGE PHENOMENA

1.1. Conduction due to Field-intensified Ionisation

The breakdown of a gas gap - be it relatively hot gas in a circuit breaker or cold gas in a homogeneous field gap - is largely determined by the ionization and attachment properties of the gas which, in turn, depend strongly on the ratio of field strength to density E/N, as this ratio determines the energy which electrons can take up from the applied electric field. To simplify the consideration of basic processes involved in low current density conduction in a gas, let us concentrate initially on homogeneous field gaps; this is justified here as such gaps will, on the one hand, be used to acquire the gas discharge data required for calculations of breakdown voltage and, on the other hand, as one will strive for a fair degree of field

---

*) Presented at the Symposium by T. H. Teich.
**) On leave from UMIST, Manchester.

homogeneity in SF_6-insulated systems, not counting surface imperfection which we shall have to consider later.

In any gas space, we shall always have charge carriers present due to their generation by cosmic radiation and radioactivity of the construction materials. Under the influence of the applied field E, these charges q_i will move parallel to the direction of the field (Fig. 1) and will thus also produce a current I_i in the external circuit.

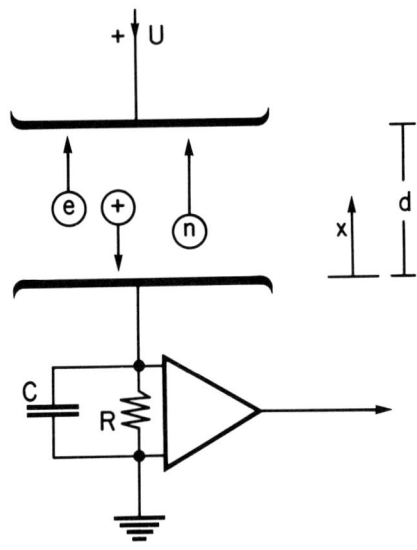

Fig. 1. Basic arrangement for the measurement of charge carrier currents; the sum of the currents of electrons (e), of positive ions (+) and of negative ions (n) appears as a voltage drop across the resistor R.

In the homogeneous field, these charges move with a constant mean velocity, the drift velocity v_i, and the instantaneous value of the current component I_i due to each species of charge carrier can be calculated by equating the mechanical and the electrical work done in moving the charges. The mechanical work is

$$dW_m = Fds = n_i q_i Eds$$

where ds is a line element along the charge carrier path and n_i the number of charge carriers of species i with a charge q_i. The mean distance covered by $n_i q_i$ in time dt is $v_i dt$, so that we may also write (as $E \parallel v_i$)

$$dW_m = n_i q_i E v_i \cdot dt. \tag{1}$$

The electrical work done is

$$dW_e = UI_i dt = I_i Ed \cdot dt \qquad (2)$$

when U is the voltage across the gap and d the gap spacing.

Equation (1) and (2) gives

$$I_i(t) = n_i(t) q_i v_i/d = n_i(t) q_i/T_i \qquad (3)$$

where T_i is the total transit time of the charge carriers of species i. Such an equation (3) will now hold for each species of charge carrier, for instance, for electron current, for positive and negative ion current. The total current which one can measure in the external circuit is then the sum of these individual currents and it reflects the number of charge carriers in the gap at the particular instant in time.

The change in the number of charge carriers is determined by formation and removal processes; among the latter is the arrival of charge carriers at the electrodes. Ignoring migration, one can find the electron number by means of very simple differential equations. The number of free electrons is increased by ionization:

$$dn_{e+} = dn_+ = \alpha n_e dx = \alpha n_e v_e dt \qquad (4)$$

The number of new electrons equals the number dn_+ of newly formed positive ions; n_e is the total electron number and v_e the electron drift velocity; $dx = v_e dt$ is the distance covered by electrons in field direction. α is the ionization coefficient, defined as the number of ionizations (positive ions formed) per electron per unit path length in the direction of the electric field.

The formation of negative ions, on the other hand, reduces the number of free electrons. This attachment of electrons to neutral molecules can be quantitatively described by means of an attachment coefficient η, representing the number of attachments taking place per electron and per unit path length in the direction of the electric field. The incremental loss of electrons by negative ion formation can thus be described by

$$-dn_n = dn_{en} = -\eta n_e dx = -\eta n_e v_e dt. \qquad (5)$$

The net change in the electron number is the sum of (4) and (5):

$$dn_e = (\alpha-\eta) n_e(t) v_e dt. \qquad (6)$$

For $n_e(0)$ electrons starting simultaneously from the cathode at time t = 0, (6) is easily integrated and gives the time-dependent electron number

$$n_e(t) = n_e(0) \exp((\alpha-\eta) v_e t) \qquad (7)$$

for the first electron transit period, $0 \leq t \leq T_e$.

Integration of (4) and (5), using n_e from (7), gives essentially similar exponential waveforms for the positive and negative ion numbers:

$$n_+(t) = \frac{\alpha}{\alpha-\eta} [\exp((\alpha-\eta)v_e t) - 1] \qquad (8)$$

$$\text{for } 0 \leq t \leq T_e$$

$$n_n(t) = \frac{\eta}{\alpha-\eta} [\exp((\alpha-\eta)v_e t) - 1] \qquad (9)$$

After the passage of the first electron front all charge carriers of the avalanche have been formed (we ignore feedback at this stage) and changes in current are now due to the removal of charge carriers by their arrival at the electrodes. When the value of $(\alpha-\eta)$ is positive and appreciable, most ions will have been formed close to the anode; thus negative ions, which migrate towards the anode, will disappear quickly, while the number and thus the current of positive ions will initially remain almost constant and will only decay appreciably once positive ions arrive in large numbers at the cathode (Fig. 2).

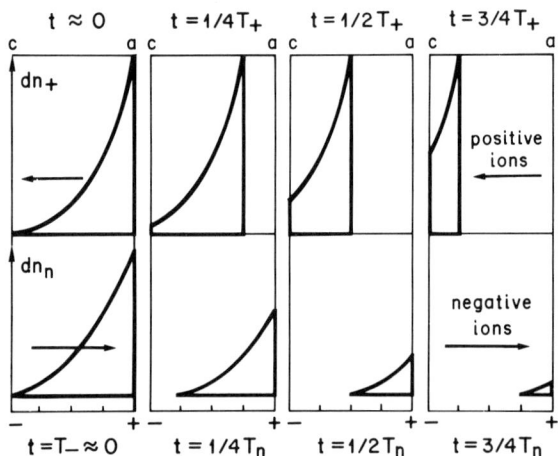

Fig. 2. Distribution of positive and negative ions in the discharge gap at different instants of time.

Equations for the charge carrier numbers can once more be derived by integrating (4) and (5), taking n_e from (7):

$$n_+(t) = \int_{v_+ t}^{d} n_e(0)\alpha \exp((\alpha-\eta)x)dx = \frac{\alpha}{\alpha-\eta} n_e(0) \{\exp((\alpha-\eta)d - \exp((\alpha-\eta)v_+ t\} \qquad (19)$$

$$\text{valid for } T_e \leq t \leq (T_+ + T_e)$$

$$n_n(t) = \int_0^{d-v_n t} n_e(0) \, \eta \exp((\alpha-\eta)x)dx = \frac{\eta}{\alpha-\eta} n_e(0) \{\exp((\alpha-\eta)(d-v_n t)) - 1\} \quad (11)$$

valid for $T_e \leq t \leq T_n$

The electron number is zero after one electron transit time T_e. To obtain the charge carrier currents, one applies (3) to equations (7) to (11); as the ion transit times T_+ and T_n are usually more than a hundred times longer than T_e, the maximum electron current is also appreciably higher than the ion current. A typical current pulse shape is shown - in principle - in Fig. 3. We can recognize the fast rise, typically about 100 ns, of the electron current which falls back to zero after one transit. The ion currents persist much longer, typically for about 30 μs for a 20 mm gap. From a record of the current pulse shape it is now possible to determine the values of the most important discharge parameters:

v_+ : positive ion drift velocity from the duration of the positive ion current;

$\alpha-\eta$: 'effective' ionization coefficient from the decay of positive ion current, time constant $([\alpha-\eta]v_+)^{-1}$ according to (10);

v_n : negative ion drift velocity from the decay of the negative ion current, time constant $([\alpha-\eta]v_n)^{-1}$;

α/η : ratio of ionization and attachment coefficient from the relative levels of positive and negative ion current after one electron transit time T_e; the value of the positive ion current at T_e is found by extrapolating back from the ion plateau;

α, η : determined separately from $\alpha-\eta$ and α/η;

v_e : electron drift velocity from the duration T_e of the electron current or from its rise time constant $([\alpha-\eta]v_e)^{-1}$, according to (7);

$n_o \equiv n_e(0)$: primary electron number form the absolute levels of current.

Actual current recordings for SF_6 are shown in Figs. 4, 5, 6.[1] At high values of the field strength to density ratio E/N we have $\alpha \gg \eta$ and therefore only negligible negative ion current (Fig. 5); at lower E/N the positive and negative ion currents become quite comparable (Fig. 6), while at even lower E/N, approaching the condition $\alpha = \eta$, phenomena not included in the model outlined above make themselves felt in the current pulse. Figure 7 shows a rise of current well after one electron transit time. This delayed current rise is usually attributed[2] to instability of the negative ions formed: The attached electron becomes detached again and once more takes part in ionization. Computer simulation of the process[3] seems to indicate that only a small proportion of the negative ions formed lose their electron again, and it may be reasonable to assume a rapid stabilization of initially unstable negative ions;[4] this assumption was made for the current simulation Fig. 8[5] which reproduces very well the current pulse shape of Fig. 7.

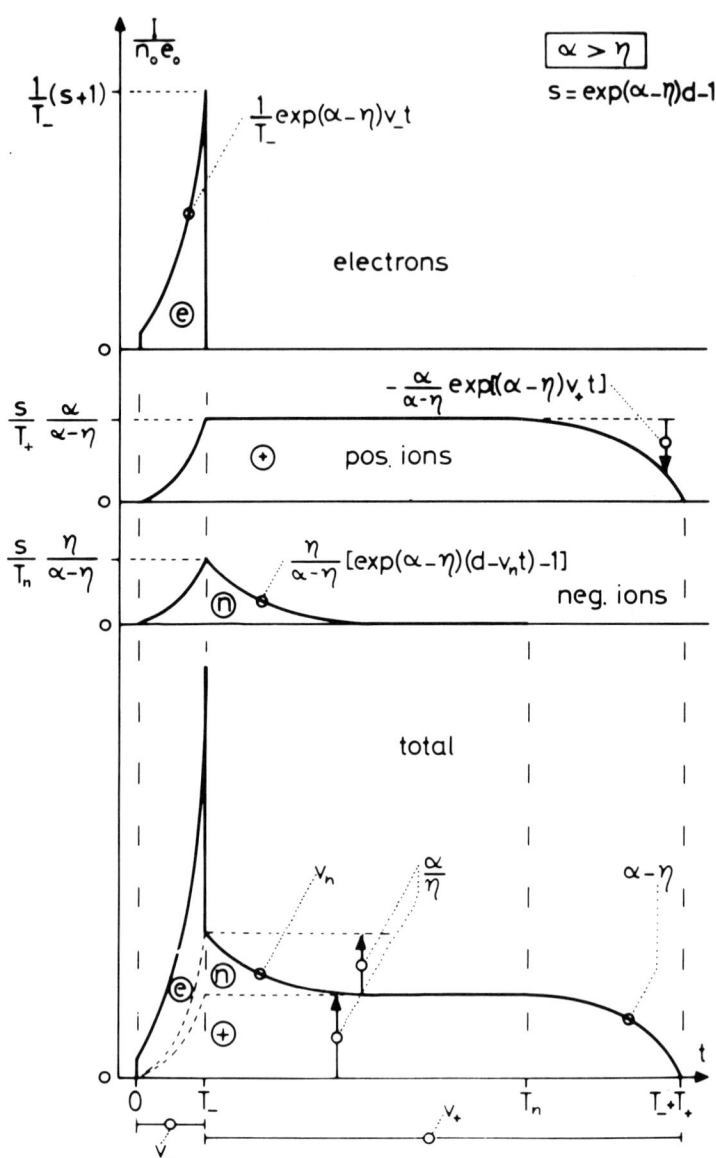

Fig. 3. Current pulse of an electron avalanche in a homogeneous field discharge gap. The quantities which can be determined from particular parts of the pulse are indicated against these. The time scale up to $T_e \equiv T_-$ has been stretched for the sake of clarity.

DIELECTRIC STRENGTH 275

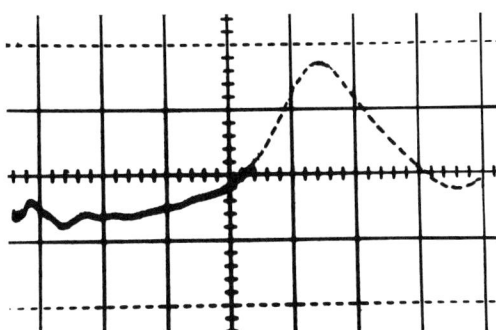

Fig. 4. Avalanche oscillogram for SF_6, (8 torr, 140 V/cm·torr, 6 µA/division, 20 ns/div., d = 2.4 cm) showing largely electron current; a timing pulse is superimposed at the beginning of the current trace to mark the instant of primary electron generation by an ultraviolet flash lamp pulse.

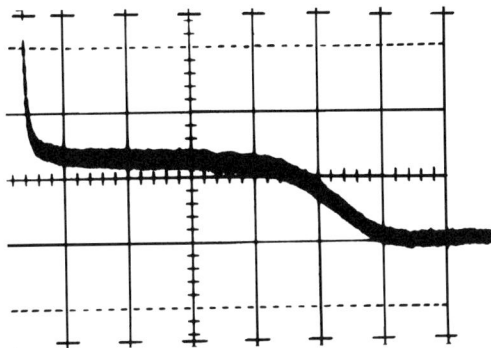

Fig. 5. Avalanche oscillogram showing predominantly positive ion current preceded by an electron current peak (SF_6, 0.6 torr, 550 V/cm·torr, 600 nA/division, 2 µs/div., d = 1.7 cm).

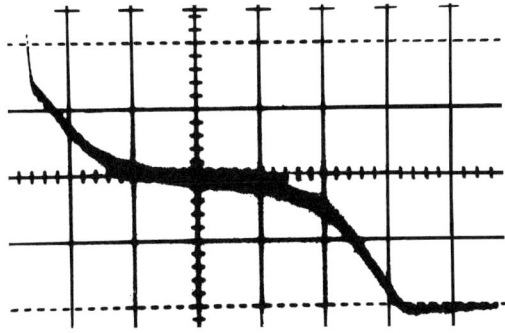

Fig. 6. Avalanche oscillogram showing positive and negative ion currents (compare with Fig. 3). (SF_6, 7 torr, 180 V/cm·torr, 300 nA/div., 1 µs/div., D = 0.55 cm).

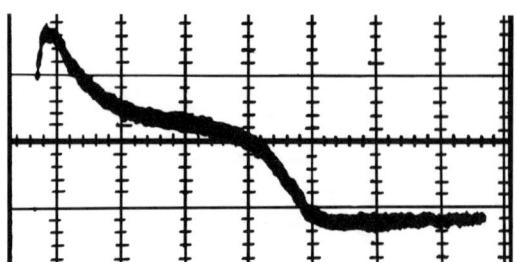

Fig. 7. Oscillogram showing a delayed current rise well after one electron transit time (ca. 100 ns), so that the maximum of the ion current is only reached after approx. 1 μs. (SF_6, 60 torr, 125 V/cm·torr, 300 nA/div., 5 μs/div., $n_o \sim 10^4$ electrons).

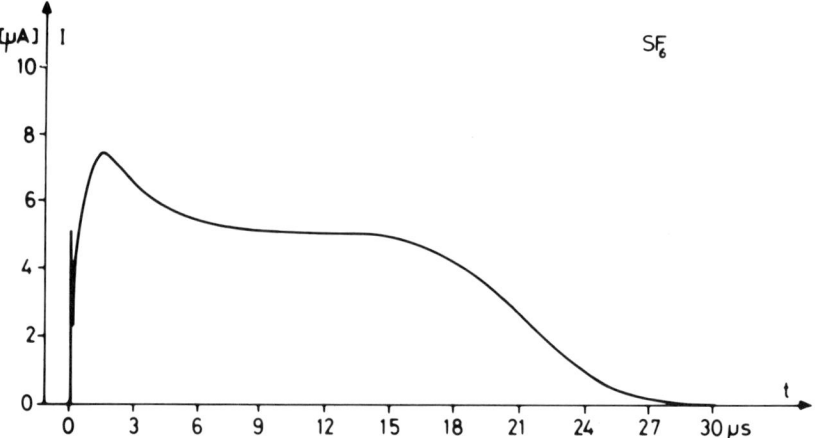

Fig. 8. Computer simulation of the current oscillogram of Fig. 7 (natural lifetime of negative ions against detachment τ = 5 μs; lifetime against stabilization B = 150 ns).

Due to the high noise levels of wide-band amplifiers, sensible time-resolved measurements are obviously only possible with fairly large numbers of charge carriers in each generation; for conditions under which $(\alpha-\eta)d < \sim 12$, these large numbers can only be obtained by means of high values of n_o. Sufficiently large numbers of starting electrons n_o within a time interval short compared with an electron transit time can, however, be generated by means of a very high power ultraviolet flash source such as a Q-switched ruby laser with frequency doubling;[6] with such an electron source, time resolved measurements can be extended into the range of low E/N used in technical insulation for which attachment dominates over ionization, $\alpha < \eta$. - Figures 9 to 11 give a view examples of data determined by means of time-resolved current measurement. Figure 9 shows $\alpha-\eta$ for CO_2; measurements in this electronegative

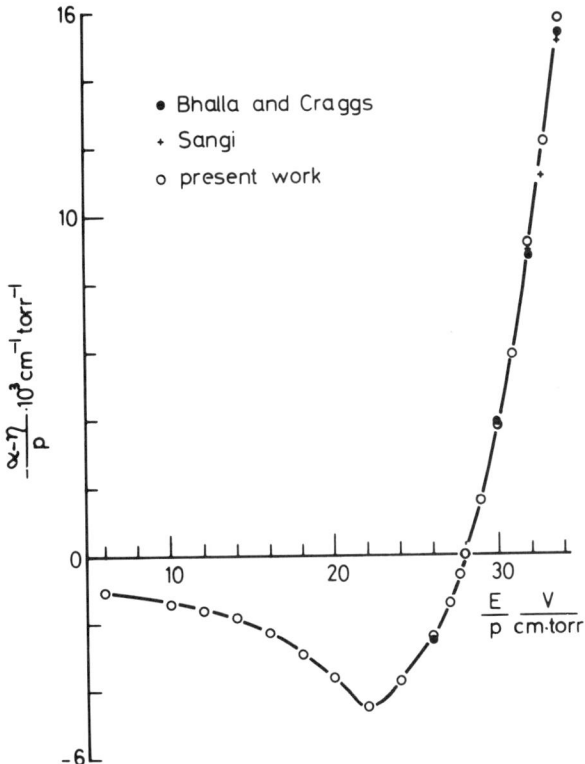

Fig. 9. Values of α-η obtained from time-resolved current measurement in CO_2. We have α = η for E/p ~ 28 V/cm·torr, but measurements extend to below one quarter of this minimum value for breakdown.[7]

gas were extended to E/N values of less than one quarter of the minimum value for breakdown[7] by generating primary electron numbers of the order of 10^8. A plot of α-η in SF_6 up to very high values of E/N is shown in Fig. 10[1] and for the region of α ~ η in Fig. 11.[7]

1.2. Discharge and Breakdown Criteria

Technical interest in an insulating gas will concentrate on its breakdown properties, and means of predicting the breakdown voltage will be wanted. Such predictions may be based on the application of relatively straightforward breakdown criteria containing basic gas data such as the ionization and attachment coefficients. Obviously, low values of the effective ionization coefficient α-η will contribute to keep down current growth, but effects not yet considered so far are involved in promoting breakdown. Electrons will often produce collisional excitation of gas molecules even when their energies are well below the ionization energy; this leads to emission of light which

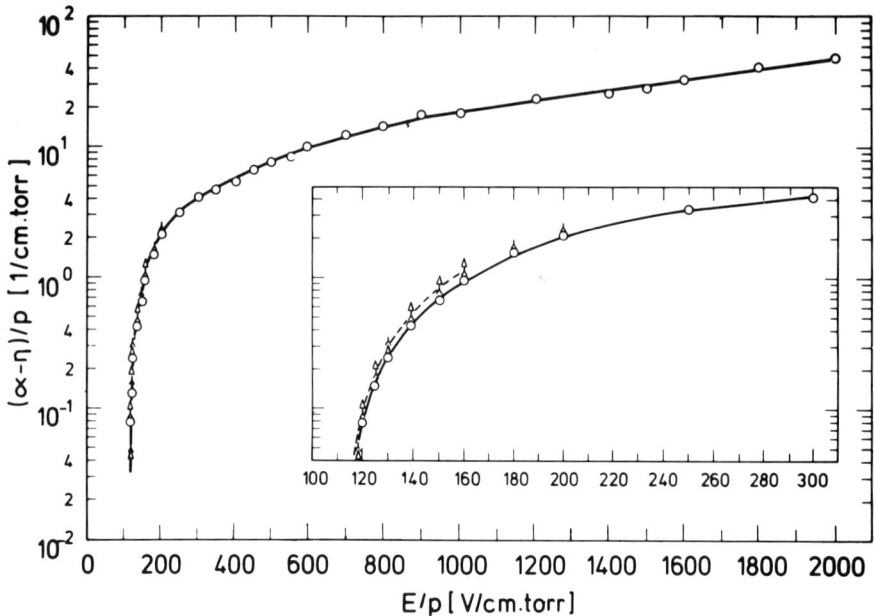

Fig. 10. α-η in SF_6 (-o-[1]).

may, in turn, produce further electron emission from the cathode. Thus avalanches may have successors, manifesting themselves as a sequence of electron current maxima spaced about one electron transit time apart (Fig. 12). When every electron originally leaving the cathode secures the occurrance of a successor, by ionization leading to gas amplification and by excitation leading to emission of ultraviolet light, then the discharge can maintain itself; for the case of fairly homogeneous fields, this state of the discharge is the (complete) electrical breakdown. To quantify such feedback processes, which may come about by photons, by positive ions or by metastable molecules incident upon the cathode, one has defined a feedback coefficient γ which denotes the number of secondary electrons produced at the cathode per ionization (i.e. per positive ion) in the discharge.

The effect of cathodic feedback can be represented as an increase in the number of primary electrons at the cathode. Let n_{oo} denote the number of primary electrons due to external causes and n_o the total primary electron number including the effects of feedback. The number n_+ of positive ions generated due to n_o primary electrons is

$$n_+ = \int_0^d n_o \alpha \exp((\alpha-\eta)x)dx = \frac{\alpha}{\alpha-\eta} n_o \{\exp((\alpha-\eta)d) - 1\}.$$

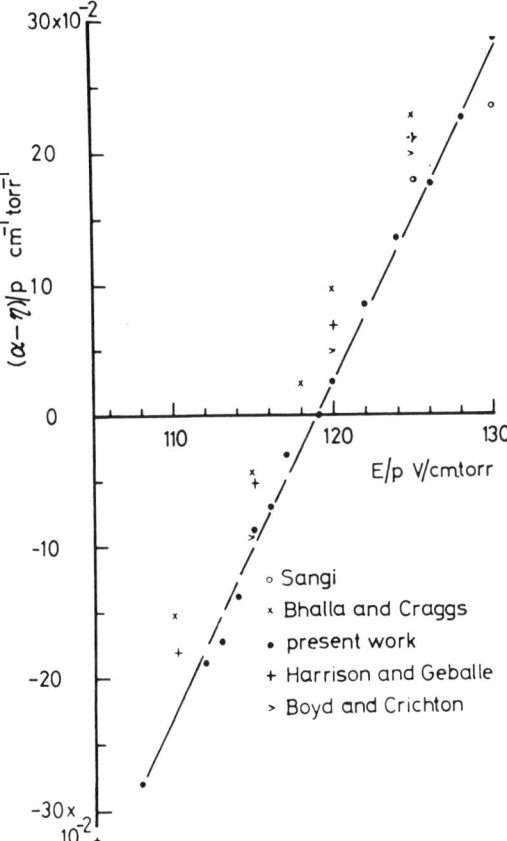

Fig. 11. α-η in SF_6 for the region of E/N for which $\alpha \underset{\sim}{\sim} \eta$.[7]

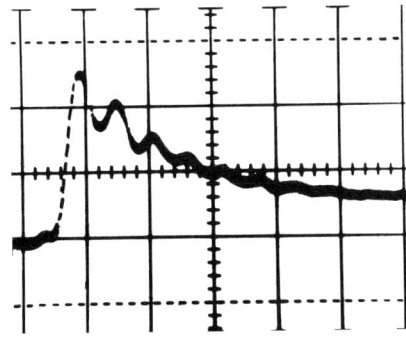

Fig. 12. Current oscillogram of an avalanche with photo successors in SF_6 (150 μA/div., 50 ns/div).[1]

In every generation there will be γn_+ secondary electrons for these n_+ positive ions, so that the total number of electrons at the cathode will be given by

$$n_o = n_{oo} + \gamma \frac{\alpha}{\alpha-\eta} n_o \{\exp((\alpha-\eta)d) - 1\}$$

and thus

$$n_o = \frac{n_{oo}}{1 - \gamma \frac{\alpha}{\alpha-\eta} \{\exp((\alpha-\eta)d) - 1\}}.$$

Obviously n_o will tend to infinity when $\gamma \alpha/(\alpha-\eta) \{\exp((\alpha-\eta)d)-1\}$ approaches unity; then the current will also become infinite, that means, it will only be limited by the external circuit. This gives us the 'Townsend breakdown criterion'

$$\gamma \{\exp((\alpha-\eta)d) - 1\} \geq \frac{\alpha-\eta}{\alpha} \qquad (12)$$

which can often be used as a basis for calculations of breakdown voltages. Let us consider here a simplified form which applies when $\exp((\alpha-\eta)d) \gg 1$, $(\alpha-\eta)/\alpha \sim 1$. Then we may write instead of (12)

$$\exp((\alpha-\eta)d) \geq 1/\gamma \qquad (13)$$

or

$$(\alpha-\eta)d = -\ln\gamma = K. \qquad (13a)$$

The expression on the left of (13) is the 'gas amplification', the number of electrons in an avalanche per starting electron. For SF_6, $-\ln \gamma = K$ is of the order of 14...16 at very low pressures and, for well founded reasons, increasing moderately with gas density (Fig. 13).

Another breakdown criterion, which presupposes a different mechanism of breakdown development - to be elucidated below -, is formally very similar to (13) and will also yield quite equivalent breakdown data; this 'Streamer Criterion' is

$$n_o \exp((\alpha-\eta)d) \geq N_c, \quad N_c = 3 \cdot 10^8 \ldots 10^9 \ldots \qquad (14)$$

or for $n_o = 1$:

$$(\alpha-\eta)d = \ln N_c = \ldots 18 \ldots 20 \qquad (14a)$$

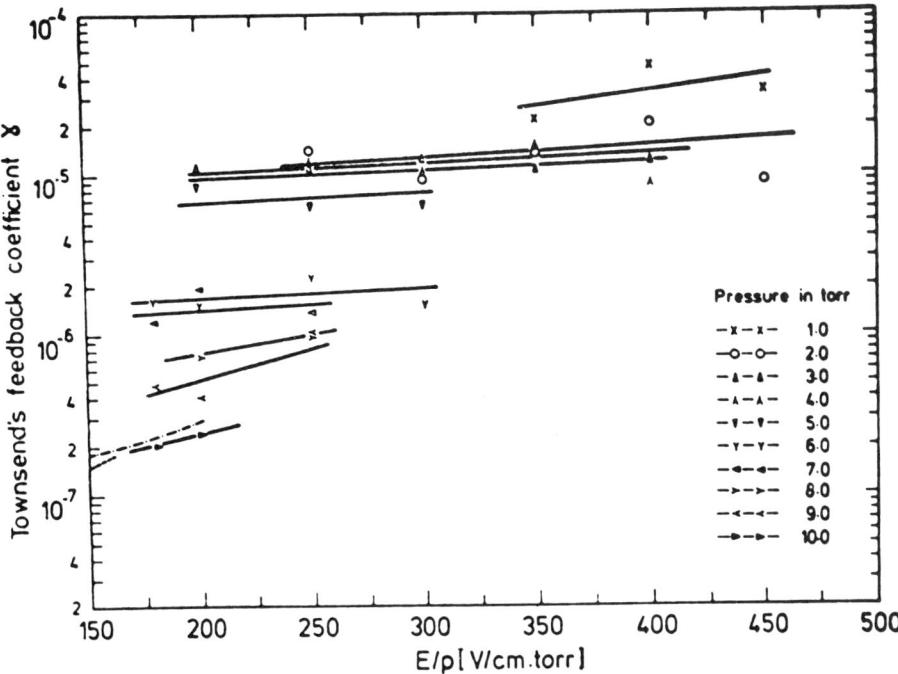

Fig. 13. Feedback coefficient γ for SF_6.[1] γ decreases markedly with increasing gas density.

This obviously means: breakdown will occur when a certain critical number of charge carriers in an avalanche head is exceeded; this type of breakdown development is thus space charge controlled from a relatively early stage onwards.

Both criteria can be expanded to apply to inhomogeneous fields; for the case $n_o = 1$ (one primary electron), the left sides of (13) and (14) comprise merely the gas amplification; the right sides are critical numbers N_c. Notwithstanding modifications which we shall have to introduce later we write a breakdown criterion for inhomogeneous fields as

$$\exp \int_0^d (\alpha-\eta)dx \geq N_c \text{ or } \int_0^d \{\alpha(x) - \eta(x)\} dx \geq \ln N_c = K \quad (15)$$

The models on which the two original criteria are based may just lead us to slightly different values of K, quantities which are difficult to determine with great accuracy and which do not seem to be strongly dependent upon most discharge conditions. In first approximation, we may thus consider K as constant.

Let us briefly describe breakdown by action of the streamer mechanism: In the growing electron avalanche, charge separation will occur; the electron front moving towards the anode leaves behind positive charges, and thus a space charge field will be superimposed upon the applied field (Fig. 14). In particular, the electric field

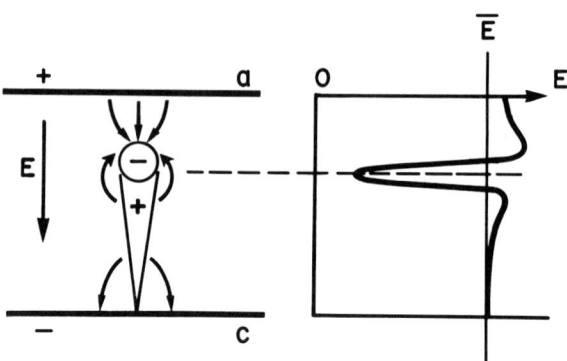

Fig. 14. Space charge field of an electron avalanche. On the right, a plot of field strength along the axis of the avalanche. $\bar{E}$ is the applied field strength which would prevail in the absence of space charge.

strength is enhanced in front of the avalanche head, leading to intensified ionization and to an increase in the propagation velocity (Fig. 15). The increased field also promotes the emission of high energy ultraviolet radiation ($\lambda \leq 100$ nm) capable of ionizing the gas (under some conditions, detachment of electrons from negative ions by photons may be involved), thus generating new primary electrons in the gas which then produce new electron avalanches in the enhanced field region well behind the avalanche head; thus they bring about a backward growth of ionization (Fig. 16). One can show this impressively by means of streak photography, which can be arranged to give us a record of the luminosity along the axis of the discharge versus time. The darker zone observed (Fig. 17) behind the hypothetical position of the avalanche head very probably corresponds to the region of reduced field strength indicated in Fig. 14.

Once this form of discharge development is initiated anywhere in the gap, complete or at least partial breakdown will occur. We should therefore modify our discharge criterion (15) to the form

$$\text{Max} \left(\int_{\substack{\text{any section} \\ \text{of the gap}}} \{\alpha(x) - \eta(x)\} \, dx \right) \geq \ln N_c = K \tag{16}$$

Partial or complete breakdown will occur once the integral on the left exceed the value K for any limits chosen along the 'most favourable' field line between the electrodes.

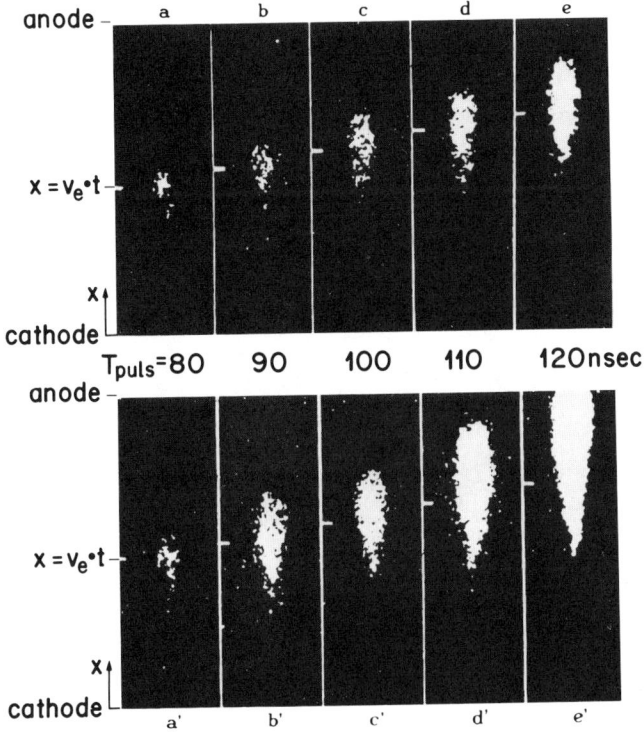

Fig. 15. Acceleration of the avalanche head under the influence of its own space charge. Without space charge influence, the head progression would be strictly proportional to electron drift velocity v_e and time t, as marked by $x = v_e \cdot t$. (Image intensifier records of avalanches in N_2 + 10 % CH_4, $p_{tot} \sim$ 100 torr; gap distance 3 cm; E/p = 59 V/cm·torr. Voltage pulse of duration T_{puls}, 19 % overvoltage. For a-e: $\ln n_o \sim 5$. For a'-e': $\ln n_o \sim 6$. See K. H. Wagner, Ref.[8], Fig. 1).

Obviously, the quantity of greatest influence upon the breakdown voltage is the effective ionization coefficient α-η. As shown in Fig. 10, for SF_6 this coefficient has been measured, at room temperature, over a wide range of values of E/p, the ratio field strength to pressure. Compared with ionization, attachment is negligible at high values of E/p; over the E/p range from 200 to 2000 V/cm·torr, the effective ionization coefficient α-η is practically identical with α and (α-η)/p can be represented by a straight line of slope 21/kV; at lower E/p, near the critical value of about 118 V/cm·torr for which α equals η, the slope steepens to about 27.7/kV (Fig. 11). Neglecting the change of this steepness for larger values of E/p, we may thus use the linear dependency:

$$(\alpha-\eta)/p = C \ [E/p - (E/p)_o] \tag{17}$$

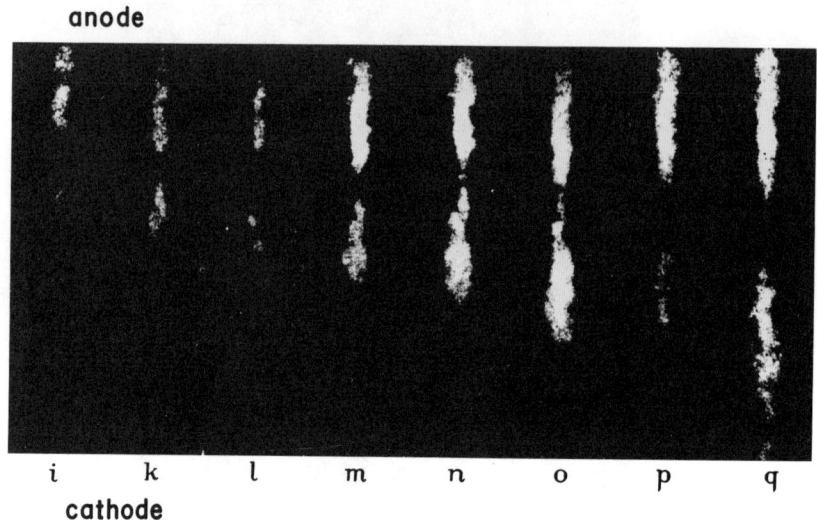

Fig. 16. Transition to cathode directed propagation of ionization ('cathode directed streamer'). Similar conditions as in Fig. 15. (K. H. Wagner,[8] Fig. 25). T_{puls} = 240 ns.

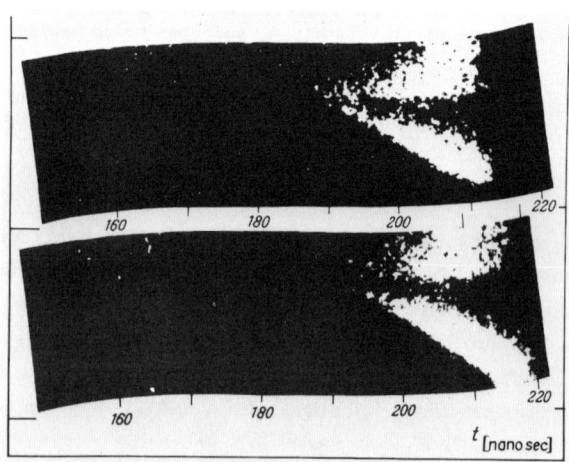

Fig. 17. Streak record of streamer development in N_2 + 10 % CH_4. Similar conditions as in Fig. 15 (K. H. Wagner,[8] Fig. 26).

with

$$C = 27.7 \ (kV)^{-1};$$

$$(E/p)_o = 88.5 \ kV/cm \cdot bar);$$

Temperature: 20° C;

$$60 \leq E/p \leq 150 \ kV/cm \cdot bar.$$

This dependency is very widely used today and deviates only little from data published elsewhere. It should be noted, however, that for values $E/p \gtrsim 150$ kV/cm·bar the measured ionization coefficients are smaller and not in agreement with Fig. 10.

The breakdown criterion (16) can be re-written for SF_6 with the help of (17) for $(\alpha-\eta)$:

$$\underset{\substack{\text{any section} \\ \text{of the gap}}}{\text{Max}} \left(\int [E(x)/p - (E/p)_o] dx \right) \geq K/(p \cdot C). \tag{18}$$

For homogeneous fields, a very simple relationship between breakdown voltage U_d and the product of gap spacing d times p results, as $U = E \cdot d$:

$$U_d = (E/p)_o \ dp + K/C. \tag{19}$$

Under high pressure conditions, for SF_6 the value of K will be in the order of 18 ... 20. Then the term K/C is well below 1 kV and can be neglected. The breakdown voltage under these homogeneous field-configuration and $U_d \gg 1$ kV is therefore

$$U_d/kV = 88.5 \ pd/(bar \cdot cm). \tag{20}$$

This Paschen relationship would give a linear increase in breakdown voltage with pressure. Discharge properties depend, however, on gas density, as the density determines the free path length. SF_6 does not behave like an ideal gas if pressure of more than some bars are taken into consideration. As the density increases more than linear with the gas pressure and depends furtheron on gas temperature,[9] U_d will increase also more than linear with pressure, contrary to[20]. For a temperature of 20° C, this breakdown voltage is about 10 % higher for p = 10 bar and 22 % for p = 15 bar and should not be neglected.

Contrary to this, deviations from Paschen relationship, that means, lower breakdown voltages, have been widely reported in the literature,[10] and various explanations have been offered. Extensive investigations seem to prove beyond doubt that these 'deviations' can mainly be interpreted as the effect of local disturbances of the electric field by imperfections of the electrodes.

2. EXPERIMENTAL DIELECTRIC STRENGTH

All experimental results published elsewhere indicate lower breakdown voltages in comparison to[19], if the field strength calculated by the macroscopic dimensions of the electrode system used, is higher than 150 ... 250 kV/cm. Especially in a plane-plane-configuration the increase of breakdown voltage with p·d is much more smaller than $(E/p)_o$ and typically about 30 kV/cm·bar only, if this critical field strength to pressure ratio is passed. Furtheron, the following effects are in general associated with such breakdown measurements, if these deviations of Paschen relationship occur:

- Even with originally fine polished electrodes a 'conditioning effect' takes place in the historical sequences of breakdown voltages, characterized by several low breakdown values in the beginning of a sequence. After multiple sparking finally a good stabilization of the breakdown voltages takes place. It can be shown, however, that this final value may strongly be influenced by the stored energy within the test circuit, which is converted in this gap during breakdown.
- Appreciable pre-breakdown currents are often observed also in homogeneous field configurations, for which with high gas pressures no 'corona breakdown' should occur.[12]

Explanations for this phenomenas offered in the literature are based upon field emission of electrons, leading to field enhancement at the cathode and initiating breakdown; upon statistical properties of the ionization processes; upon detachment of electrons from negative ions and finally, upon the possibility that conducting particles in the gas or protrusions at the electrode surfaces will lower the breakdown voltage.

As it could be assumed that due to the steep increase of the ionization coefficients with field strength even in very small regions of enhanced fields the breakdown criterion (18) may be fulfilled, purposeful experiments were carried out with relatively ideal, smooth and clean electrodes, which had not suffered breakdowns before, in a wide range of high pressure, by R. Baumgartner([11]; some preliminary results in[10]). Under such circumstances no significant deviations from equ. (20) are found at pressures up to 12 bar with plane electrodes having a highly stressed area of about 2 cm^2 (Fig. 18) and up to 15 bar with sphere electrodes (Fig. 19). For these investigations, an experimental system with very low stored energy was used to keep electrode damage by sparking to a minimum, and in general only the first and only breakdown of the carefully polished and cleaned electrodes was used for the calculation of breakdown strength E_d. To test the influence of the methods used for cleaning the electrodes, a multi-station 'electrode carrousel' permitted rapid change of electrodes without changing the gas in the test vessel and to provide equal test conditions. The gas was well filtered to exclude all dust and particles from the gas as far as possible. Such clean electrode systems showed no measureable

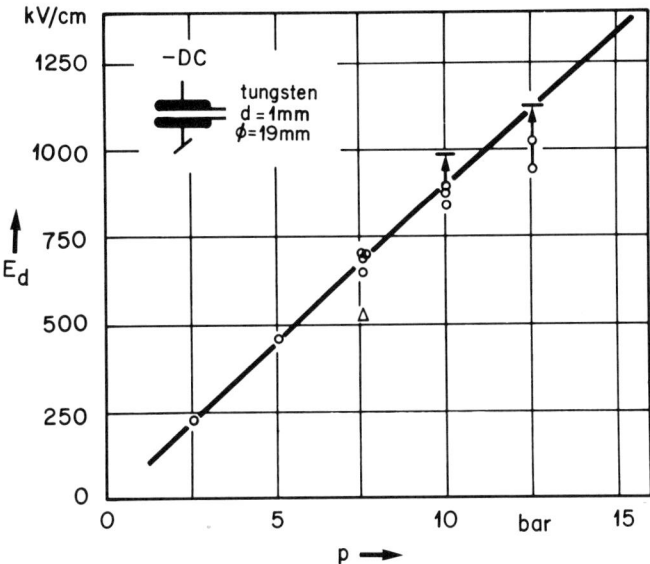

Fig. 18. Breakdown field strength between parallel-plane tungsten electrodes in dependence upon pressure.[11] The solid line is the Paschen curve.[20]
о First breakdown of the gap
♂ First breakdown with final value after conditioning with sparks
△ First breakdown with very small oil contamination of electrode; 1 μA prebreakdown current.

prebreakdown currents. However, the introduction of only traces of oil on the electrodes led to an appreciable prebreakdown current and a breakdown voltage well below the value predicted for a clean gap (see Fig. 18).

This investigations confirmed not only the linear dependence of dielectric strength in SF_6 with gas pressure, but also with gas density. The dashed line in Fig. 19 takes this density-dependence into account, whereas the full line represents the breakdown criteria according to equ. (18) by assuming an ideal gas. In both cases, these calculated curves have regard to the field distribution of $E(x)$ within this gap.

Such results can only indicate the big influence of any kind of 'contamination' to the breakdown strength of a gap. An experimental confirmation of the influence of surface irregularities, however, can be made by provision of controlled protrusions on one electrode of a parallel plate system, for which the field distribution $E(x)$ can either be calculated or approximated with good accuracy. The results of measurements can then be compared with calculations applying the streamer breakdown criterion.

In Fig. 20 the calculated, monotonously falling field strength starting from the tip of a spherical protrusion at one plane electrode ($x = 0$) is shown. The falling field

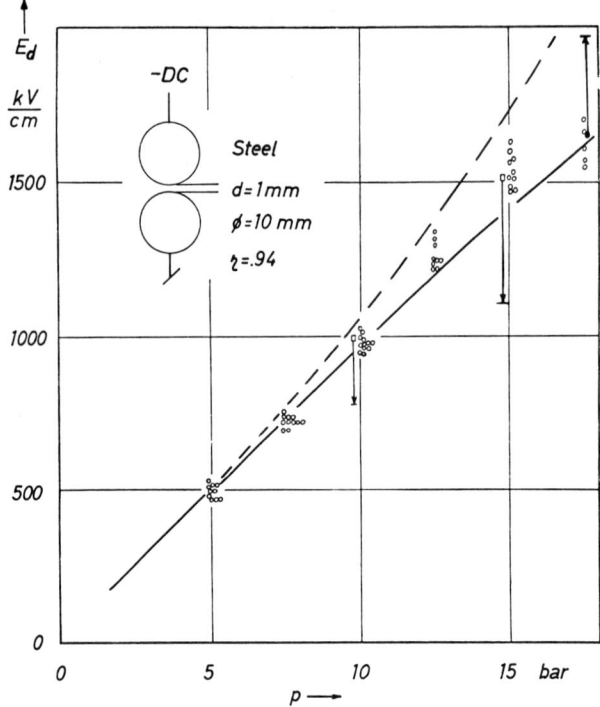

Fig. 19. Breakdown field strengths in a sphere-sphere discharge gap,[11] made of steel; calculated values with and without density correction.
◯ First breakdown of the gap
⌷ First breakdown and deconditioning after 5 sparks
● First breakdown and conditioning for a single gap (p = 17.5 bar).

strength in this 'microscopic' field reaches after a distance, which is about equal to the height of this protrusion (2·R), the 'macroscopic', constant field strength E_{hom}. This field distribution may be roughly approximated by two straight-line segments. The horizontal segment will represent the macroscopic field strength, the falling segment will originate at a point corresponding to the maximum field strength (tip of the protrusion) and meet the horizontal segment at a suitably chosen distance x_{st} for which the integral under the curve is about the same as for the real field distribution, if SF_6 is taken into account.

The breakdown criterion (18) can now easily be applied to this simplified field configuration. With the monotonously falling field strength, the integration has now only to be extended from the protrusion tip to a critical distance x_{crit} at which α equals η, that means, at which E/p reaches $(E/p)_0$; however, particularly at low pressures, this may nevertheless involve integration over the entire gap. The maxi-

DIELECTRIC STRENGTH

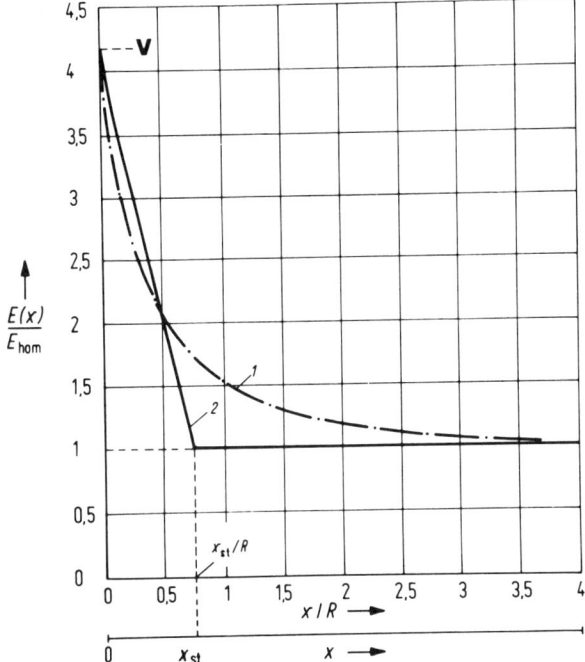

Fig. 20. Field disturbance introduced by a small spherical protrusion in a parallel plate electrode system. Distance x measured from the tip of the protrusion. The theoretical field curve (1) has been replaced by an approximation (2) for a simplified calculation of breakdown voltage.

mum value which the integral in (18) may have without breakdown occurring is obviously inversely proportional to pressure; this means that the curve representing field strength over pressure in dependence upon distance x has to be compressed towards lower E/p values with increasing pressure in order to keep the integral below its critical limit (Fig. 21); the result is a lower macroscopic field strength which solely governs the breakdown or discharge voltage U_d. With further increasing, but very high pressures, the breakdown field strength is limited by the critical field strength to pressure ratio $(E/p)_o$ divided by the field intensification factor v.

The equation for the breakdown voltage resulting from the application of the breakdown criterion (18) to the simplified model as shown in Fig. 20/21 is

$$U_d = (E/p)_o pd/v + (v-1)Kd/(Cv^2 x_{st}) \{1 + \sqrt{(2(E/p)_o vCpx_{st}+(v-1)K/((v-1)K)}\}; \quad (21)$$

this equation is valid for conditions under which the critical amplification exp K is attained within a distance x_c which is smaller than the extent x_{st} of the field dis-

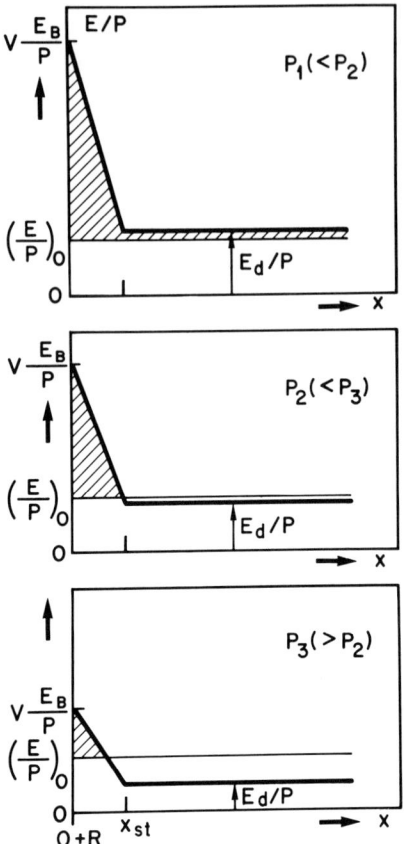

Fig. 21. Sketch showing the rapid decrease of breakdown field strength E_d/p with increasing gas pressure ($p_1 < p_2 < p_3$). The hatched area represents the value of the integral (18) taken to the critical upper limit of x at which E/p equals $(E/p)_0$.

turbance (see Fig. 21 for p_2 and p_3). - The breakdown field strength according to (21) in dependence upon px_{st}, the product pressure times length of the field disturbance, is shown in Fig. 22. The plot shows clearly that small field disturbances can be tolerated at low pressure, but that they will severely impair insulation performance at high pressures. This model explains also the dependency of measured Paschen curves - $U_d = f(p \cdot d)$ - from the gap distance d, as in (21) the second term does not depend from the product (pd) only.

The breakdown criterion (18) cannot predict, whether in any gap a direct breakdown will occur or not. Also in macroscopic, strongly inhomogeneous field configurations with SF_6-pressures below some bar this criterion indicates the inception field

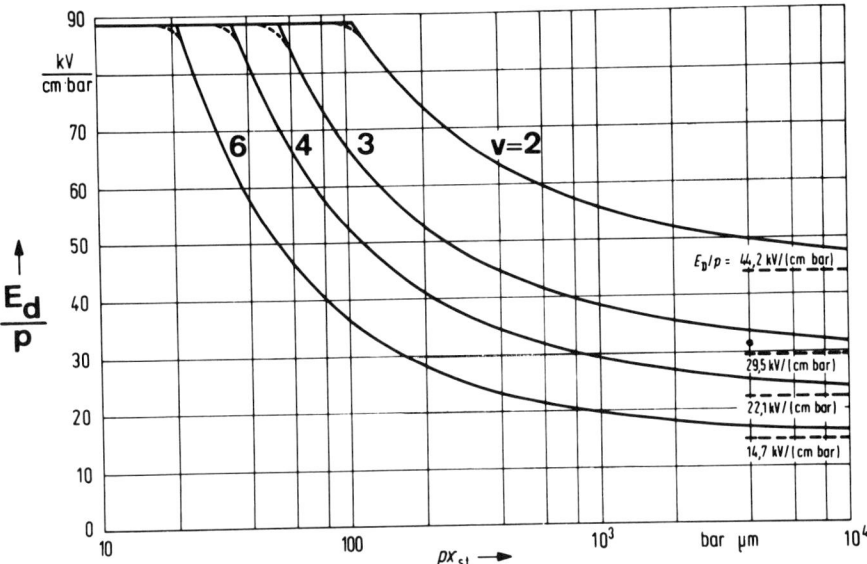

Fig. 22. Predicted breakdown field strength according to (21) in dependence upon $p \cdot x_{st}$, with field intensification factor v as a parameter (see also Fig. 20).

strength or inception voltage only. A steady state or pulse corona increases the breakdown voltage and produces a 'corona breakdown'.[12] For a homogeneous macroscopic field disturbed by protrusions within 'microscopic' field regions similar phenomena will occur; no systematic investigations concerning the relationship between these transition region and all related parameters are, however, made until now. Therefore, Fig. 23 will also contribute to this phenomenon: Here the measured values of corona inception or breakdown field strength E_d are presented together with calculated curves for very precisely manufactured, hemiellipsoidal protrusions within a plane-to-plane electrode system. The calculated curves - according to equ. (18) - are based upon the real field distribution E(x) for this type of a protrusion; however, the same field intensification factor v as defined in Fig. 20 is used for the different shapes. It can be seen that direct breakdown occurs with not too high values of v and high pressures only, whereas in all other cases corona or partial discharge (PD) will be formed. In all cases there is an excellent agreement between the calculation and measurements. With these small gaps there is no difference in the behaviour with direct and with alternating voltages. With impulse voltages, especially at moderate pressures, breakdown voltages may be higher, at least while there is no strong ultraviolet radiation of the gap.

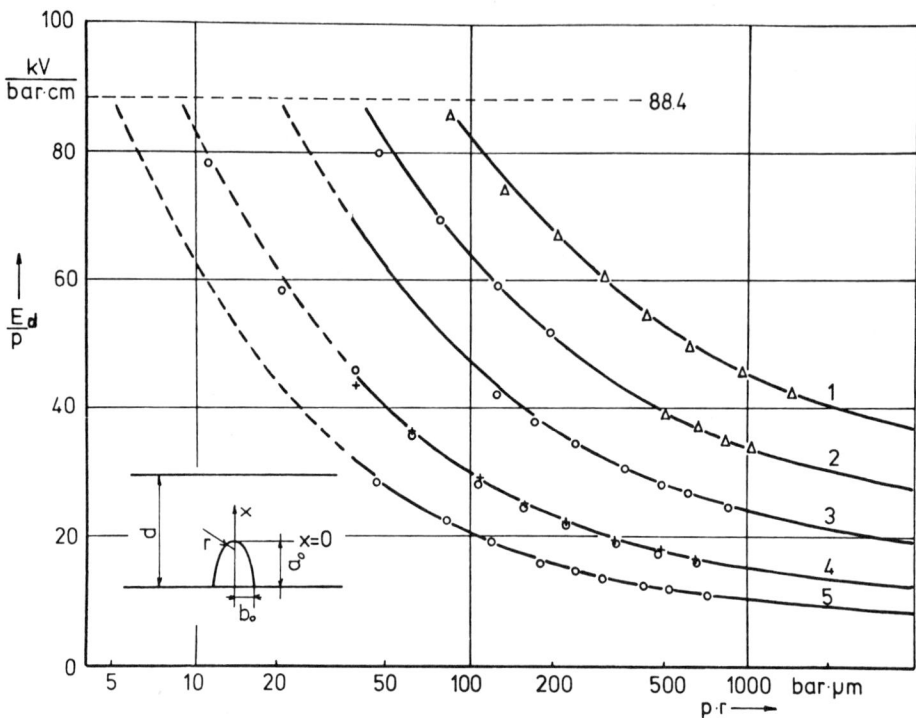

Fig. 23. Breakdown or inception field strength as predicted (curves) and measured in dependence of p·r for hemi-ellipsoidal protrusions.
Δ   Measurements; direct breakdown
o   Measurements; PD-inception, neg. DC
+   Measurements; PD-inception, AC 50 Hz   } UV-irradiation

Curve	1	2	3	4	5
Intens. factor v	3	4,1	5,8	9,2	13,2
Radius of curvature r/μm	500	345	250	167	125
Material	Steel	W	W	W	W
Height of protr. a_o/mm	0,5	0,725	1	1,5	2

OUTLOOK

Is there any hope to improve the insulation performance of this gas? The quantity to which the breakdown criteria are sensitive is the effective ionization coefficient $\alpha-\eta$. The ionization coefficient α, which is exceptionally high in SF_6 at the critical $(E/p)_0$ = 118 V/cm·torr, namely $\alpha/\eta \sim$ 1/cm·torr, can only be reduced by reducing the electron temperature, that means by introducing a process of effective energy transfer from electrons to other particles. On the other hand, one will aim to increase the attachment rate; once more, this could be achieved by a reduction in the electron energy, as the cross section for attachment in SF_6 has its maximum at very low electron energies; this manifests itself in the decrease of η with increasing E/p. The alternative to this is the introduction of a gas with appreciable attachment cross sections at higher energies. The investigation of gas mixtures would therefore lead to improved breakdown behaviour.

Furtheron, from the point of view of the use of SF_6 in circuit breakers and associated systems we shall have to ask ourselves to what extent the models and results presented here can still be applied. To the best of the knowledge of the authors, there is so far little information available on the temperature dependence of the ionization and attachment coefficients of SF_6 and fragments of this molecule. Recently, attempts have been made[13] to calculate such coefficients from published cross sections of SF_6 and the electron energy distribution in dependence upon E/N and the temperature, taking the decomposition of SF_6 at temperatures above 1500 K into account.[14] At temperatures below this value, the calculated data for α/N and η/N versus E/N show no temperature dependence. Such a result is to be expected for the ionization coefficient α. The attachment coefficient η or at least the stability of negative ions formed might conceivably be influenced significantly by vibrational excitation. A significant change in the relative composition of the negative ion population over the temperature range 273 ... 500 K has been observed by Fehsenfeld.[15] Furthermore, the lifetimes of negative ions in SF_6, which may strongly influence electron detachment, are also covering a fairly wide spectrum indicating different excited states of the ion the populations of which might be influenced by temperature. If these effects lead to an appreciable change of the effective attachment has still to be investigated experimentally.

We may find fairly unfavourable conditions for the avoidance of breakdown in a circuit breaker after current interruption. Apart from large numbers of residual charge carriers, we have reduced gas density because of the higher temperature. Furthermore, electrodes will neither remain perfectly smooth nor perfectly clean, so that there is no chance to achieve the high breakdown strengths possible in clean apparatus with good electrode finish, not even when we can discount the problems of elevated gas temperature.

ACKNOWLEDGEMENTS

The authors wish to thank Mr. R. Baumgartner for making the results of his measurements available to them. They are also indebted to workshop personnel both at Manchester and at Zürich for their help and fine workmanship in the construction of apparatus used, and to Miss B. Münst for her great help with the preparation of the manuscript and illustrations. Part of the work at Manchester was supported by the Science Research Council (Grant B/RG/11094). Part of the work at ETH Zürich was supported by the Swiss National Fund (Project 2.570.71).

REFERENCES

1. B. Sangi, 'Basis discharge parameters in electronegative gases'. Ph.D. Thesis, UMIST, Manchester, 1971
2. H. Schlumbohm, Z. Naturforsch. 16a (1961) 510
3. M. A. A. Jabbar, 'Simulation of discharge processes in electronegative gases'. Ph.D. Thesis, UMIST, Manchester, 1974
4. F. C. Fehsenfeld, J. Chem. Phys. 53 (1970) 2000
5. K. A. Sattar, 'Computer simulation of homogeneous field discharge'. M.Sc. Dissertation, UMIST, Manchester, 1974
6. D. W. Branston and T. H. Teich, Nature 244 (1973) 504
7. D. W. Branston, 'Generation of electron swarms and avalanches in gases by high intensity laser light pulses'. Ph.D. Thesis, UMIST, Manchester, 1973
8. K. H. Wagner, Z. Phys. 189 (1966) 465
9. W. H. Mears, E. Rosenthal and J. U. Sinka, J. Phys. Chem. 73 (1969) 2254
10. For instance, see the literature quoted and reviewd by: R. Baumgartner, ETZ-A 98 (1977) 369; R. Baumgartner, ETZ-A 97 (1976) 177
11. R. Baumgartner, 'Untersuchungen über die Gültigkeit des Aehnlichkeitsgesetzes in Schwefelhexafluorid'. Dissertation ETH Zürich Nr. 5997 (1977)
12. S. Sangkasaad, 'Dielectric Strength of Compressed SF_6 in Nonuniform Fields'. Dissertation ETH Zürich Nr. 5738 (1976)
13. B. Eliasson, E. Schade, 'Electrical breakdown of SF_6 at high temperatures (< 2300 K)'. Proceedings of the XIIIth International Conference on Phenomena in Ionized Gases, Berlin (East), September 12-17, 1977
14. W. Frie, Z. Physik 201 (1967) 269
15. F. C. Fehsenfeld, J. Chem. Phys. 53 (1970) 2000

DISCUSSION
(Chairman: A. N. Bird, SECV)

J. J. Lowke (University of Sydney)

I want to raise two points. There has been a lot of controversy about the question of breakdown, regarding whether one should use the Townsend or the streamer breakdown criterion. Now, I propose that for the circuit-breaker engineer neither of these breakdown criteria need be considered, but rather a very much simpler one; namely, $\alpha = \eta$. The justification for this is as follows: If you have an initial number of electrons n_0 they are amplified in the case $\alpha-\eta > 0$ according to:

$$n_e = n_{eo} \exp (\alpha-\eta)d \qquad (1)$$

If n_e goes up to 10^8 you get space charge, the avalanche proceeds and you get breakdown. On manipulating this breakdown criterion we obtain

$$(E-3.5\cdot 10^{-15} \text{ V/cm}^2 \cdot N) \text{ d} = 0.7 \text{ kV}, \qquad (2)$$

where N is the gas-particle density. The value of E/N for $\alpha = \eta$ (critical field strength E_c) is taken[1] to be $E_c/N = 3.5\cdot 10^{-15}$ V/cm^2.

In a circuit breaker we have a typical pressure of 10 bar, and therefore a high particle density. This has the consequence that an extremely small relative increase of E above E_c already fulfills criterion (2). I might mention that this is in agreement with the paper by Schade and Eliasson.[2]

My second comment applies to the case of nonuniform fields. Recent experimental work[3] has shown that after the initial breakdown process, as discussed above, there is a phase where the discharge has the form of a glow. The major property of the glow is that the electron temperature is very much higher than the gas temperature. For such a glow discharge you get another simplification if you have an attaching (electronegative) gas such as SF_6. Figure 1 shows results from a calculation[4] for

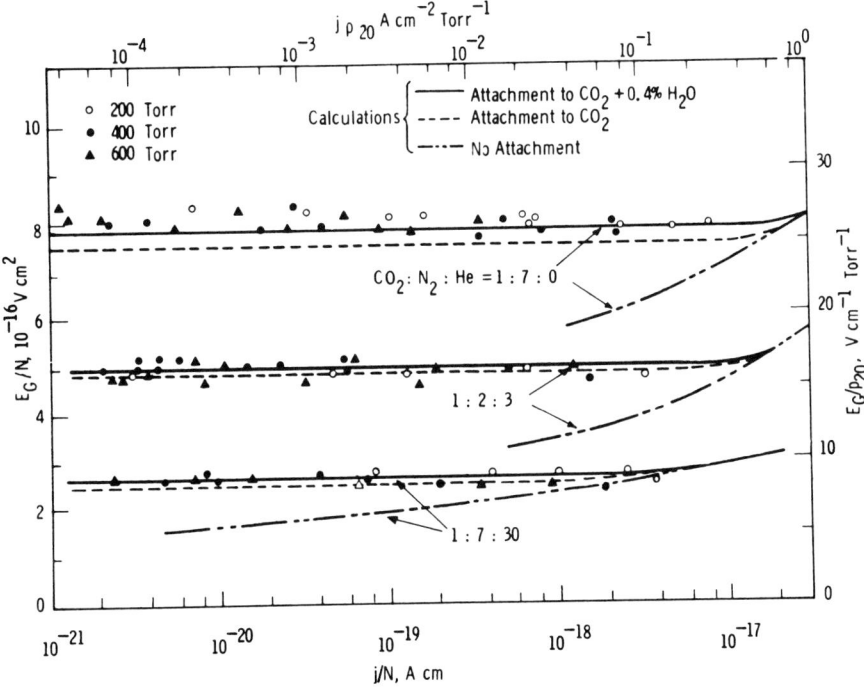

Fig. 1. Values of E/N in the glow stage of breakdown for three CO_2:N_2:He laser mixtures. Experimental values at three values of pressure are indicated by data points. Curves represent results of numerical calculations with varying conditions of attachment.[4]

laser mixtures which are also attaching (CO_2), therefore it's directly applicable to circuit breakers. It is found that theoretical and experimental V-I characteristics are in good agreement if it is assumed that during the glow phase the discharge operates at a value of E/N such that the ionization coefficient equals the attachment coefficient. The values of E/N are constant over three orders of magnitude of current density and are also independent of electrode spacing and gas pressure.

Now in the circuit breaker I think the following will happen. Local perturbations produce a high field strength and as a result a corona discharge will be initiated near the electrodes if the field is above the critical value. The discharge will proceed and set up a uniform E/N by space-charge effects. The overall voltage, however, may be such that while it is possible to have local regions with $E > E_c$, the voltage is insufficient for E to be $> E_c$ at all axial positions. The discharge will then terminate, only to restart when the space charges are finally absorbed by the electrodes. On the other hand, if the voltage is sufficient to maintain $E > E_c$ at all axial positions, complete breakdown occurs. If this were to be the case it would be a dramatic conceptual simplification for us circuit-breaker engineers.

K. Ragaller (Brown Boveri)

As it is shown in J. Kopainsky's lecture there is a discharge current flowing prior to breakdown due to the rest ion density. The axial conductance profile generated by these ions determines the axial field variation. The conditions shown in Fig. 1 by J. Lowke would correspond in a circuit breaker already to breakdown because there the voltage is imposed which means an unlimited current for the VI characteristic of Fig. 1. (Comment subsequent to meeting.)

H. Noeske (General Electric)

In a circuit breaker we have two additional influences which you did not mention. One is the temperature, which can still be very high due to the arc the breaker has just cleared. The second is the flow. Can you say something about the influence of these two effects on the physics of breakdown?

T. Teich

As far as the temperature dependence of breakdown is concerned there are two different influences. The first is a change in particle density N. This is a major effect since $E_c \sim N$. The second is a temperature dependence of the ionization and attachment coefficients. There are some physical effects, such as the vibrational excitation of the attachment molecules, which can cause a significant reduction of the attachment coefficient at higher temperatures. Above the temperature where dissociation sets in you change, in addition, the composition of the gas. Unfortunately, there are not enough basic data available to make reliable calculations of α and η in this regime.

As far as the influence of the flow is concerned, I would say that the discharges are mainly determined by the drift velocity of the electrons which is much higher than the flow velocities.

REFERENCES

1. M. S. Bhalla and J. D. Graggs, Proc. Roy. Soc. 80 (1962) 151
2. E. Schade and B. Eliasson, Proc. 13th International Conference on Phenomena in Ionized Gases, September 1977, Berlin (East)
3. M. C. Cavenor and J. Meyer, Aus. J. Phys. 22 (1969) 155
4. L. J. Denes and J. J. Lowke, Appl. Phys. Lett. 23 (1973) 130

RADIATIVE ENERGY TRANSFER IN CIRCUIT BREAKER ARCS

J. J. LOWKE
School of Electrical Engineering
University of Sydney, Australia

SUMMARY

Computer studies have revealed that in high-current gast-blast arcs, energy losses at the arc centre are dominated by the radiation losses so that the arc is approximately isothermal. However, most of this radiation is in the far ultra violet region of the spectrum and is re-absorbed in the outer cold region of the arc. Thus, the electrical energy is largely expended in producing arc plasma rather than lost as radiation. Formulae which result for arc diameter as a function of current are in reasonable agreement with experiment. The model for radiation transfer provides a mathematical foundation for the Cassie arc model that for high-current gas-blast arcs the arc diameter is approximately proportional to the square root of the current and arc voltage and temperature are largely independent of current.

Radiation processes are also of primary importance in determining the onset of dielectric breakdown. Recent theoretical investigations have verified that in the breakdown process a conducting channel is propagated by space charge and radiation effects of $\exp(\alpha d) \sim 10^8$ where α is Townsends ionization coefficient and d is the transit distance of an avalanche. Furthermore, absorption of radiation from the arc before current zero will produce an outer mantel of hot gas which will not rapidly cool by conduction because of its large radius. Because α/n increases rapidly with E/n, breakdown voltages are significantly reduced by any warm gas remaining in the inter electrode region after current zero; n is gas number density, E the electric field.

1. INTRODUCTION

Our understanding of the complex processes occurring in circuit breakers, in particular the influence of radiation, has been considerably extended in recent years through the use of computers. It has been concluded from theoretical calculations that the dominant energy loss occurring at the center of high current arcs is from radiation losses in the vacuum ultra violet region of the sprectrum. Such radiation is not directly observable, because its wavelength is so small that it photoionizes the cold gas surrounding the arc.

A second example of where computer analyses have yielded new insights is in the understanding of what is called dielectric breakdown in circuit breakers. The basic processes of space charge distortion, avalanche growth due to ionization and the propagation of the conducting electrical channel in the initial breakdown process have been predicted theoretically as a function of position and time for the simple case of plane parallel gaps. Streak photographs have yielded information of avalanche growth on a nano second time scale. The detailed spatial and temporal dependence of the gross phenomena on photoionization cross-sections and photo emission processes at the cathode can be investigated in a unique way by comparison of the experimental results with computer predictions.

In the present paper we describe recent progress in understanding radiation processes occurring in circuit breakers. In Section 2 the problem of accounting for radiation transfer is discussed, together with approximate methods that are of use to the circuit breaker engineer. In Section 3 our approximate model for radiation transfer is applied to high-current gas-blast arcs to enable predictions to be made of arc diameter, temperature and electric field as a function of axial position and arc current. The radiation model is also used to make predictions of properties of arcs in clogged circuit breakers, where the arc fills the throat of the circuit breaker nozzle.

In Section 4 recent developments are described of our understanding of radiation processes that occur during the process of dielectric breakdown, when a new conducting channel is formed after current zero in gas which is no longer ionized. The influence of the absorption of radiation from the arc before current zero is crucial. Resultant hot gas that is produced may remain between the electrodes after current zero and critically reduce the voltage required to produce breakdown.

2. RADIATION TRANSFER

The problem of representing the radiation transfer within an arc is most complex, due to two main difficulties. Firstly, radiation emission spans the wavelength range of from 100 A° to 10,000 A°, and consists of hundreds of lines plus continuum radiation. Moreover the spectral absorptivity can range over 7 orders of magnitude, and is a strong function of temperature as well as frequency. Secondly, absorption of radiation can be a dominant process in many regions of the arc. Determining the amount of absorption at any position is complex because it depends on the radiation intensity at that position and thus on the temperature of all other regions of the arc. In this section we discuss in turn the radiative transfer equation, processes contributing to the spectral absorptivity, and exact and approximate methods of calculating radiation transfer.

2.1. Radiative Transfer Equation

The radiation intensity $I_\upsilon(\vec{r},\vec{n})$ which is the radiative power density per unit of radiation frequency per unit solid angle at position $\vec{r}$ in direction $\vec{n}$, is defined by the radiative transfer equation[1] which is:

$$\vec{n} \cdot \nabla I_\upsilon = \varepsilon_\upsilon - I_\upsilon K_\upsilon \tag{1}$$

ε_υ is the emission coefficient at radiation frequency υ and K_υ is the spectral absorptivity. This equation states that the increment in I_υ per unit distance in the direction of observation $\vec{n}$ is increased by the local emission ε_υ but decreased by the absorption $I_\upsilon K_\upsilon$. For the conditions of local thermal equilibrium which apply to a good approximation over at least the central regions of the temperature profile,[2] $\varepsilon_\upsilon = B_\upsilon K_\upsilon$, which is Kirchhoffs law.[3] Thus equation (1) becomes

$$\vec{n} \cdot \nabla I_\upsilon = B_\upsilon K_\upsilon - I_\upsilon K_\upsilon \tag{2}$$

and the radiation properties of both emission and absorption are determined by $K_\upsilon(\upsilon,T)$. The effect of stimulated emission[4] can be accounted for by replacing K_υ by $K_\upsilon(1-\exp(-h\upsilon/kT))$, but for the high temperatures of electric arcs this factor is negligible except in the infra-red region of the spectrum, h is Planck's constant, k is Boltzmann's constant and T the temperature.

2.2. Radiative Processes Contribution to Spectral Absorptivity

2.2.1. Continuum Radiation

The determination of values of $K_\upsilon(\upsilon,T)$ for any arc plasma represents a major investigation in itself. Contributions to K_υ arise from both continuum and line radiation. Contributions to continuum radiation arise from electron transitions between free-free, bound-free and negative ion energy states. Free-free electron transitions are dominant in the infra-red region of the spectrum, but the resulting radiation is generally a negligible fraction of the total radiation. Bound-free absorption is significant below 1500 A° and at low temperatures is particularly important in causing self absorption of radiation energy in the region of cold gas immediately surrounding the arc. For this process to occur, the photon energy $h\upsilon$ must be sufficient to photoionize the initial electron state of energy E^a; i.e.,

$$h\upsilon \geq E^\infty - \Delta E^\infty - E^a$$

where h is Planck's constant, E^∞ is the ionization potential of the atom or ion and ΔE^∞ is the lowering of the ionization potential due to the space charge of the plasma.

Generally $\Delta E^\infty = Ze^2/\rho_D$ where Z is the net charge, e is the electrostatic charge and ρ_D is the Debye length.[4] Thus K_υ as a function of υ is a series of steps, for as υ is increased, higher energy states of a given species can be photoionized, thus increasing K_υ at each threshold frequency.

Experimental values of the photoabsorption cross-sections are frequently not available in the literature. However, theories have been developed to estimate these absorption cross-sections. The quantum defect method of Burgess and Seaton[5] as modified by Peach[6] has been used[7] to calculate these cross-sections at their threshold for S and F, appropriate to SF_6 arcs. Values of K_υ for nitrogen are given by Wilson and Nicolet.[8] Generally K_υ beyond each threshold frequency varies as $1/\upsilon^3$ as with hydrogenic atoms.

There is additional continuum radiation in arcs of the circuit breaker gases air and SF_6 due to the formation of negative ions; i.e., O^- and N^- ions in air and F^- ions in SF_6. Equilibrium concentrations[9] of F^- ions exceed electron concentrations in SF_6 for temperatures of less than 4000 K, and the percentage of F^- rises to almost 0.01 % of the total particle concentration at 9000 K. Absorption cross-sections of F^- radiation are given by Popp[10,11] and for O^- and N^- by Boldt,[12] and Morris.[13] There is doubt as to whether the additional continuum radiation in nitrogen is due to the N^- ion, and a recent publication attributes the additional radiation to overlapping lines.[14]

2.2.2. Line Radiation

Although continuum radiation is a significant contribution to radiation for both air,[15] and sulfur hexafluoride[16] arcs, calculations indicate that line radiation dominates over continuum radiation. At 1 atmosphere, line radiation is about an order of magnitude higher than continuum radiation,[9] but this ratio decreases as the pressure increases.

Evaluation of absorptivities from line radiation is fraught with difficulties.

(i) The number of lines involved for SF_6 for example is almost 2000. Even including only the strongest of these lines means accounting for several hundred lines.

(ii) Most of the significant lines lie in the ultra violet portion of the spectrum where experimental data[17] on line strengths is sparse. However, as sulphur and fluorine atoms are expected to obey L-S coupling rules, line strengths can be calculated using formula given by Griem[18] and Biedenharn[19] et al.

(iii) The effective contribution of a line to radiation is critically dependent on the degree of line broadening. Experimental values of half widths of N II lines are given by Hunt and Sibulkin[20] but generally line broadening parameters are difficult to obtain and for many lines, particularly in the ultra violet, experimental information is unavailable. Theoretical estimates are possible using a first order electron impact analysis[18] for neutral lines, for which results are in reasonable agreement with experiment. However, for the all important ion lines only semi empirical theories are available.[21] Further difficulties arise because although broadening is usually dominated by Stark broadening, the other broadening mechanisms[4] of resonance broadening, Van der Waal broadening and Doppler broadening need to be taken into account. For the very strong resonance lines the central absorptivity is over 1000 cm^{-1} so that at these frequencies there is negligible radiation flux for a major part of the line profile because of self absorption of radiation. However, the central absorptivity of a line is directly dependent on the broadening parameter, which thus represents a major determinant of the amount of radiation contributed by a strong line.

(iv) The perturbations on the electron energy levels produced by the high electron density also causes shifts in the line centre of spectral lines. These shifts will be a function of temperature because the electron density is a strong function of temperature. Furthermore, many of the broadened lines overlap and there are uncertainties in the theory of the combined profiles of overlapping lines which may depart significantly from the normal Lorentz profile

$$K_\upsilon = \frac{e^2 N f \omega / mc}{(\upsilon - \upsilon_c)^2 + (\omega)^2}$$

where υ_c is the frequency at the line centre, ω is the half width of the line at half its peak value, m is the electronic mass, c the velocity of light, e the electronic charge, f the oscillator strength and N the number density of the absorbing particle.

2.3. Calculations of Temperature Profiles

Despite the complexities of analysing spectral absorptivities of a whole spectrum, such an analysis has been done by Hermann and Schade[15] for a 350 A arc in nitrogen with a central temperature of 26,000 K. Very good agreement was obtained between derived and theoretical and experimental radiation emission coefficients. In Fig. 1 are shown derived[15] values of K_υ at 20,000 K at a pressure of 20 atmospheres for wavelengths ranging from 700 Å to beyond 15 μ. It is seen that K_υ varies from 0.1 to over 1000 cm^{-1}.

It is obviously highly desirable to make approximations in dealing with the complex radiation processes that occur. Yet in making approximate calculations, one needs to also make exact calculations for any given model of $K_\upsilon(\upsilon,T)$ with which to compare the approximate calculations. Furthermore, we wish to not just analyse a

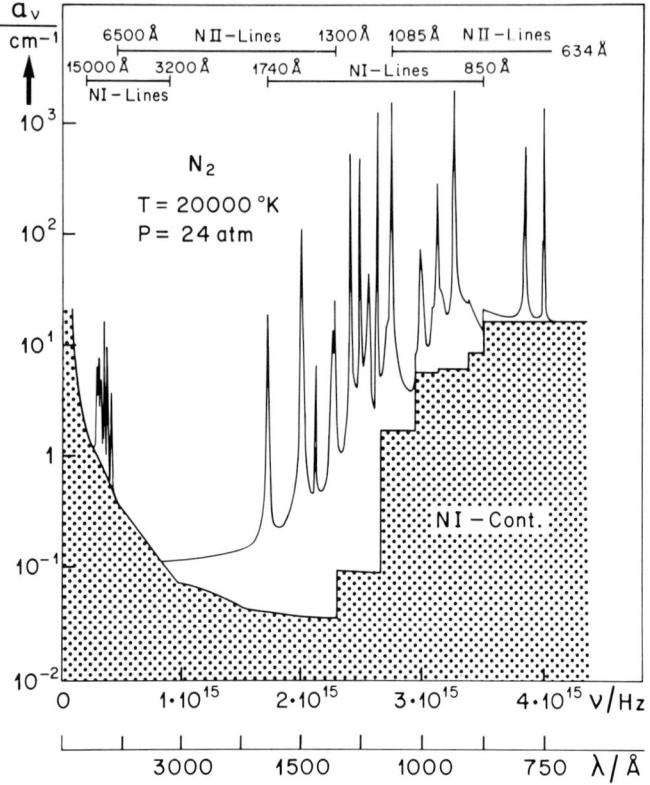

Fig. 1. Spectral absorptivity of nitrogen at 20,000 K and a pressure of 20 atmospheres, taken from Ref.[15].

given experimental arc temperature profile in terms of energy transfer from conduction and convection, but to be able to derive the temperature profile for any given current and arc radius for any given values of $K_\nu(\nu,T)$ together with the thermal conductivity $\kappa(T)$ and electrical conductivity $\sigma(T)$. In Section 2.3.1 we give the theory of such a procedure and in 2.3.2 discuss various approximate procedures that have been attempted for dealing with radiation transfer.

2.3.1. Exact Calculations

The radiative transfer equation (2) can be integrated after multiplying by the integrating factor $\exp \tau$ where $\tau = \int K_\nu dx$ to obtain

$$I_\nu = \exp(-\tau) \int_0^\tau B_\nu \exp(\tau') \, d\tau',$$

which on substituting $\tau' = \tau-\tau''$ simplifies to

$$I_\upsilon = \int_0^\tau B_\upsilon \exp(-\tau'')d\tau''. \tag{3}$$

The radiation intensity $I_\upsilon(\vec{r},\vec{n})$ is a scalar quantity dependent not only on position $\vec{r}$, but also on the direction of the line of sight, defined by $\vec{n}$, which is determined by the two angles θ and ϕ of spherical co-ordinates placed at $\vec{r}$; θ is the angle the line of integration makes with the axis of the arc, and ϕ is the azimuthal angle the line makes with the radial direction. In the determination of the arc temperature profile it is the radiation flux density $\vec{F}_R$ which is important where $\vec{F}_R$ is defined by

$$\vec{F}_R(\vec{r}) = \int \vec{F}_{R\upsilon} d\upsilon = \iint_{4\pi} I_\upsilon(\vec{r},\vec{n})\vec{n}\, d\Omega d\upsilon \tag{4}$$

where I_υ at $\vec{r}$ is integrated over all angles and radiation frequencies, $d\Omega$ being an increment of solid angle. The unit vector $\vec{n}$ within the integral causes the expression to give the net flux of radiation rather than just summing all radiation intensities in different directions.

For arcs of cylindrical symmetry, uniform in the axial direction, temperatures along a line of sight defined by θ, ϕ can always be related to temperatures along the line $\pi/2$, ϕ, i.e., a line through $\vec{r}$ along a plane perpendicular to the axis of the arc. As a consequence[22] the integral over θ in equ. (4) can be eliminated. Then from equs. (3) and (4)

$$F_R = -4 \int_{\upsilon=0}^\infty \int_{\phi=0}^\pi \int_{S=0}^{P(\phi)} K_\upsilon(s')B_\upsilon(s')G_1[\tau(s')]\cos\phi\, d\upsilon d\phi ds' \tag{5}$$

where $\tau(s') = \int_0^{s'} K_\upsilon(s'')\, ds''$

and $G_1(x) = \int_0^{\pi/2} \sin\theta \exp(-x/\sin\theta)d\theta$

The integration from O to $P(\phi)$ is along a straight line in a plane perpendicular to the axis of the arc from a point O at radius r where $S' = 0$ to P at the edge of the arc. ϕ is the angle OP makes with the radial direction; S' and S'' indicate distances from O along OP. The function G_1 is evaluated as a function of x and tabulated for use in any arc calculation. Its use makes it unnecessary to do any integration with respect to θ in equation (5).

The temperature profile of an arc of given radius R and current I is determined by the energy balance equation relating heat input from the electric field to heat losses by conduction and radiation i.e.,

$$\sigma E^2 = \nabla \cdot \vec{F}_C + \nabla \cdot \vec{F}_R \qquad (6)$$

where $\vec{F}_C = -\kappa \nabla T$ is the conduction flux density, σ is the electrical conductivity and κ is the thermal conductivity. In cylindrical co-ordinates this equation becomes

$$0 = \sigma E^2 + \frac{1}{r}\frac{d}{dr}\left(r\kappa \frac{dT}{dr}\right) - \frac{1}{r}\frac{d}{dr}(r F_R) \qquad (7)$$

where E is related to I by Ohms law i.e.,

$$E = I / \int 2\pi r\, \sigma\, dr. \qquad (8)$$

Equation (7) with E and F_R determined by equations (8) and (5) is a second order differential equation with T(r) as the only unkown. The two boundary conditions required are that dT/dr = 0 at r = 0 and T = T_w the wall temperature at r = R. This equation can be solved numerically using an iterative relaxation method[23] for any given values of the material functions $\kappa(T)$, $\sigma(T)$ and $K_\upsilon(\upsilon,T)$. The basis of the relaxation method is to successively modify any initial arbitrary temperature profile T(r) using the time dependent form of equation (7). By putting $\partial T/\partial t$ on the left hand side of equation (7), an incremental temperature ΔT for any time step Δt is determined for all radial positions by evaluating the RHS of equation (7). This iterative procedure using a simple explicit numerical scheme is stable for sufficiently small Δt where Δt is chosen by trial. The conduction term causes instabilities if $\Delta t > \Delta r^2/2\kappa$ where Δr is the radial step size.[24]

The numerical effort required in determining F_R from equation (5) is prohibitively large if it is desired to treat hundreds of radiation lines, each of which would require say 30 frequency points. However, the equation has been solved for two useful cases. Firstly, a model of $K_\upsilon(T)$ has been used[25] representing the radiation of high pressure arcs in air as fourteen continuum bands of equal width in frequency, extending to 400 Å. Secondly, calculations have been performed[23] for sodium arc lamps, where the discharge is dominated by radiation from the merged D lines, which are represented by 24 frequency bands. From these exact solutions for model $K_\upsilon(T)$ values, assessments can be made of the validity of approximate methods. The qualitative features of the effects of ultra violet and resonance radiation can also be determined.

2.3.2. Approximate Calculations

For arcs in circuit breakers we are mainly interested in determining the temperature of the arc, so that all that is required is ∇F_R for equation (6). The angular and frequency dependence of I_υ in equation (2) is of only incidental interest. So we integrate equ. (2) over all angles to obtain

$$\nabla \cdot \vec{F}_{R\upsilon} = 4\pi B_\upsilon K_\upsilon - 4\pi J_\upsilon K_\upsilon \tag{9}$$

where J_υ is the average radiation intensity given by

$$J_\upsilon = (1/4\pi) \int I_\upsilon d\Omega \tag{10}$$

and $F_{R\upsilon}$ is defined by equations (4) and (5).

Equation (9) is known as the first moment equation of equation (2). It is useful to obtain the second moment equation by multiplying equation (2) by $\vec{n}$ and again integrating over all angles. The term in $B_\upsilon K_\upsilon$ becomes zero because $\int_{4\pi} \vec{n}\, d\Omega = 0$ and the equation becomes

$$\int \vec{n} (\vec{n} \cdot \nabla I_\upsilon)\, d\Omega = -\vec{F}_{R\upsilon} K_\upsilon \tag{11}$$

The left hand side of this equation becomes equal to $4\pi\nabla J_\upsilon/3$ for several quite different angular approximations of $I_\upsilon(\theta,\phi)$ and equation (11) becomes[25]

$$\vec{F}_{R\upsilon} = \frac{-4\pi}{3K_\upsilon} \nabla J_\upsilon \tag{12}$$

Equation (12) is used extensively in astrophysics and is given various names such as the Eddington approximation,[3] the diffusion approximation,[26] the Schuster-Schwarzschild approximation [27-28] or the differential approximation.[29] We shall refer to equation (12) as the diffusion approximation as it describes diffusion of radiative energy by $\vec{F}_{R\upsilon} = -D\nabla E_N$ where $E_N = 4\pi J_\upsilon/c$ is the energy density of photons and $D = \ell c/3$ is the classical diffusion coefficient from kinetic theory where $\ell = 1/K_\upsilon$ is the mean free path of photons.

2.3.2.1. Diffusion Approximation

Calculations[25] have been made comparing results of the diffusion approximation with exact calculations for the model of the continuum $K_\upsilon(T)$ for air at high pressure, and also for various values of K_υ constant with T, but ranging from 0.01 to 1000 cm^{-1}. In all cases agreement between exact and approximate values of $F_{R\upsilon}$ and J_υ is generally less than 20 %, as shown in Fig. 2 and this accuracy is adequate for circuit breaker engineers. However, in this authors view, nobody has as yet succeeded

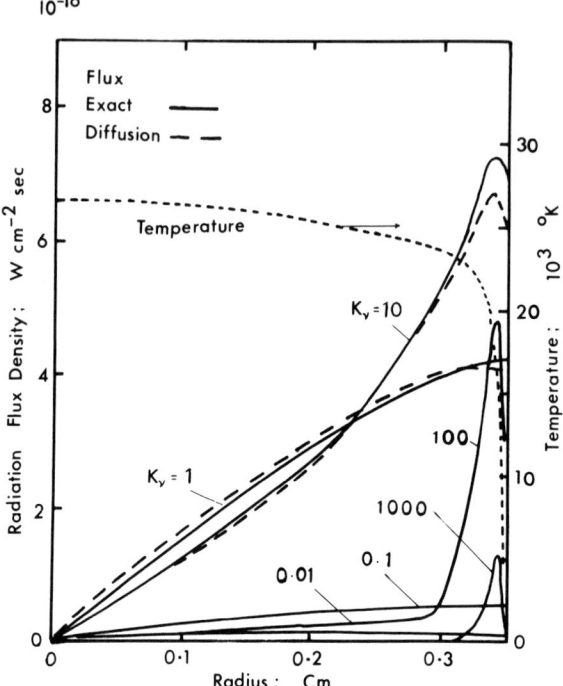

Fig. 2. Comparison of the radiation flux density calculated using the diffusion approximation and exact calculations, from Ref.[25].

in devising an efficient practical scheme in using the equations to justify their extensive use.

For a given temperature profile, i.e., with B_υ and K_υ known as a function of r, we need to solve equations (9) and (12) in the two unknowns $F_{R\upsilon}$ and J_υ. The problem has split boundary conditions[25] in that at r = 0, $dJ_\upsilon/dr = 0$ and at r = R, $J_\upsilon \sim 0.2\ F_{R\upsilon}$, so that solutions cannot be found from a straight forward integration from one boundary to the other. Furthermore, equation (9) becomes stiff[30] for large optical depths. These difficulties can be overcome and solutions obtained, but generally it is more straight forward to determine the exact values of $F_{R\upsilon}$ from equation (5).

2.3.2.2. Planck and Rosseland Mean Absorptivities

An enormous simplification would be made if one could use effective average absorption coefficients, that could be evaluated as a function of temperature just once for the whole spectrum, and then used for any temperature profile. But in this authors view this method leads to gross errors.

If we integrate equations (9) and (12) over all radiation frequencies we obtain

$$\nabla \cdot \vec{F}_R = 4\pi(\varepsilon - K_1 J) \qquad (13)$$

and

$$\vec{F}_R = -(4\pi/3K_2)\nabla J \qquad (14)$$

where

$$\varepsilon = \int B_\upsilon K_\upsilon d\upsilon,$$

$$K_1 = (1/J) \int J_\upsilon K_\upsilon d\upsilon, \qquad (15)$$

$$\frac{1}{K_2} = \int \frac{1}{K_\upsilon} \nabla J_\upsilon d\upsilon / \int \nabla J_\upsilon d\upsilon, \qquad (16)$$

$$J = \int J_\upsilon d\upsilon \text{ and } \vec{F}_R = \int \vec{F}_{R\upsilon} d\upsilon.$$

The problem is that the absorptivities K_1 and K_2 can only be determined knowing the weighting factors which involve knowing J_υ. The Rosseland mean K_R is similar to K_2 and can be evaluated as a function of temperature for it assumes $J_\upsilon = B_\upsilon$. K_R is defined by

$$\frac{1}{K_R} = \int \frac{1}{K_\upsilon} \frac{dB_\upsilon}{dT} d\upsilon \Big/ \int \frac{dB_\upsilon}{dT} d\upsilon$$

As $\nabla J_\upsilon = (dJ_\upsilon/dT)\nabla T$, K_R and K_2 are identical if $J_\upsilon = B_\upsilon$, which is true in the optically thick limit.

The Planck mean K_P is defined by

$$K_P = \int B_\upsilon K_\upsilon d\upsilon / \int B_\upsilon d\upsilon$$

and is used to obtain ε using $\varepsilon = K_P \int B_\upsilon d\upsilon$. In the optically thin limit when $J_\upsilon < B_\upsilon$ the term in K_1 can be neglected in equation (13) which can then be used to determine F_R. In the optically thick limit $J_\upsilon = B_\upsilon$ and equation (14) can be used to determine F_R with $K_2 = K_R$.

Unfortunately most radiation transfer occurs for values of K_υ between the very low values of K_υ when the arc is optically thin, and the very high values of K_υ when the arc is optically thick, see Fig. 2. In the region intermediate between the thick and thin limits, very significant errors can occur, in that $J_\upsilon \neq B_\upsilon$, particularly near the edge of the arc. Values of K_1 and K_2 for example can differ by four orders[25] of magnitude from K_R to K_P. Methods of using combinations of K_P and K_R have been advocated,[1] but are untried for the very large temperature gradients that occur at the edge of circuit breaker arcs.

2.3.2.3. Equivalent thermal conductivity

It has frequently been suggested[31,32] that radiation transfer might be represented as an effective increase in the thermal conductivity when K_υ is high. From equ. (12)

$$F_{R\upsilon} = -\frac{4\pi}{3K_\upsilon}\frac{dJ_\upsilon}{dT}\frac{dT}{dr} = -\kappa_r \frac{dT}{dr}$$

where $\kappa_r = -(4\pi/3K_\upsilon)dB_\upsilon/dT$ if $J_\upsilon = B_\upsilon$, and κ_r is an effective thermal conductivity due to radiation.

In this authors view the approximation is not valid because J_υ is not even approximately equal to B_υ for values of K_υ at which radiation transfer is significant. For example[33] in a calculation for radiation at 500 Å in an arc in SF_6, at the arc centre $K_\upsilon \sim 0.1$ cm^{-1} and J_υ is about an order of magnitude less than B_υ. At the edge of the arc $K_\upsilon \sim 30$ cm^{-1} and J_υ is orders of magnitude above the value of B_υ corresponding to the local temperature. The distance required for J_υ to reach B_υ for an isothermal plasma is $\sim 1/K_\upsilon$. It is because the temperature gradients are so steep in the outer region of the arc that the temperature varies by thousands of degrees in the distance $1/K_\upsilon$ and B_υ, which is very sensitive to temperature, is in general not approximately equal to J_υ.

2.3.2.4. Net Emission Coefficients

We now discuss an approximation[33] which in this authors view does give a useful method, of determining the principal arc properties of central temperature and electric field for a given current.

For any temperature profile the net emission of radiation at any radial position for any frequency υ can always be represented by $\nabla \cdot F_{R\upsilon} = 4\pi(\varepsilon_\upsilon - J_\upsilon K_\upsilon)$ as in equation (9) where $\varepsilon_\upsilon = B_\upsilon K_\upsilon$. Our approximation is to use a net emission coefficient ε_N (W cm^{-3} ster^{-1}) defined by

$$\varepsilon_N = \int \varepsilon_{N\upsilon}\, d\upsilon = \int (B_\upsilon K_\upsilon - J_\upsilon K_\upsilon)\, d\upsilon \tag{17}$$

where J_υ is the radiation intensity at the centre of an isothermal spherical plasma of radius R. J_υ in equation (17) is given from equation (3) by

$$J_\upsilon = B_\upsilon[1-\exp(-K_\upsilon R)]$$

where R is the arc radius.

On substituting for J_υ in equation (17) we obtain

$$\varepsilon_{N\upsilon} = B_\upsilon K_\upsilon \exp(-K_\upsilon R) \tag{18}$$

Values of $\varepsilon_N = \int \varepsilon_{N\upsilon} d\upsilon$ can be evaluated from equation (18) as a function of T and R for any set of $K_\upsilon(\upsilon,T)$. As the integration over frequency is only done once for each value of R, complex spectra including hundreds of lines can be treated. Temperature profiles can then be derived using equation (6) but with $4\pi\varepsilon_N(T)$ for the radiation term $\nabla \cdot \vec{F}_R$. Values of $\varepsilon_N(T)$ are only weakly dependent on arc radius R.

The approximate method yields central arc temperatures which are in good agreement with central arc temperatures obtained from exact calculations.[33] Central arc temperatures are most important in determining arc conductance for a given current or arc radius. Temperatures near the outer edge of the arc may be significantly in error, because self absorption of radiation by the cold outer arc region is not accounted for in this method.

Values of ε_N determined with the aid of equation (18) differ only by ~10 % from values evaluated from more complex formulae calculated on the basis of J_υ being evaluated at the centre of an isothermal cylinder instead of a sphere. Then[33] $\exp(-K_\upsilon R)$ in equation (18) is replaced by $G_1(K_\upsilon R)$.

3. PROPERTIES OF HIGH CURRENT ARCS IN CIRCUIT BREAKERS

On the basis of the previous theoretical work, it is possible to derive properties of arcs of currents of thousands of amps at pressures of the order of 10 atmospheres, in either air or SF_6.

3.1. General Properties Determined by Radiation

Approximate radial temperature profiles can be derived using equation (7) which neglects effects of axial convection except that use of the equation effectively assumes that axial convection defines a radius R. Results obtained using[25] the approximate model for continuum radiation are listed below.

(i) At the arc centre, energy losses are almost completely by radiation. Thus the central temperature of the arc is determined simply by

$$\sigma E^2 = \nabla \cdot F_R \sim 4\pi\varepsilon_N \qquad (19)$$

This result is not violated by later assessments that have been made of the effect of losses by turbulence[34] or axial convection[35] at the arc centre, for large currents.

(ii) Arcs are approximately isothermal. From solutions of equation (7), conduction is only important at the outer edge of the arc so that temperature profiles are very flat. Experimental results of temperature profiles in the region upstream of the nozzle throat have confirmed[36] this expectation. For isothermal profiles

$$I = \sigma E \cdot \pi R^2 \qquad (20)$$

and this equation, together with equation (19), means that if any two of the variables I, E, T and R are known, the other two can be determined, provided $\varepsilon_N(T,R)$ and $\sigma(T)$ are known for a given pressure. Families of E-I characteristics can be drawn for various arc radii and central temperatures. The curves of Fig. 3 were calculated[16] from the ε_N values derived for SF_6 at 10 atmospheres pressure.

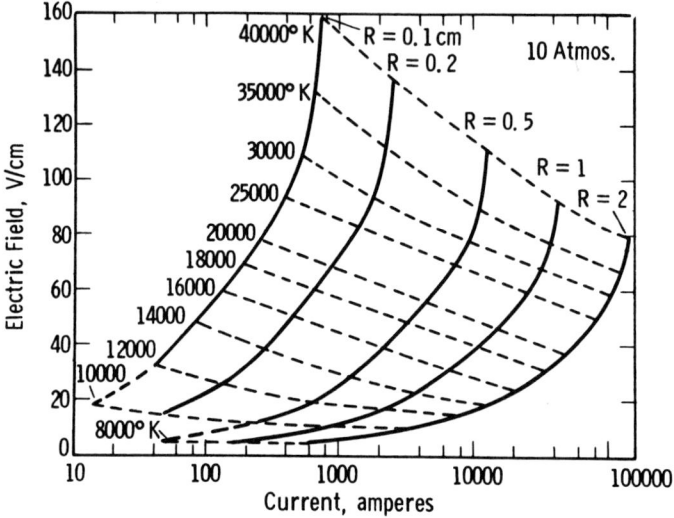

Fig. 3. E-I characteristics calculated from net emission coefficients and $\sigma(T)$ values of SF_6 at a pressure of 10 atmospheres, from Ref.[16].

(iii) The effect of radiation on derived temperature profiles is shown in Fig. 4 for a 2800 A arc in air.[25] Most of the radiation emitted from the arc centre is in the ultra violet region of the spectrum. Furthermore line radiation dominates over continuum radiation. These facts are illustrated in Fig. 5 showing calculated[16] emission coefficients for SF_6 at 1 atmosphere and the contribution of continuum and radiation > 2000 Å compared with total radiation. The curve labelled 'No self absorption' is for R = 0 or simply $\int B_\upsilon K_\upsilon d\upsilon$. Results indicate that self absorption of radiation is significant even within the isothermal region of the arc.

The 80 % or more of radiation that is generally in the ultra violet region of the spectrum is expected to be absorbed by the cold gas immediately surrounding the arc. This absorption is a result of the very high absorptivities at short wavelengths at low temperature. Thus we have a situation where radiation losses are dominant at the arc centre, yet in general most radiation does not leave the total arc column.

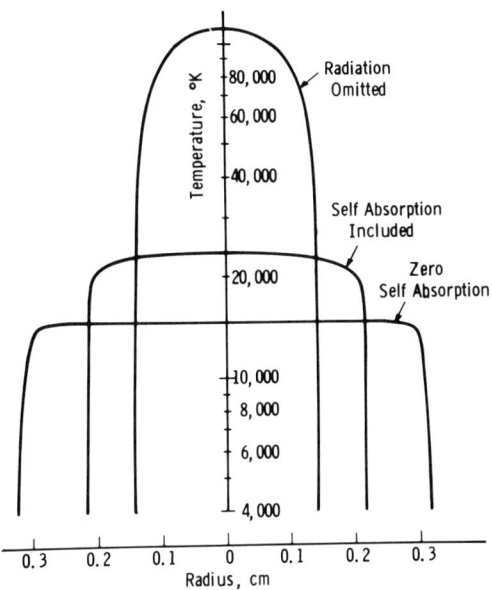

Fig. 4. Influence of Radiation on an Arc in air at 30 atmospheres of 2800 A and E = 140 V/cm; from Ref.[25].

(iv) Constricted Arc Modes are Possible.

For arcs in both SF_6 and air, in the temperature range of 10,000 K to 20,000 K the calculated net radiation emission coefficient increases more rapidly with temperature than σ increases with temperature. The time dependent energy balance equation applied to the arc centre, with the conduction term omitted, is from equation (6)

$$\rho C_p \frac{\partial T}{\partial t} = \sigma E^2 - 4\pi\varepsilon_N \tag{21}$$

ρ is the density and C_p the specific heat. For a given electric field E this equation gives a stable solution with a central temperature T_o defined by $\sigma E^2 = 4\pi\varepsilon_N$ and energy balance effects will bring the central temperature back to T_o if there are any small perturbations in the central temperature from T_o.

However, it is quite possible that for some temperature ranges ε_N/σ decreases with temperature. For example, ε_N could fall with temperature due to a reduction in neutral line radiation because of a reduction of the density of neutral particles with increase in temperature.

If ε_N values are such that $4\pi\varepsilon_N/\sigma$ values are given by Fig. 6, for some values of E or I there are three instead of one solution of the energy balance equation.[37]

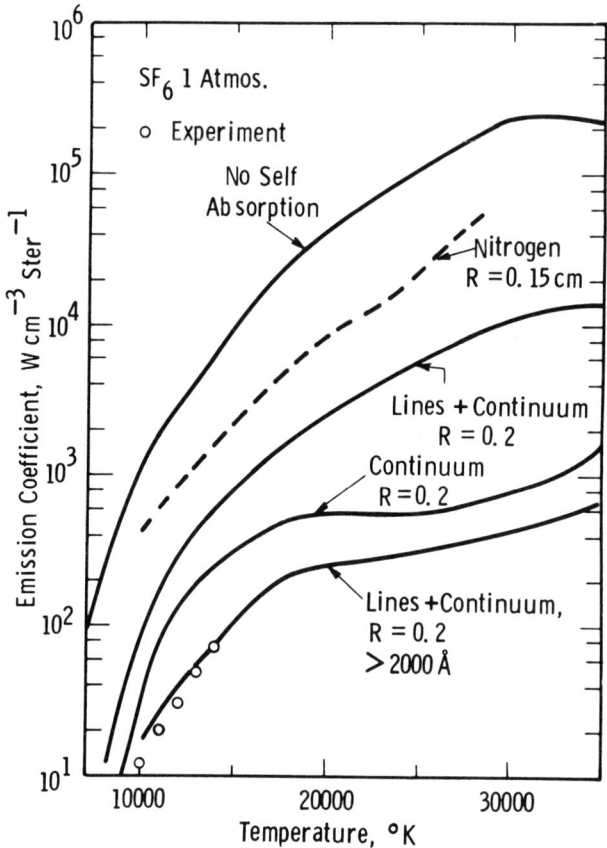

Fig. 5. Components of the total radiation emission coefficient for SF_6 at 1 Atmosphere; from Ref.[16].

For E = 180 volt/cm, in addition to the normal isothermal solution of central temperature 24,600 K, two other solutions exist of central temperature 41,000 K and ~100,000 K. The second solution would be unstable for normal power supplies as is shown by the derived E-I characteristics of Fig. 7. The temperature profiles of the other two solutions are not isothermal, but have a high temperature core as they are conduction dominated at the arc centre. There is no solution with a central temperature of 38,000 K, where σE^2 also equals $4\pi\varepsilon_N$, see Fig. 8, because such a central temperature would be unstable from equation (21).

The predictions of Figs. 7-8 only indicate the possibility of temperature profiles at high current having a high temperature core. The figures are not actual predictions as the model of absorptivities does not represent line radiation and self absorption of radiation in the outer region of the arc in a realistic way. Furthermore the predictions are for wall stabilized arcs rather than arcs stabilized by forced convection.

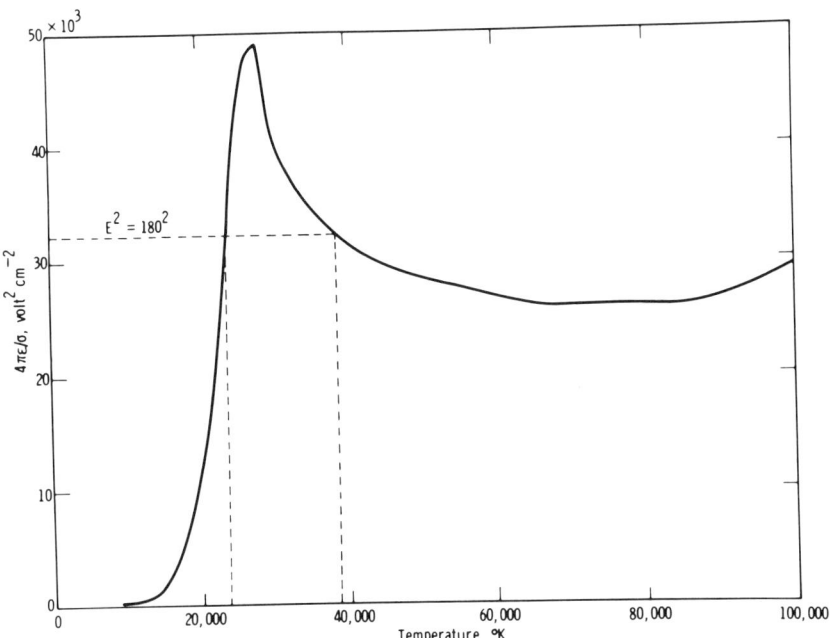

Fig. 6. Values of $4\pi\varepsilon_N/\sigma$ for which multiple solutions occur; from Ref.[37].

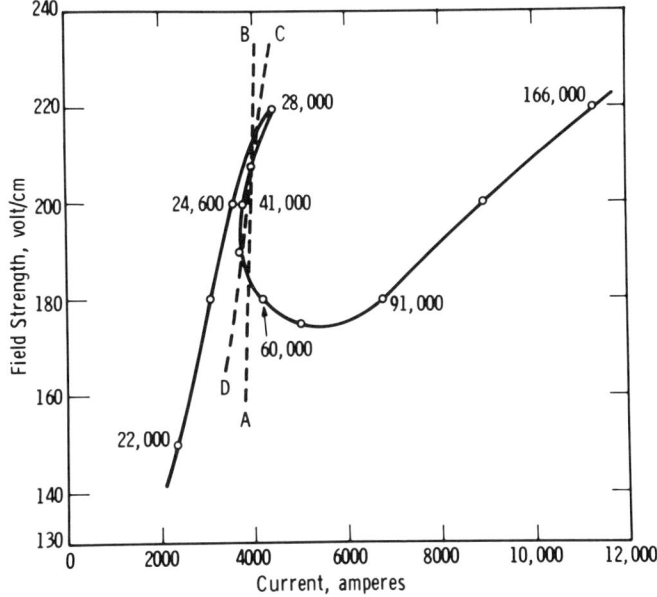

Fig. 7. Derived E-I characteristics using ε_N values of Fig. 6 in region of multiple solutions for arc radius of 0.2 cm; from Ref.[37].

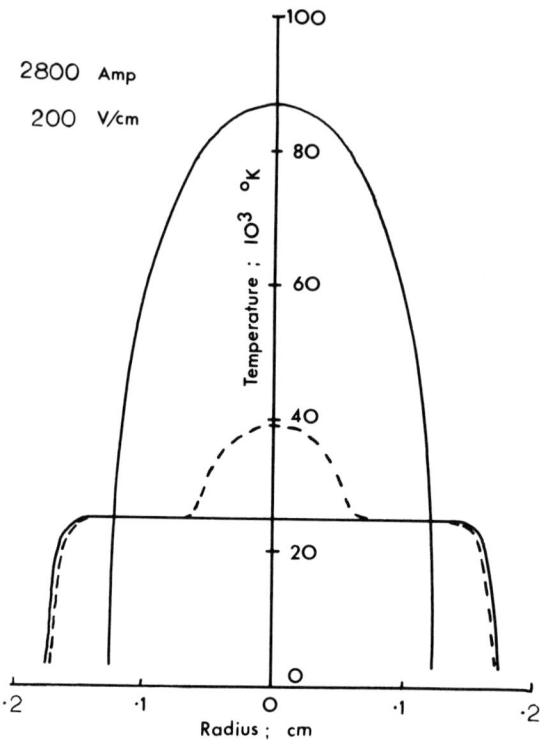

Fig. 8. Three solutions all for 2800 A and E = 200 V/cm at 30 atmospheres pressure.

However, it is of interest that changes in arc mode similar to the above qualitative predictions have been observed by Airy[38] for arcs of > 20,000 A in SF_6 in a gas blast nozzle. These authors attributed the constriction to the magnetic pressure. Effects due to the onset of metal vapor occuring in the arc make it difficult to make a conclusive interpretation.

3.2. Radius of Arcs in Forced Flow

As a large fraction f of the radiative losses within the arc are in the ultra violet portion of the spectrum, it is expected that this radiation will be absorbed in a small region of gas that surrounds the arc probably of the order of a mm. This absorption helps the cold gas which is drawn into the arc to attain the plasma temperature. Thus the bulk of the electrical energy expended in the arc goes into producing arc plasma, i.e.,

$$\sigma E^2 A f = \frac{\partial}{\partial z}(\rho h v_z A) \tag{22}$$

where A is the area of cross-section of the arc, h is the enthalpy and v_z the velocity of plasma. Equation (22) states that the input of electrical energy per unit length of the arc that goes into producing ultra violet radiation, results in an increase in the flux of plasma energy $\rho h v_z A$. The fraction 1-f of input energy would be lost as radiation to the walls of the circuit breaker.

Equation (22) enables an approximate but useful formula[35] for arc area to be derived by integration, if we assume $\rho h v_z$ and σf is independent of z and substitute $E = I/\sigma A$. Thus,

$$A = \sqrt{\frac{2zf}{\rho h v_z \sigma}} \cdot I \tag{23}$$

from which arc radius R can be derived using $\pi R^2 = A$. As R depends on $(2zf/\rho h v_z \sigma)^{1/4}$, R is insensitive to the approximation that $\rho h v_z$ is constant, or to the rather arbitrary division of radiation into a fraction f which is in the ultra violet.

Usually circuit breaker engineers are particularly interested in the arc radius at the nozzle throat, where v_z is approximately sonic. In Fig. 9 are shown[35] various experimental investigations of arc radius at the throat, compared with results from equation (23), and it is seen that there is approximate agreement between theory and experimental for both air and SF_6.

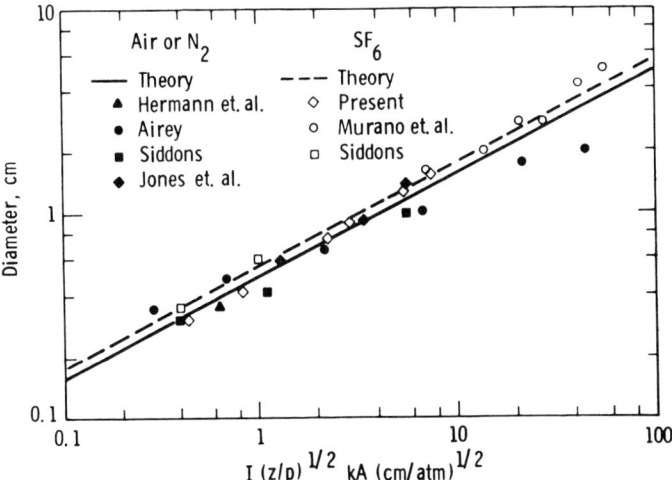

Fig. 9. Measurements of arc diameter at the nozzle throat compared with results from equation (23); from Ref.[35].

For the purpose of predicting the throat diameter required for any given current in a circuit breaker, a significant safety margin would be required above the radius predicted from equation (23) to prevent clogging. The mantle of hot gas which surrounds the arc core that is produced from turbulence, conduction and radiative absorption is of unknown radius, and will inhibit gas flow.

3.3. Cassie Arc Model

A further consequence of the radiative model of the arc in terms of emission coefficients is that arc temperature and electric field distributions, which are a function of axial position, can be shown[35] analytically to be independent of arc current, and arc area shown to be directly proportional to current, provided ε_N is independent of R.

The axial distribution, of central arc temperature $T(z)$ is determined by the energy balance equation, where z is the axial co-ordinate. The energy balance equation appropriate to the arc centre is

$$\sigma E^2 = \rho C_p v_z \partial T / \partial z + U \qquad (24)$$

where $U = 4\pi\varepsilon_N$ is the net emission coefficient in watts/cc, and consists mainly of ultraviolet radiation. Losses of energy by radial and axial conduction are omitted, which will be a good approximation only at high currents.

The approximation is again made that the arc is isothermal with radius and of cross-sectional area $A(z)$. This arc area is largely determined by the energy balance equation integrated over the arc cross-section i.e.,

$$\sigma E^2 A = \frac{\partial}{\partial z}(\rho h v_z A) + U_t A \qquad (25)$$

This equation is similar to equation (22) except that we represent losses of radiation from the whole arc column by U_t. U_t will consist largely of radiation in the visible region of the spectrum.

It is assumed that the axial Mach number distribution $M(z)$ is determined by the nozzle shape through the standard gas dynamic equation

$$\left(\frac{Q}{Q^*}\right)^2 = \frac{1}{M^2}\left[\frac{2}{\gamma+1}\left(1 + \frac{\gamma-1}{2}M^2\right)\right]^{(\gamma+1)/(\gamma-1)}. \qquad (26)$$

$Q(z)$ is the area of cross section of the nozzle, Q^* the area of the nozzle at the throat and γ the isentropic gas constant given by $pV^\gamma = $ const. where V is the volume. Then $v_z = cM$ where c is the sonic velocity appropriate to the local temperature through $c(T)$ but otherwise is independent of current.

A similar gas dynamic relation relates p(z) with M(z), i.e.,

$$p_o/p(z) = [1 + \frac{1}{2}(\gamma-1)M^2]^{\gamma/(\gamma-1)} \tag{27}$$

where p_o is the upstream tank pressure and p(z) is the axial pressure distribution.

We can now eliminate E from equations (24), (25) using $E = I/\sigma A$, which is Ohms' law. Then A from equation (24) is given by

$$A = I/\sigma^{1/2}[\rho C_p v_z(\partial T/\partial z) + U]^{1/2} \tag{28}$$

and can be used to eliminate A from equation (25). In the resulting equation the current I cancels leaving a second order differential equation with temperature as the only unknown as σ, p, C_p, h, v_z, U_t and U are all determined by temperature. It is assumed that U and U_t are independent of A. Thus the axial temperature distribution T(z) is independent of current, provided that the two required boundary conditions are independent of current. A first boundary condition is that the temperature at the upstream electrode, T(z=0), is the melting point of the electrode, and so independent of current. For a second boundary condition, $\partial T/\partial z$ from equation (24) is given by

$$\frac{\partial T}{\partial z} = (I^2/A^2\sigma - U)/\rho C_p v_z \tag{29}$$

and we need to assume that A_o, the value of arc cross section at the electrode, is proportional to I.

Having shown that T(z) is independent of current the remaining properties follow easily. From equation (24) E is a function of T(z) and is thus independent of current. Then Ohms law gives $A = I/\sigma E$ so that arc area is proportional to current. Calculated[35] values of T, E, R, and p as a function of z are shown as a function of z for the 2000 A arc of Hermann et al.[34] in Figs. 10 and 11. Experimental results are shown as points, and it is seen that there is reasonable agreement between theory and experiment. Predicted plasma velocities are however a factor of two higher than the single experimental measurement.

Interesting relations also follow regarding the dependence of arc properties on pressure. From equation (26), M and thus v_z are independent of pressure if γ is independent of pressure. The differential equation in T obtained on eliminating A and E from equation (24) and (25) is independent of pressure if U, U_t and ρ are proportional to pressure. Thus, T(z) should also be approximately independent of pressure although from the second boundary condition we need to assume $A_o \propto 1/\sqrt{p}$. Numerical solutions of the equations confirm that T(z) is largely independent of p and from equation (24) $E \propto p^{1/2}$ and from Ohms' law $A \propto 1/p^{1/2}$.

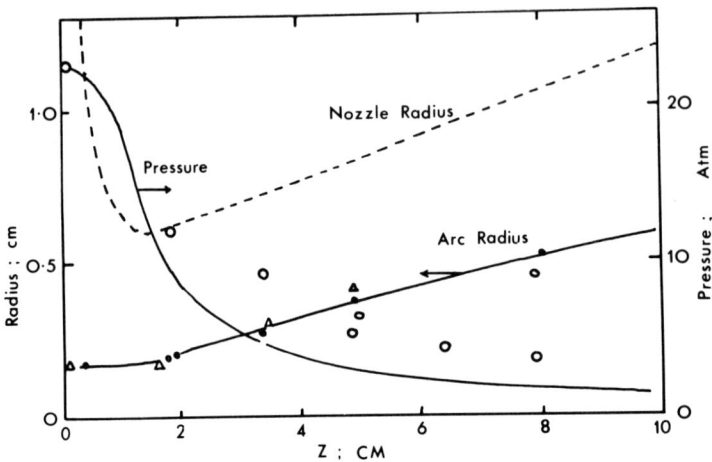

Fig. 10. Theoretical predictions of simple model compared with experimental points of Ref.[34] for 2000 A arc in N_2; from Ref.[35].

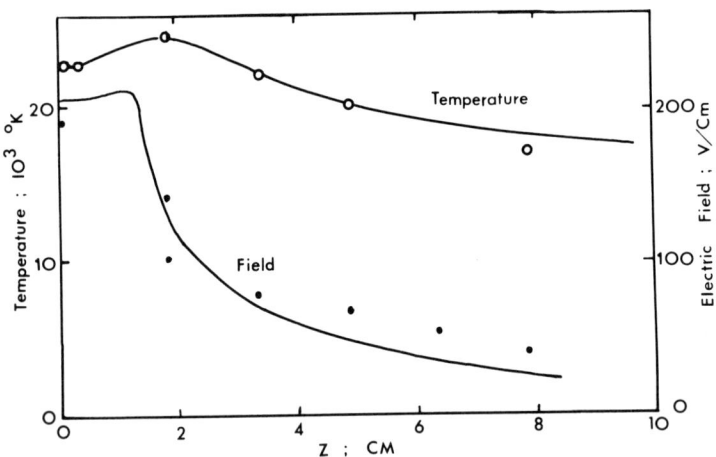

Fig. 11. Theoretical predictions of simple model with experimental points of Ref.[34] for 2000 A arc in N_2; from Ref.[35].

3.4. Ablation Stabilized Arc in Clogged Mode

For very high currents, the arc nozzles of circuit breakers become 'clogged'. Then the arc diameter becomes equal to the diameter of the nozzle, there is no longer a significant flow of cold gas between the arc and the nozzle, and the nozzle becomes ablated by the arc. Because of the restricted mass flow and also because of the production of ablated material within the nozzle, the plasma pressure rises very rapidly to become of the order of or even higher than the upstream tank pressure. In this case although plasma flow still proceeds at the downstream exit of the nozzle at sonic velocity, within the nozzle there may be a stagnation point and a region of considerably increased pressure.

The model of radiation transfer again enables a simple analysis to be made. Calculations of Kovitya et al.[39] indicate that the high axial flow rates tend to even out axial temperature variations so that to a first approximation the temperature can be taken to be uniform axially. Thus the energy balance equation (24) at the arc centre reduces to $\sigma E^2 = U$ or

$$j^2 = U\sigma \tag{30}$$

where current density $j = \sigma E$.

The approximation is again made that the arc is isothermal with radius. The radiation produced by the input electrical energy will be absorbed by the nozzle, and will raise the temperature of the surface above its boiling point to produce additional plasma at the arc temperature. It is assumed that losses of energy by conduction through the solid nozzle wall to the external surroundings are small compared with energy losses from the increased enthalpy flow that results from the ablated material. Thus the integrated energy balance equation similar to equation (25) is

$$\sigma E^2 A = \partial(\rho h v_z A)/\partial z$$

where A is the area of cross-section of the arc, it being assumed that the arc fills the nozzle. For the simple case of a cylinder where A is independent of z this equation reduces to

$$j^2/\sigma = \partial(\rho h v_z)/\partial z \tag{31}$$

Integrating equation (31) from the stagnation point within the nozzle where $z = 0$ to the downstream exit of the nozzle where $z = L$ we obtain $j^2 L/\sigma = \rho h c$ where c is the velocity of the plasma at the exit, which should be close to sonic velocity. Then substituting $j^2 = U\sigma$ from equation (30) we obtain

$$3L/4 = \rho\, h\, c/U \tag{32}$$

where the factor 3/4 arises because of the axial pressure variation.[39]

If the material functions $\rho(T)$, $h(T)$, $c(T)$ and $U(T)$ are known, equation (32) effectively determines the plasma temperature which is seen to be independent of current. Furthermore if ρ and U are both proportional to pressure, equation (32) defines a temperature independent of plasma pressure. Having thus determined the temperature, the relation $j^2 = U\sigma$ from equation (30) defines the pressure. As $U \propto$ pressure p, the pressure in the nozzle is proportional to the square of the current.

Figure 12 is taken from the more detailed calculations of Ref.[39] and shows calculated plasma pressures and temperatures as a function of current density and tube length. The material functions in this figure should be for ablated nozzle material, but instead are for air[40] so that the figure gives just a first order estimate of parameters.

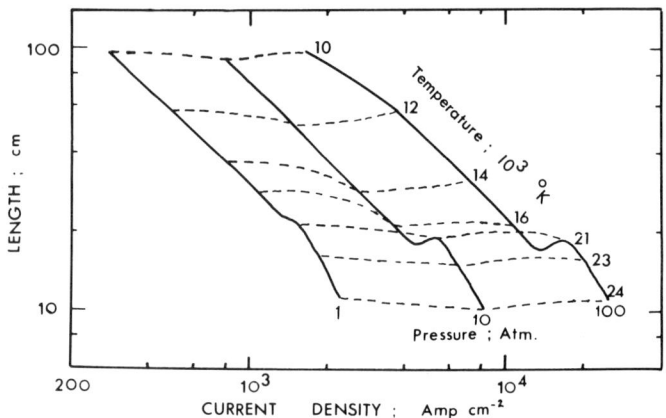

Fig. 12. Plasma pressure and temperature in a clogged cylindrical nozzle as a function of current density and nozzle length; from Ref.[39].

4. INFLUENCE OF RADIATIVE ENERGY TRANSFER ON DIELECTRIC BREAKDOWN IN CIRCUIT BREAKERS

Probably the most important question for a circuit breaker is whether there is arc reignition after current zero. We skip over the mechanism of thermal reignition, where turbulence rather than radiation processes are probably more important, and discuss dielectric reignition. In this process electrical breakdown occurs in times of 100 µs or more after current zero when voltages of 100 kV or more are applied across the electrodes. The physics of dielectric breakdown is quite different from thermal breakdown in that for the late times of dielectric breakdown a plasma no longer exists, and a new conducting channel has to be formed by electron ionization processes.

There has been a significant increase in our understanding of the mechanism of breakdown in the past few years, and the associated radiation processes have been clarified. This breakdown mechanism is discussed in the paper of T. Teich and W. Zaengl in this volume. In this section we shall discuss specifically the role of radiation processes before current zero and their influence on the size of the channel of hot gas in which dielectric breakdown occurs.

As is mentioned in the paper of K. Ragaller and K. Reichert of this volume, a mantle of hot gas forms around the arc plasma before current zero. From Fig. 17 of their paper the diameter of this mantle is of the order of 5 mm even 100 μs after current zero.

A simple order of magnitude estimate shows that the extent of this thermal boundary layer of the arc cannot be produced by thermal conduction. The thermal diffusivity α of SF_6 at 20 bar is 10^{-5} m^2/s at the estimated temperature of the boundary layer of 2000 K. Thus, the approximate thickness of the layer produced by thermal conduction alone is $\sim \sqrt{\alpha t} < 0.1$ mm for times of the order of 100 μs, appropriate to the time for arc plasma to travel ~10 cm at 2000 K. This conduction thickness is very much smaller than the observed thicknesses.

The thermal boundary layer of Fig. 17 is observed under conditions of a stable arc in laminar flow. Thus the influence of turbulence can here be neglected and the only physical process which can supply heat fast enough to produce such a thick thermal boundary layer is radiation. Although turbulence and flow instabilities can produce a still larger radius of the thermal mantle in a circuit breaker, the radius caused by radiation is most important in that it determines the lower limit to the extent of the mantle. In the paper of J. Kopainsky of this volume an analysis of this channel decay is given and related to the dielectric limit of a circuit breaker.

Fairly detailed experimental investigations [41-43] have been made of the photoionization cross-sections of SF_6 at room temperature as a function of wavelength for the ultraviolet region of the spectrum. Results are summarized in Fig. 13 which gives the absorption coefficient in cm^{-1} for a pressure of 1 atmosphere as a function of wavelength. It is seen that continuous absorption begins at about 1500 Å, the absorption coefficient being 1 cm^{-1} at 1420 Å and rises to over 1000 cm^{-1} for wavelengths below 900 Å.

The results of Fig. 13 indicate a sharp division between radiation of wavelengths shorter than 1250 Å, which would be absorbed in a distance of less than 1 mm, and radiation of wavelength greater than 1400 Å which would be lost to the wall. However, it would be highly desirable to have further investigations of the absorption coefficient for wavelengths greater than 1400 Å. It is in this region that a significant fraction of the radiative flux exists, for example see Fig. 14 calculated[7] for the central region of a 100 A arc, but accounting for continuum radiation only.

Some limit can be placed on the radius of the combined arc plus hot sheath channel. In Fig. 9 reasonable agreement is obtained between experimental results of arc radius R and values calculated from equation (23) with f = 1 which assumes that no energy is lost by radiation from the arc core. But from equation (23) R varies as $f^{1/4}$ and a value of f = 0.5 would only change the theoretical values of R by 20 %.

In a manner similar to the derivation of equation (23), we can derive an expression for the additional area A_2 of a hot sheath (see Fig. 15). Thus, we write an equation similar to equation (22)

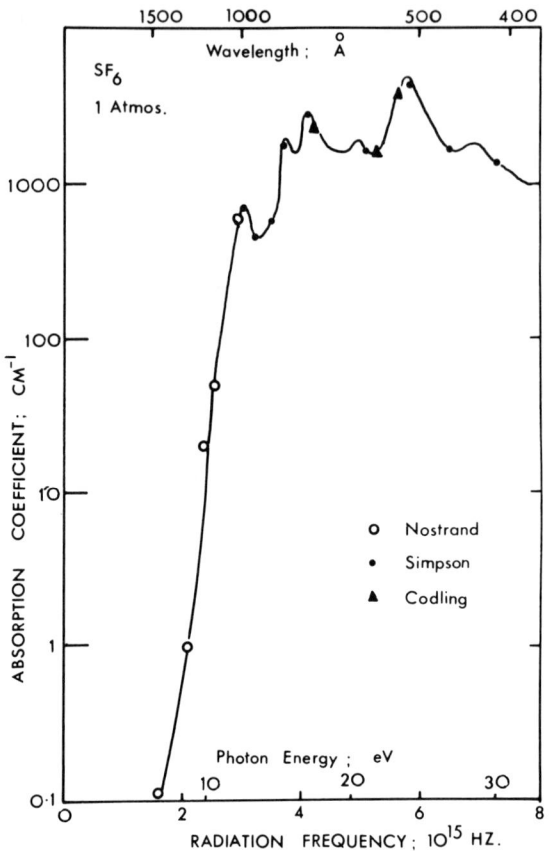

Fig. 13. Measured absorption coefficients in SF_6 at 1 atmosphere pressure as a function of wavelength from Refs.[41-43].

$$\sigma_1 E^2 A_1 f_2 = \frac{\partial}{\partial z}(\rho_2 h_2 A_2 v_{z2})$$

where f_2 is the fraction of input energy absorbed in the sheath and the subscripts 1 and 2 refer to the arc temperature and sheath temperature respectively. On eliminating $\sigma_1 E^2 A_1$ from the above equation with equ. (22), integrating with respect to z and approximating $\rho_2 h_2 \simeq \rho_1 h_1$ we obtain:

$$A_2/A_1 = f_2 v_{z1}/f_1 v_{z2} \tag{33}$$

Thus if $f_2 \simeq f_1 \simeq 0.5$, then as the sonic velocity v_{z1} at 10,000 K is about 5 times the sonic velocity v_{z2} at say 2000 K equation (33) would give $A_2 \simeq 5A_1$. Thus the ra-

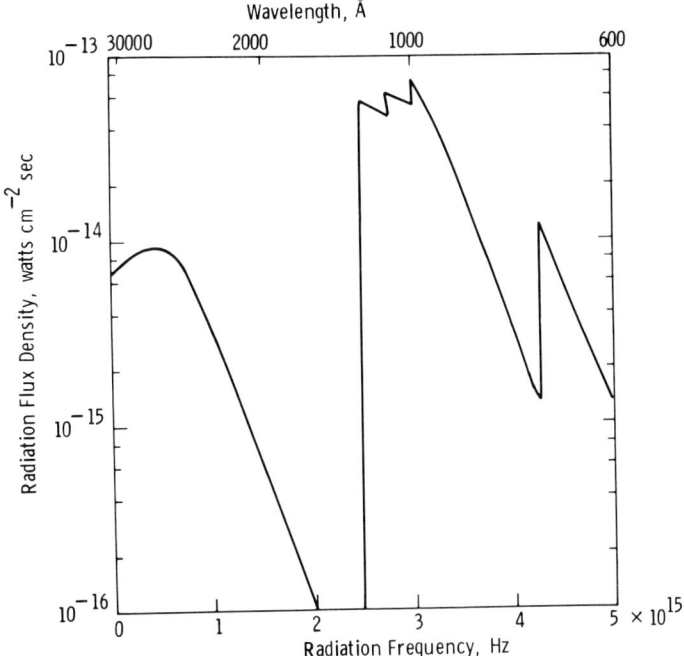

Fig. 14. Calculated radiative flux near the centre of a 100 A arc from continuum radiation; from Ref.[7].

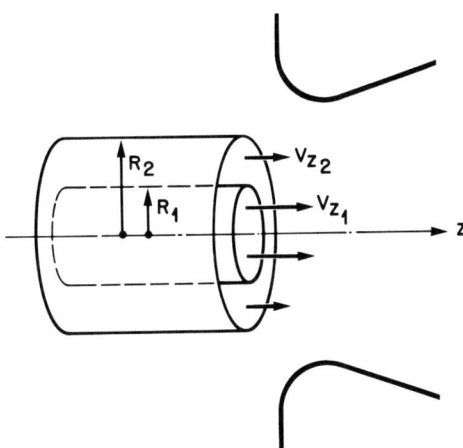

Fig. 15. Arc and thermal arc boundary layer in the region upstream of the nozzle.

dius of the hot gas plus arc core would be ~ 2.5 times the radius of the conducting core. Although it is unlikely that f_2 is as high as 0.5, the calculation demonstrates that radiation absorption can be the determining influence in the radius of the hot gas remaining when the large system voltages are applied across the circuit breaker. The final temperature is determined by convection and radial conduction which are critically dependent on the initial diameter, as is discussed in the following paper by J. Kopainsky.

REFERENCES

1. D. H. Sampson, Radiative contributions to energy and momentum transport in a gas, p. 25, Interscience, New York (1965)
2. J. B. Shumaker, JQSRT 14 (1974) 19
3. L. H. Aller, Astrophysics, p. 223, Ronald Press, New York (1963)
4. H. R. Griem, Plasma Spectroscopy, McGraw Hill, New York, 1964
5. A. Burgess and M. J. Seaton, Mon. Not. Roy. Astr. Soc. 120 (1960) 121
6. G. Peach, Memoirs R. Astron. Soc. 71 (1967) 13
7. J. J. Lowke and R. W. Liebermann, J. Appl. Phys. 42 (1971) 3532
8. K. H. Wilson and W. E. Nicolet, JQSRT 7 (1967) 891
9. J. J. Lowke and R. W. Liebermann, J. Appl. Phys. 43 (1971) 1991
10. H. Popp, Z. Naturforsch. 22a (1967) 254
11. H. Popp, Physics Reports 16 (1975) 169
12. G. Bold, Z. Phys. 154 (1959) 319 and 154 (1959) 330
13. J. C. Morris, R. U. Krey and R. L. Garrison, Phys. Rev. 180 (1969) 167
14. W. D. Barfield, JQSRT 17 (1977) 471
15. W. Hermann and E. Schade, JQSRT 12 (1972) 1257
16. R. W. Liebermann and J. J. Lowke, JQSRT 16 (1976) 253
17. W. L. Wiese, M. W. Smith and B. M. Miles, Atomic Transition Probabilities, NBS NSRDS-NBS, 22, Vol. 2 and NSRDS-NBS, 4, Vol. 1
18. H. R. Griem, Phys. Rev. 128 (1962) 515
19. L. C. Biedenharn, J. M. Blatt and M. E. Rose, Rev. Mod. Phys. 24 (1952) 249
20. B. L. Hunt and M. Sibulkin, JQSRT 7 (1967) 761
21. H. R. Griem, Phys. Rev. 165 (1968) 258
22. C. H. Church, R. G. Schlecht, I. Liberman and B. W. Swanson, AIAA 4 (1966) 1947
23. J. J. Lowke, JQSRT 9 (1969) 839
24. R. D. Richtmyer and K. W. Morton, Difference methods for initial valve problems, Interscience, New York (1967)
25. J. J. Lowke and E. R. Capriott, JQSRT 9 (1969) 207
26. W. Finkelnburg and H. Maecker, Handbuch der Physik, Vol. 22, p. 254 (Edited by S. Flügge), Springer Verlag, Berlin (1956)
27. R. Viskanta, Advances in Heat Transfer, Vol. 3, p. 175, Edited by T. F. Irvine and J. P. Hartnett, Academic Press, New York, 1966
28. A. Schuster, Astrophys. J. 21 (1905) 1
29. S. C. Traugott and K. C. Wang, Int. J. Heat Mass Transfer 7 (1964) 269
30. C. F. Curtiss and J. O. Hirschfelder, Proc. Natn. Acad. Sci. USA 38 (1952) 235
31. M. F. Hoyaux, Arc Physics, Springer Verlag, New York (1968)
32. H. W. Emmons, Phys. Fluids 10 (1967) 1125
33. J. J. Lowke, JQSRT 14 (1974) 111
34. W. Hermann, U. Kogelschatz, L. Niemeyer, K. Ragaller and E. Schade, J. Phys. D 7 (1974) 1703, also IEEE Trans. PAS 95 (1976) 1165
35. D. T. Tuma and J. J. Lowke, J. Appl. Phys. 46 (1975) 3361, also J. J. Lowke and H. C. Ludwig, J. Appl. Phys 46 (1975) 3352
36. W. Hermann, U. Kogelschatz, K. Ragaller and E. Schade, J. Phys. D 7 (1974) 607
37. J. J. Lowke, J. Appl. Phys. 41 (1970) 2588
38. D. R. Airey, R. E. Kinsinger, P. H. Richards and J. D. Swift, IEEE Trans. PAS 95 (1976) 1
39. P. Kovitya, J. J. Lowke and A. D. Stokes, Proc. 13th Intern. Conf. Phenomena in Ionized Gases, Berlin, 1977, p. 515

40. K. A. Ernst, J. G. Kopainsky and H. Maecker , IEEE Trans. Plasma Science PS 1 (1973) 4 and J. M. Yos, AVCO Technical Memorandum RAD-TM-63-7
41. E. D. Nostrand an A. B. F. Duncan, J. Amer. Chem. Soc. 76 (1954) 3377
42. J. A. Simpson, C. E. Kuyatt and S. R. Mielczarek, J. Chem. Phys. 44 (1966) 4403
43. K. Codling, J. Chem. Phys. 44 (1966) 4401

DISCUSSION
(Chairman: W. Bötticher, Technical University, Hannover)

K. Ragaller (Brown Boveri)

For circuit-breaker design it would be of interest to reduce the temperature and size of the thermal arc boundary layer. Do you think it will be possible to predict from basic radiative transport data what needs to be done, for example, with respect to gas mixtures, to achieve this effect?

J. J. Lowke

The problem of representing in detail this radiation with so many lines, all as a function of wavelength and temperature, is extremely difficult. What one would need is probably something with a very high absorption coefficient. Then, of course, you would absorb the radiation in a much smaller region, ionize it, and hopefully keep the radius of the thermal arc layer very much smaller.

D. M. Benenson (State University of New York at Buffalo)

Do you have any suggestions as to ways in which one might be able to scale, or develop similarity criteria for, the arc including radiation?

J. J. Lowke

We derived that the net emission coefficient is roughly proportional to the pressure. This has a number of important consequences: the electric field scales with the square root of the pressure, the temperature is independent of current, the arc area is proportional to current, etc. It isn't exact, but I think it's a good approximation.

W. Bötticher

Has anyone observed photoionization outside of the arc due to absorption of ultraviolet light emitted from the arc? There may be some radiation which reaches outer regions and produces there electrons and ions by photoionization. The charge carriers could stay there and influence the late stage of dielectric breakdown. The configuration is very similar to the photoionization processes used in TEA lasers.

INFLUENCE OF THE ARC ON BREAKDOWN PHENOMENA IN CIRCUIT BREAKERS

J. KOPAINSKY
Brown Boveri Research Center, Baden, Switzerland

SUMMARY

After the thermal interruption regime a zone of hot gas $300 < T < 5000$ K is present in the circuit breaker. This zone is contracting and cooling down simultaneously. In the peak-regime the axial temperature is of the order of 2000 K. The development of the temperature distribution with time is investigated.

The decaying hot zone still contains charge carriers (mainly positive and negative ions). The ion concentration is generated by rather complex processes. It is large enough to be able to change the external electric field by space charges. As a consequence, a small current prior to breakdown occurs and the breakdown voltage is reduced.

The theory which is developed determines the ions which are generated by a decaying, axially blown arc. During the decay time until breakdown, recombination of ions, attachment, and recombination of atoms occurs to generally no more than diatomic molecules. The partial equilibrium concentration, as well as the kinetics of the reactions leading to equilibrium, is investigated.

1. INTRODUCTION

There are two kinds of dielectric failure in circuit breakers. In the first class the breakdown occurs in cold gas. Examples are late restrikes and breakdowns between thermally nonstressed parts. This type has been discussed in detail in the paper of T. H. Teich and W. Zaengl and will not be considered here.

The second group is concerned with breakdown in hot gas. Terminal fault and out-of-phase switching are examples for this case. The heating of the gas produces a great variety of new effects, and measurements have shown that the cold gas breakdown mechanism cannot be applied in this case. The phenomena involved in hot-gas investigations are complex and little work has been done to date. Only recently have experiments been performed, in the hot wake of a cross-flow arc which acted as heat source, and the results explained by a simple model.[1] The essential result is that the hot wake possesses an ion concentration which is sufficiently high (about 10^{11} cm^{-3}) to effect some change of the electric field applied to the gap by space charges. The changed electric field causes a lower breakdown voltage than for the charge-free gas.

Measurements of the breakdown voltage for high-voltage gas-blast circuit breakers also yield results which cannot be explained in terms of the cold-gas mechanism. The investigation of the decay process of the plasma column after current zero provides the first hint that the discrepancy might have its origin in effects similar to those occurring in the cross-flow arc experiment. Near the peak of the recovery voltage, where the breakdown appears, the hot gas has shrunk to a column less than one millimeter in diameter with an axial temperature between 2000 and 3000 K. The hot-gas column contains positive and negative ions which have survived from the high-current phase whilst electrons have been reduced to a greater degree by attachment and recombination. The ion density of roughly 10^{13} cm^{-3} is sufficient to change the local electric field by space charges to such a degree that it becomes independent of the circuit-breaker geometry.

It is the aim of the work described here to determine the size, radial extension, and temporal development of the ion column in a double-nozzle geometry which is typical of that used in high-voltage gas-blast circuit breakers. This information is essential for the development of an explanation of the breakdown mechanism for hot gases. The current state of understanding is reported in the present paper.

We discuss first the separate phenomena involved in the breakdown process, and subsequently consider the problem as a whole. In what follows we limit our considerations to the arc region between the two nozzle throats (cf. Fig. 9 of the paper of K. Ragaller and K. Reichert). In a circuit breaker, this portion of the arc normally has a cylindrical geometry. This is also true for the experiments performed in a laboratory model of the double-nozzle geometry.

2. DECAY OF THE TEMPERATURE DISTRIBUTION

By investigating the decay of an axially-blown arc without an applied recovery voltage we gain an understanding of the physical state of the hot gas column between current zero and breakdown several 100 μs later. The situation is illustrated for a model experiment in nitrogen in Fig. 1. The inner bright region is the current carrying core with a temperature between 7,000 and 20,000 K. This core is surrounded by the nonconducting thermal boundary layer (see also Fig. 16 of K. Ragaller's and K. Reichert's paper) with temperatures between room temperature and 3000 K. It is visualized by the Schlieren fringes (light and dark bends) in Fig. 1.

We consider the case of a decaying 30 kA arc in SF_6 at a pressure of 14 bar. This arc has a streak record very similar in appearance to that of the nitrogen in the laboratory investigation. Therefore we carry over the measured temperature dis-

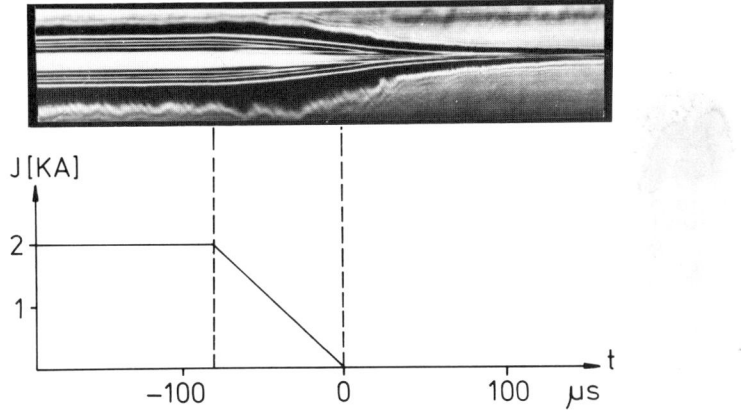

Fig. 1. Schlieren-streak picture of decaying arc.

tribution at current zero for nitrogen to this case as shown in Fig. 2. The conducting core has a diameter of about 1 mm and an axial temperature of approximately 10,000 K. The thermal arc boundary layer is a cylinder roughly 10 mm in diameter. This layer is generated by radiative absorption from the hot core as discussed in the paper of J. Lowke.

The decay process is described quantitatively by the conservation equations for mass, momentum, and energy. The simultaneous solution of the coupled differential

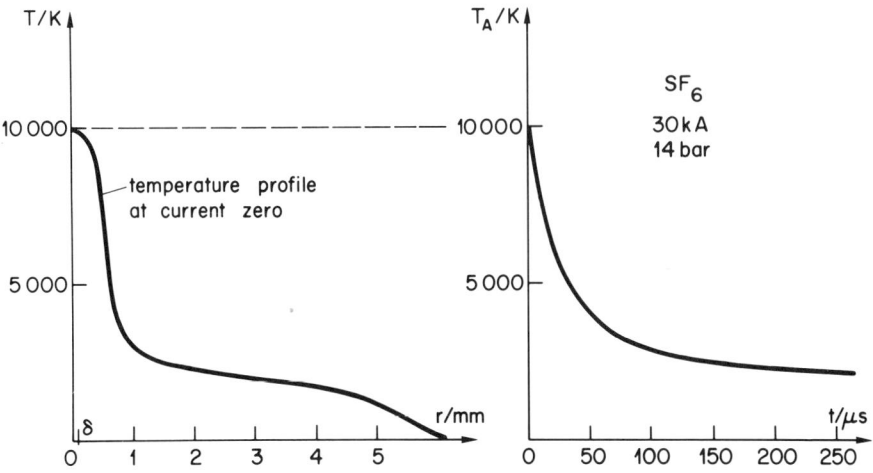

Fig. 2. Temperature profile at current zero and decay of axial temperature for an SF_6 arc.

equations is tedious, as pointed out in the paper of B. Swanson. Since a numerical solution does not easily permit an insight into the physical mechanisms involved, an approximative, analytical treatment is preferred here.

For a cylindrical arc column with sonic, or nearly sonic, conditions at the nozzle throat both the radial and axial momentum balances, as well as mass conservation, yield an expression for the radial velocity to within a good approximation:

$$v_r = c \cdot \frac{r}{2\ell} \quad (1)$$

where c is the local sound speed and ℓ the distance between the stagnation point and the nozzle throat.

In axially blown, cylindrical arcs the temperature is independent of the axial position[2]. To a good approximation, the sound speed is a function of temperature alone, and so equation (1) implies that cold gas enters the hot zone with equal radial velocities at all axial positions. Substitution of equation (1) into the energy balance yields

$$\frac{\partial T}{\partial t} - \frac{c}{2\ell} r \frac{\partial T}{\partial r} = \frac{1}{\rho c_p} \frac{1}{r} \frac{\partial}{\partial r} \left(r \kappa \frac{\partial T}{\partial r} \right) \quad (2)$$

where ρ is the density, c_p the specific heat at constant pressure, and κ the thermal conductivity.

The equation states that the axisymmetric arc temperature distribution decays after current zero by conduction and convection. Since the two transport coefficients κ and c_p show strong variations with temperature, it is advantageous to introduce the heat-flux potential

$$S = \int_0^T \kappa(\tilde{T}) \, d\tilde{T}$$

Then the energy balance equation (2) may be written as

$$\frac{\partial S}{\partial t} = \frac{c}{2\ell} \cdot r \frac{\partial S}{\partial r} + \alpha \frac{1}{r} \frac{\partial}{\partial r} \left(r \frac{\partial S}{\partial r} \right) \quad (3)$$

In this form only two, slowly varying, material functions, the thermal diffusivity $\alpha(S) = \kappa/(\rho c_p)$ and sound speed $c(S)$, enter the equation.

To obtain an insight into the physics of the decay process a hypothetical situation is considered, with constant α and c, for which an exact, analytical solution is possible:

$$S = S_o \cdot \exp(-\frac{t}{\ell/c}) \cdot \exp(-\frac{r^2}{2\delta^2}) \qquad (4)$$

where $\delta^2 = 2\ell\alpha/c$.

The radial distribution of the heat-flux potential has a bell shape with an axial value S_o at current zero (Fig. 3). The decay has a time constant equal to ℓ/c. Since the heat-flux potential is less commonly encountered than the temperature, the corresponding temperature distribution is also plotted in Fig. 3. The temperature distribution is distorted compared with the Gaussian shape of the heat-flux-potential curve as a result of the nonlinear relation between the heat-flux-potential and temperature.

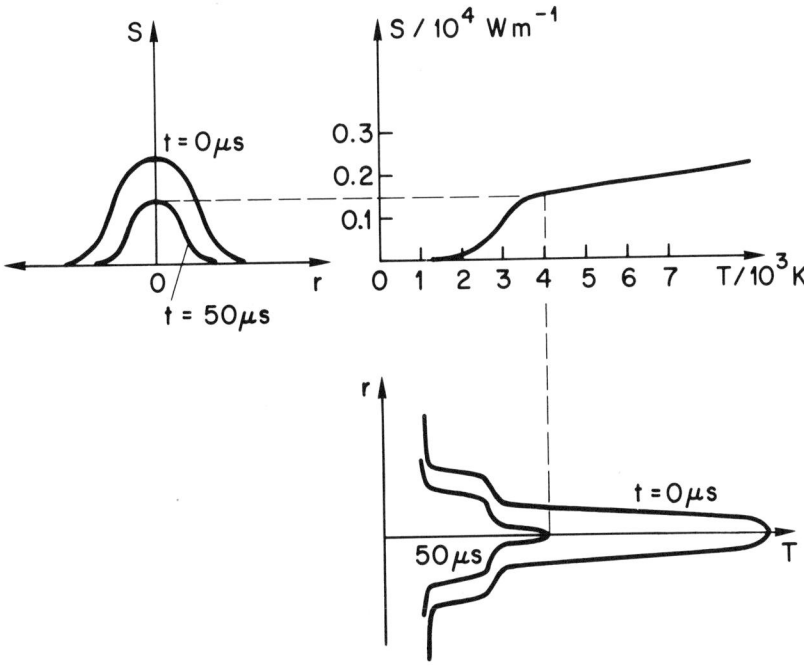

Fig. 3. Distribution of temperature and heat-flux potential for the hypothetical case of constant transport coefficients α and c.

The energy balance shows that convection and conduction are not equally important at all radial positions. Near the axis, the temperature gradient $\partial T/\partial r$ is small and conduction is the most important mechanism in this zone. The opposite is true for the outer region, where convection is dominant.

The entire hot region can be modeled in a crude way by separating it into an inner part with pure conduction and an outer zone with only convection. By defining a separation surface $r = \delta$, at a position where both mechanisms are equally strong, leads to the same expression for δ as was found quite formally in equation (4). In this way, the physical meaning of δ becomes clear.

For the outer region the isotherms are identical to particle pathlines, i.e. the isotherms are 'frozen' in the fluid and so move together with the fluid particles (Fig. 4). The fluid streams out through the nozzles with a velocity, at the nozzle throats, roughly equal to the sound speed. A fluid cylinder with its upper end at the stagnation point plane, i.e. with particles of zero axial velocity at its upper end, becomes longer and thinner due to the outflow in a similar way to a stretched elastic band. The outflow of the inner rings is faster than that of the outer ones because of the increase of sound speed with temperature.

In the absence of thermal conduction, the column would only become thinner in the course of time, whilst its axial temperature remained constant.

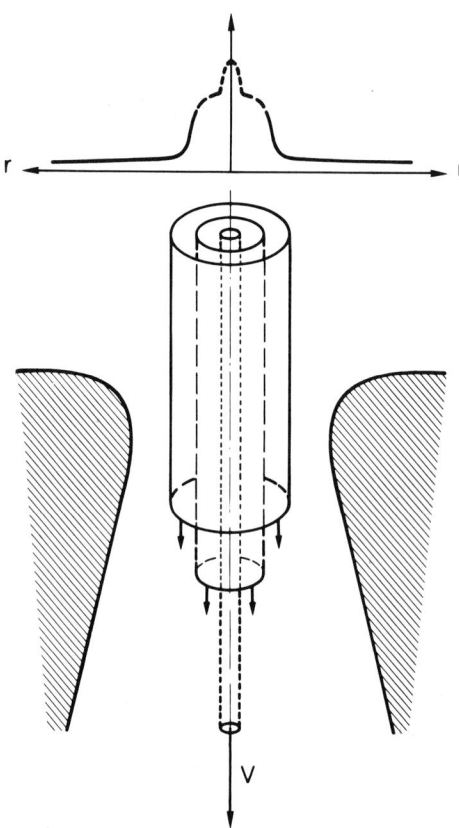

Fig. 4. Stretching and simultaneous shrinking of isothermal rings.

At r = δ the gas is sucked in across the boundary surface which, in the analytical model, remains fixed during decay. In reality δ decreases slightly due to the temperature dependence of the transport coefficients. The continuous decrease in temperature at the boundary leads to a temperature decrease of the inner, conduction-dominated region and thus also of the axial temperature.

This approximate model has been solved numerically with the actual transport coefficients for SF_6. Guided by the qualitative form of the temperature distribution given in Fig. 3, the starting profile at current zero has been approximated by the superposition of two Gaussian functions.

The numerical treatment proceeds as follows. The outer region is split into a large number of concentric isothermal rings, whilst the temperature distribution of the inner zone is approximated by a parabola.

The first result is that the radius of the separating surface, δ, is equal to roughly 0.1 mm (Figs. 2 and 4). This means that only the immediate surrounding of the axis is conduction dominated. The second result is the quantitative form of the axial temperature decay (right half of Fig. 2). Initially, the temperature decays very rapidly, and eventually approaches a value between 2000 and 3000 K, depending on the current. Near the peak of the recovery voltage the diameter of the total hot zone is of the order of 0.1 mm.

The calculations have been checked by holographic interferograms and Schlieren records for a model experiment with an N_2 arc. Quantitative evaluations of the data for the first 40 μs after current zero show satisfactory agreement with the model.

3. DECAY OF CHARGE-CARRIER CONCENTRATION

3.1. Influence of Chemical Reactions

At current zero the particle densities are, to a good approximation, those appropriate to thermal equilibrium. They have been calculated by Frie[3] and are reproduced here in Fig. 5. The conducting core of the SF_6 arc, which has a temperature of about 10.000 K, consists mainly of S, S^+, F, and electrons.

During decay, molecules and corresponding ions are built up by reactions between the four most frequently occurring components. The large number of reactions taking place can be reduced considerably by omitting all reactions with time constants greater than 100 μs. This excludes all molecules, and their ions, with three or more atoms which are generated only in small concentrations during the time of interest because of the very small collision probabilities.

As a first crude approximation it is assumed that only ions, electrons, atoms, diatomic molecules, and diatomic molecular ions are present in the hot column. To obtain a preliminary insight into the particle concentrations it is further assumed that all these particles are in equilibrium. This situation will be called partial equilibrium. The result of the calculation is shown in Fig. 6 and the details explained in Appendix 1.

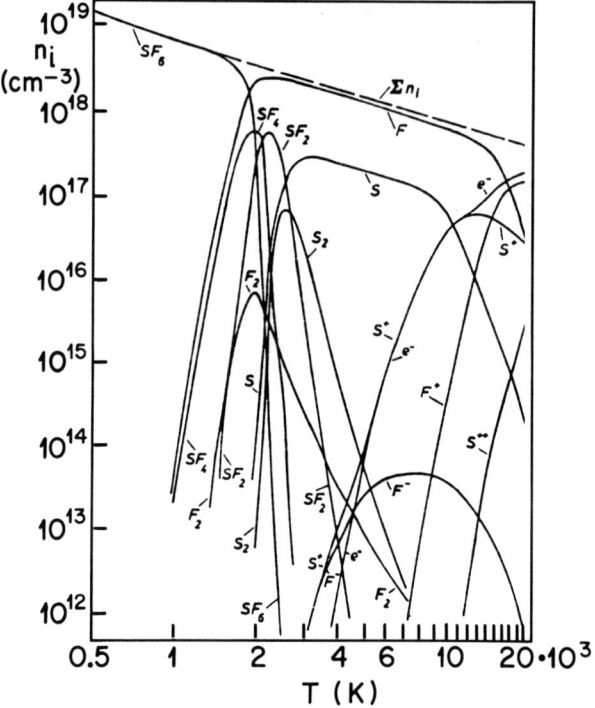

Fig. 5. Total particle densities in SF_6 according to Frie[3].

At 10,000 K, total and partial equilibrium give equal particle concentrations. The small differences between Frie's results and those given here are the result of using different data. At low temperatures, the concentration of neutral particles is in good agreement with Frie's findings. However, the presence of considerable positive and negative ion concentrations is indicated in our calculations, whereas Frie's diagram shows only small ion concentrations. The reason is that Frie included a strongly reduced number of different ions in his calculations.

At breakdown, i.e. at axial temperatures near 2500 K as obtained in section 2, the ion concentration is roughly 10^{13} cm^{-3} with S_2^+ and F^- being the main components. The electron density is several orders of magnitude lower.

As a second approximation, the kinetics of the reactions is included. This is outlined for the example of dissociative attachment of electrons to form molecular fluorine

$$F_2 + e^- \underset{d}{\overset{r}{\rightleftarrows}} F + F^- \tag{5}$$

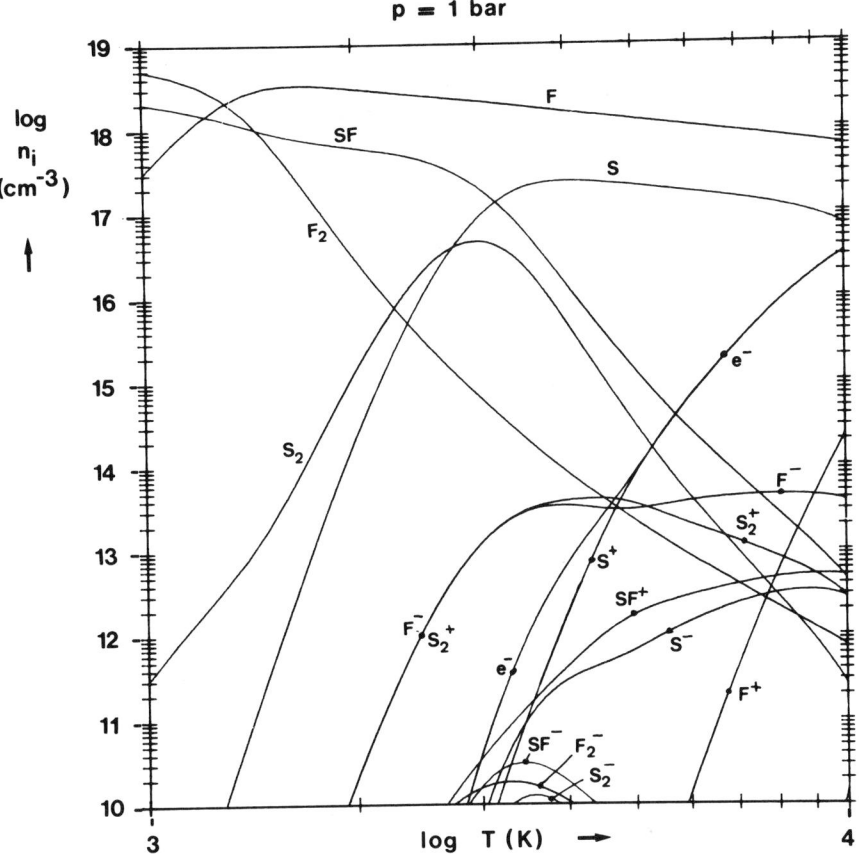

Fig. 6. Partial equilibrium densities for 1 bar.

The electron concentration is decreased according to this reaction by the forward process of dissociative attachment with reaction rate r. At the same time it is increased by the backward detachment process with rate d:

$$\frac{\partial n_e}{\partial t} = -r\, n_F\, n_e + d\, n_F\, n_{F^-} \tag{6}$$

A special case of the steady state, with the left-hand side equal to zero, corresponds to the equilibrium situation which has just been treated.

The task is now to set up all relevant reactions, find the corresponding reaction rates r and d, and solve the resulting set of differential equations. Unfortunately the available data on reaction rates is very poor and it was necessary to determine some data from simple models or extrapolation of similar, known reaction rates. In all, some twenty reactions were found to be important.

A short review of the calculation is given in Appendix 2, and the details outlined in a paper of K. P. Brand and J. Kopainsky.[4] In spite of the uncertainty in the data, the resulting densities are considered to be reliable to within an order of magnitude.

Figure 7 shows the results of the model for an arc which is decaying according to the axial temperature curve shown. After a few hundred µs the ion concentrations have decreased to approximately 10^{13} cm^{-3}, the main components being S_2^+ and F^-.

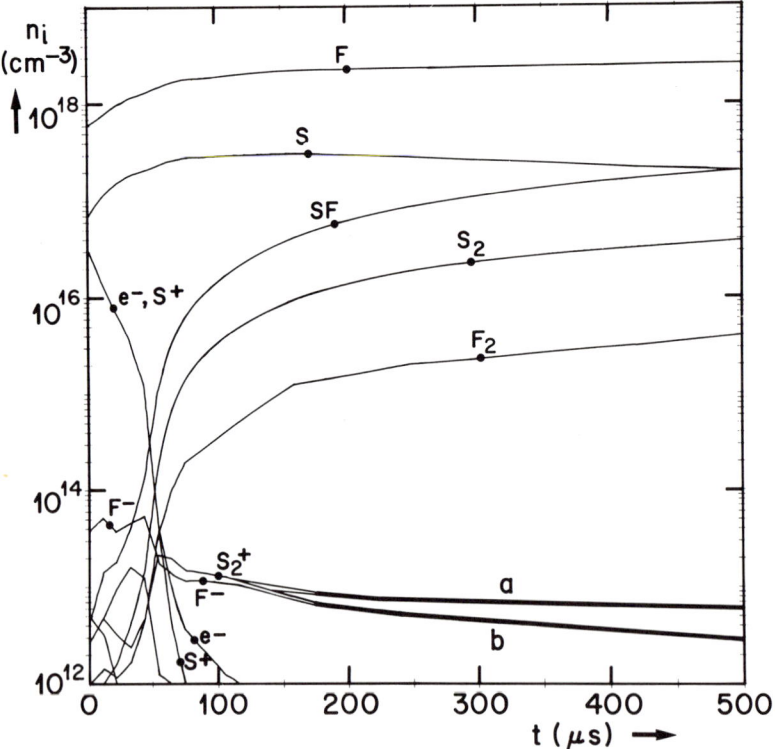

Fig. 7. Particle densities in a decaying SF_6 arc for 1 bar. Bounds for the S_2^+ and F^- concentrations arising from the unknown data for type K (see Appendix 2), neutralization reactions are estimated from calculations assuming no neutralization (curve a) and the maximum possible neutralization (curve b).

In order to reveal directly the physics of the reaction kinetics, we have plotted the relative particle densities versus time in Fig. 8, that is the ratio of actual den-

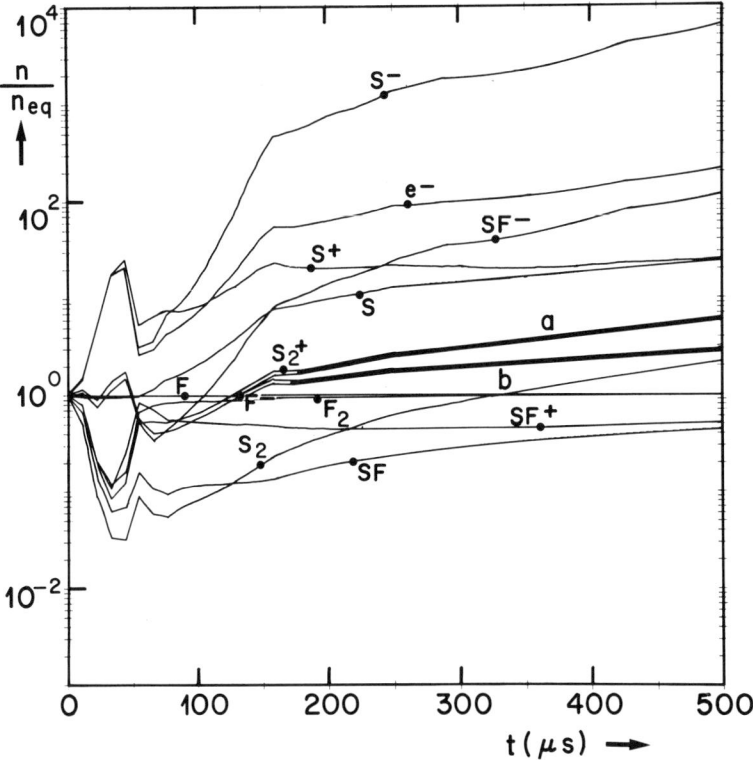

Fig. 8. Relative particle densities for situation of Fig. 7. Denominator: densities from Fig. 6.

sity to the density obtained for partial equilibrium. During the first 100 µs we find a departure from partial equilibrium, whereas at later times all components relax towards their partial equilibrium values (i.e. the curves in Fig. 8 approach the value unity). After about 100 µs, the temperature has decreased to such an extent that the reaction rates have become very slow. Even further decays in temperature lead to gradual deviations from partial equilibrium. At breakdown, all particles except the rare components S^-, e^-, and SF^- differ by less than a factor of 30 from their partial equilibrium values. The most frequently ions, S_2^+ and F^-, in particular, are very close to unity after 100 µs.

The electron density is rapidly reduced by recombination reactions resulting in a rapid decrease in electrical conductivity and Joulean heating. This is the physical reason for the transition some 10 µs after current zero from an energy-balance dominated (thermal) regime to a dielectric regime with negligible Joulean heating.

This picture should not be taken as a precise quantitative result because of uncertainties in the data used. Nevertheless, it reveals the main features of the kinetics of a decaying arc which will not change considerably with more precise data. There will still be an initial phase after current zero where strong departures from partial equilibrium appear, because of the rapid decay in temperature. In the subsequent phase, the particle densities relax towards their partial equilibrium values. Finally, a third phase after approximately 100 µs, slow deviations from partial equilibrium occur.

3.2. Ion Column

After discussing the physical phenomena of a decaying arc in the previous two parts we consider now the consequences.

The decay leads to a thin ion column which connects the electrodes along the path where the arc was previously burning (Fig. 9).

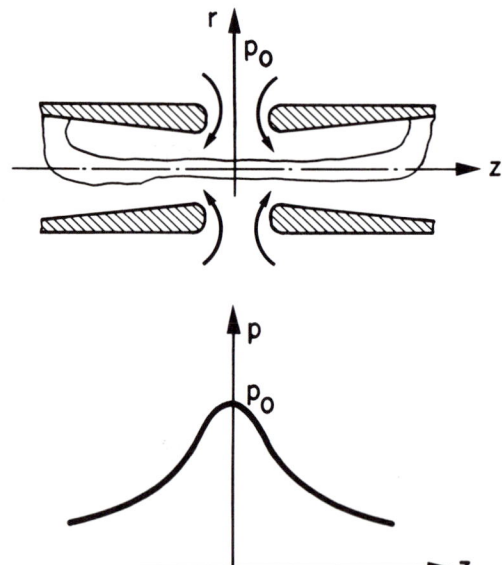

Fig. 9. Ion column. A thin rod with positive and negative ions connects the electrodes along the axis where the arc was previously burning. Below: Pressure distribution on the axis.

An important property of the ion column is that, since it contains a considerable number of ions, it may change the applied electric field in the hot zone in a manner similar to a conducting wire. This is different from the cold-gas situation where the applied electric field remains unchanged until immediately before breakdown.

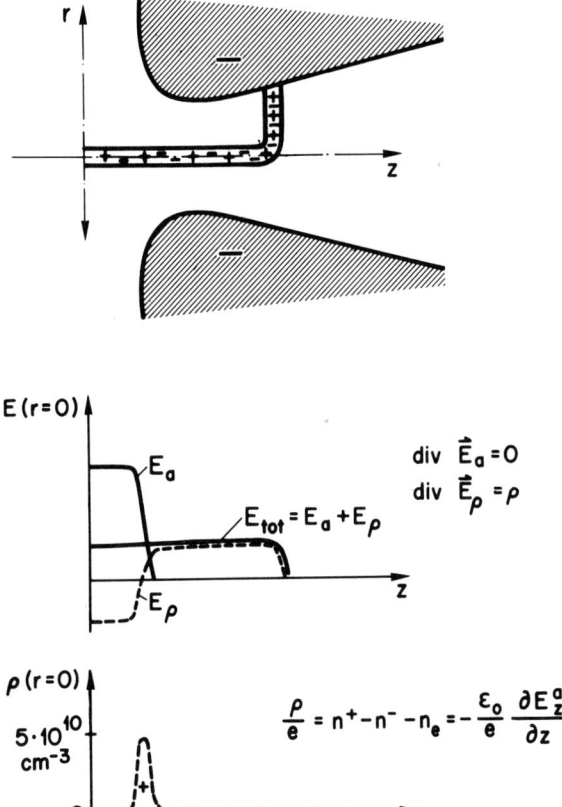

Fig. 10. Applied electric field E_a without ion column and total field E_{tot} in the presence of an ion column.

We again consider a double-nozzle geometry, specifying now that we have two metal nozzles (Fig. 10). For a cold gas the application of a voltage across the electrodes leads to an electric field on the axis E_a which is approximately constant between the nozzles and decreases to zero within the nozzles. However, in the presence of an ion column the field distribution is entirely different. Assuming for simplicity a homogeneous, cylindrical ion rod we obtain now a constant field strength E_{tot} which extends into regions which would be free of the field in a gas without space charges.

A minimum number of charged particles is required in order to establish a constant total field strength E_{tot}. If the charged particles are separated under the action of the gradient of the applied field, an influenced field E_ρ is created according to Poisson's equation. For typical circuit-breaker dimensions a minimum charge separation of $5 \cdot 10^{10}$ cm^{-3} is required for an E_ρ which is large enough to establish a constant E_{tot} close to the breakdown voltage. This means that more than $5 \cdot 10^{10}$ ions per cm^3 are needed to create an electric field in the hot zone which is determined by Ohm's law alone, as for a wire. The previous considerations show that this condition is fulfilled in decaying, axially blown arcs.

3.3. Influence of Axial Flow and Radial Diffusion

The particle densities are not only changed by chemical reactions, as discussed in the preceding section, but also by convection and diffusion. The conservation equations for the three most important charged particles are

$$\frac{\partial n_+}{\partial t} + \text{div}(n_+ \vec{v}) = \text{div } D_+ \text{ grad } n_+ + \Gamma_+ \tag{7}$$

$$\frac{\partial n_-}{\partial t} + \text{div}(n_- \vec{v}) = \text{div } D_- \text{ grad } n_- + \Gamma_- \tag{8}$$

$$\frac{\partial n_e}{\partial t} + \text{div}(n_e \vec{v}) = \text{div } D_e \text{ grad } n_e + \Gamma_e \tag{9}$$

where $\vec{v}$ is the velocity, $\Gamma_{+,-,e}$ the net production rate of +, -, e$^-$, as the sum of all reactions where the corresponding particle concentration is changed, and $D_{+,-,e}$ is the diffusion coefficient.

On the axis the equations can be considerably simplified as follows:

- For constant temperature with respect to the axial coordinate div $\vec{v} = 0$
- The high ion density creates a quasi-neutral condition: $n_+ - n_- - n_e \ll n_+, n_-$.

As shown in the last section the charge separation is small compared with the total charge density. This statement is true for a spatial resolution of the size of the Debye length, which is typically 10^{-3} mm for the situation considered here. The charge conservation requires

$$\Gamma_+ = \Gamma_- + \Gamma_e.$$

Thus we obtain for the axis of the ion column

$$\frac{\partial n_+}{\partial t} = 2D_+ \frac{\partial^2 n_+}{\partial r^2} + \Gamma_+, \tag{10}$$

$$\frac{\partial n_-}{\partial t} = 2D_- \frac{\partial^2 n_-}{\partial r^2} + \Gamma_- \tag{11}$$

$$\frac{\partial n_e}{\partial t} = 2D_e \frac{\partial^2 n_e}{\partial r^2} + \Gamma_+ - \Gamma_- \tag{12}$$

During the first 100 μs or so the chemical reactions dominate and diffusion can be neglected.

In the succeeding period the production rates approach the small rates of the partial equilibrium, viz. $\Gamma_{+,-,e} = (\partial n_{+,-,e}/\partial T)_{eq} \cdot \partial T/\partial t$ where $(\partial n/\partial T)_{eq}$ has to be taken from Fig. 6 and $\partial T/\partial t$ from Fig. 2. In this region, the effects of diffusion must be included.

It is remarkable that the convection term does not appear directly in calculating the charge-carrier concentration during the total decay period of interest. This term enters the diffusion term indirectly through the temperature and density profiles.

These considerations lead us to the following picture of a decaying, axially-blown arc after current zero without recovery voltage:

The geometry of the chamber and the filling pressure establish the axial pressure distribution (Fig. 9) which remains constant after current zero. This pressure distribution and the current amplitude in the high-current phase determine the temperature distribution at current zero. The temperature decays according to the energy balance of the most frequently occurring particles, that is the neutral particles. The charge-carrier concentrations follow the neutral particles' temperature and their densities during the first 100 μs are determined by recombination reactions, whereas later on they approach the partial, diatomic equilibrium values.

3.4. Decaying Axially-Blown Arc with Recovery Voltage

When a recovering voltage is applied to a decaying arc, two additional effects appear. Firstly, the charge carrier velocities are changed with respect to the neutral gas velocity $\vec{v}$ by the drift velocities $\vec{V}_+$, $\vec{V}_-$, $\vec{V}_e$. Secondly, the reaction rates (r,d), which were functions only of temperature in the absence of the recovering voltage, become dependent upon the electrical field strength.

The influence of the drift velocity is negligible on the axis in the case of a homogeneous column. Thus, only the changed reaction rates need to be investigated.

According to our present knowledge, the main reactions with respect to the change in carrier concentration are the following ones:

Immediately after current zero radiative electron-ion-recombination is most important:

$$S^+ + e^- \longrightarrow S + h\upsilon. \tag{13}$$

Later, dissociative attachment and mutual neutralization take over

$$F_2 + e^- \xrightarrow{r_2} F + F^- \tag{14}$$

$$F + F^- + S^+ \longrightarrow SF + F \tag{15}$$

Later, still near breakdown, collisional ionization becomes dominant

$$S_2 + e^- \xrightarrow{r_3} S_2^+ + e^- + e^-. \tag{16}$$

To give the connection between the reaction rates r_2, r_3 which have been used here, and the coefficients for ionization α and attachment η which are commonly applied in breakdown investigations (see the paper of T. H. Teich and W. Zaengl), the corresponding relations are employed:

$$\alpha = \frac{n_{S_2} \langle v\sigma_i \rangle}{v_e} = \frac{n_{S_2} r_2}{v_e} \tag{17}$$

$$\eta = \frac{n_F \langle v\sigma_a \rangle}{v_e} = \frac{n_F r_3}{v_e} \tag{18}$$

where $\sigma_{i,a}$ are the cross sections for ionization and attachment, v the electron velocity and $\langle \rangle$ indicates an average over the electron distribution function.

The breakdown voltage can now be determined by solving the reaction equations for all important reactions simultaneously with Poisson's equation for the electric field. This is a difficult task, because the reaction rates are functions not only of temperature, but also of the electric field. The available data are even more meagre than previously and it will be some time in the future before such an understanding can be completed.

Nevertheless, our present understanding of the processes involved permits a qualitative description and a hypothesis of the processes which occur.

For a given current, the axial temperature T_A first decays very rapidly and then more slowly (Fig. 11). From the diagram at a fixed time, say 100 μs after current zero, the axial temperature is found to depend upon the current. This dependence is plotted again in the temperature versus current diagram on the right side. The axial temperature yields a breakdown voltage U_{BD} which is shown in the breakdown voltage versus current curve plotted on the lower right side for the same conditions. We see that the breakdown voltage decreases slightly with rising current up to a critical value I_0. This is a consequence of the slow increase of the axial temperature. Above I_0 the increase in the axial temperature is more pronounced resulting in a stronger decrease in the breakdown voltage. This leads us to the typical form of breakdown

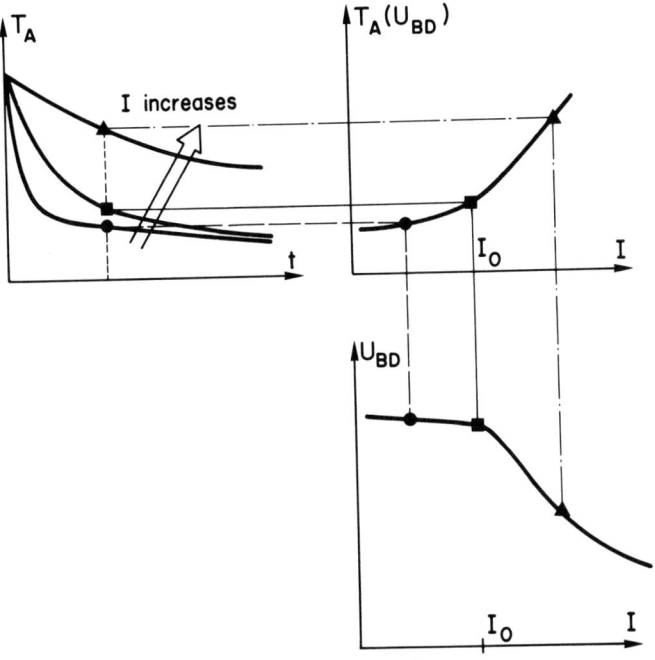

Fig. 11. Qualitative connection between decay of axial temperature T_A and breakdown voltage U_{BD}. For details see text.

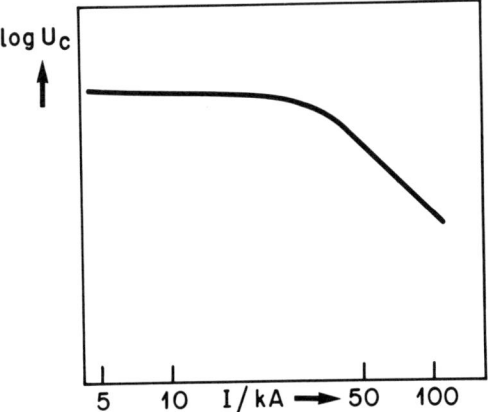

Fig. 12. Double-log plot of U-I-breakdown characteristic of Fig. 11.

characteristic found from circuit-breaker measurements which is shown in a double-log plot (Fig. 12).

Much work remains to be done for the quantitative prediction of the breakdown voltage in circuit breakers, but essential steps have been performed and they have revealed a promising way to proceed towards this goal.

The paper is a survey of the research activities of K. P. Brand, B. Eliasson, J. Kopainsky, U. Kogelschatz, L. Niemeyer, K. Ragaller, E. Schade at the Brown Boveri Research Center in the field of dielectric breakdown.

APPENDIX 1: PARTIAL EQUILIBRIUM DIAGRAM

The equilibrium-density diagram, calculated following Frie,[3] requires the knowledge of binding energies and partition functions of all components.

The necessary atomic data for atoms and ions are ionization energies, energy terms, and quantum numbers which have been taken from the tables of Moore.[5,6,7]

The binding energies of molecules and molecular ions (energy of dissociation ED, attachment EA, ionization IP) have been collected from different references as listed in Table 1.

The molecular partition functions are calculated according to Herzberg[8] with the following material coefficients: vibrational frequency ω_e, anharmonicity constant $\omega_e x_e$, rotational constant B_e, vibrational-rotational interaction constant α_e, internuclear distance r_e, symmetry number σ, and the two nuclear spins I_1 and I_2. For missing data the values have been deduced from isoelectronic molecules. Table 1 gives a list of the data employed for both relevant and isoelectronic molecules, including the literature references. As a help for orientation the electron configuration is also given.

The result is plotted in Fig. 6.

Explanations to Table 1

Atomic data for the calculation of the molecular partition functions. Isoelectronic pairs: (F_2^+, FO), (SF^-, ClF), (SF^+, PF), (SCl, S_2^-), (S_2^+, SP). Data taken from[10,15]. SCl-data from SCl-compounds.[20]

The values deduced from isoelectronic molecules have been given a number index in parantheses with the following significance:

1: from ED and ω_e, assuming a Morse potential
2: both ω_e and α_e are assumed to have the same value as the isoelectronic molecule
3: from α_e, ω_e, and $\omega_e x_e$, assuming a Morse potential
4: from the formula for a rigid rotation $[r_e(A)/r_e(B)]^2 = B_e(A)/B_e(B)$
5: average of the S-Cl bond lengths in various molecules[20]
6: From[17], $\omega_e - 2\omega_e x_e = 0.065$ eV and ED, assuming a Morse potential
7: from B_e, ω_e and $\omega_e x_e$, assuming a Morse potential
8: estimated from $\alpha_e(F_2)$, $\alpha_e(FO)$ and[11]
9: assumed to have the same value as $B_e(SF)$
10: average of data from[12] and [16]

TABLE 1

Molecule	F_2^-	F_2	F_2^+	SF^-	SF	SF^+	S_2^-	S_2	S_2^+
Config.	$^3\Sigma_u^+$	$^1\Pi$	$^2\Pi$	$^1\Sigma^+$	$^2\Pi$	$^3\Sigma$	$^3\Sigma$	$^3\Sigma_g^-$	$^2\Pi$
ω_e (cm^{-1})	510[9]	923.1[10]	1073[11]	784.29(2)	914[12]	846.75(2)	529(6)	724.66[10]	743.5(2)
$\omega_e x_e$ (cm^{-1})	6.10(1)	16.04[10]	9.00[11]	10.32(1)	5.12[12]	5.77(1)	2.13(6)	2.852[10]	2.63(1)
B_e (cm^{-1})	0.50[9]	0.8938[10]	1.200(3)	0.361(3)	0.553[13]	0.553(9)	0.2455(4)	0.2948[10]	0.3299(3)
α_e (cm^{-1})	0.0073(7)	0.022[10]	0.014(8)	0.004329(2)	0.004[12]	0.00456(2)	0.0013(7)	0.0016[10]	0.0016(2)
r_e (Å)	1.91[9]	1.41[10]	1.22(4)	1.979(4)	1.599[13]	1.599(9)	2.07(5)	1.889[10]	1.786(4)
ED (eV)	1.29[14]	1.65[9]	3.19[15]	1.81[2]	3.5[16]	3.8[12]	4.04[15]	4.38[15]	6.46[15]
EA (eV)	–	3.08[14]	–	–	2.1(10)	–	–	1.663[17]	–
IP (eV)	–	15.83[18]	–	–	10.09[16]	–	–	8.3[19]	–
σ	2	2	2	1	1	1	2	2	2
I_1, I_2	1/2, 1/2	1/2, 1/2	1/2, 1/2	0, 1/2	0, 1/2	0, 1/2	0, 0	0, 0	0, 0

Explanations see page before.

APPENDIX 2: CHEMICAL REACTIONS

The reactions which occur are mainly two- and three-body collision reactions. The reaction partners are electrons, atoms, ions, diatomic molecules, diatomic molecular ions, and photons.

Two-body collision reactions with photons are described schematically by

$$A + B \underset{d_{II}}{\overset{r_{II}}{\rightleftharpoons}} AB + h\nu \qquad (1)$$

The difference in energy and momentum which appears during recombination is carried off by radiation. The concentration of component A decreases by recombination (recombination rate r_{II}) and increases by decomposition (decomposition rate d_{II}) of the molecule AB:

$$\frac{\partial n_A}{\partial t} = -r_{II} n_A n_B + d_{II} n_{AB} n_{ph} \qquad (2)$$

For equilibrium, which is denoted by the index 'eq', the left-hand side of equation (2) is zero and a relation between r_{II} and d_{II} holds (law of mass action)

$$\frac{d_{II}}{r_{II}} = \frac{n_A n_B}{n_{AB} n_{ph,eq}} = f(T) \qquad (3)$$

where $f(T)$ is a function of temperature T.

As a first approximation, equation (3) can also be applied to irreversible processes, eq. (2).

The determination of the equilibrium photon concentration $n_{ph,eq}$ requires the solution of the radiation-transport equation which necessitates a considerable calculation effort. Therefore, only the two limiting cases, optically thin ($d_{II} = 0$; i.e. the recombination radiation leaves the ion cylinder completely) and optically thick are considered. For the latter case the approximate representation

$$d_{II} n_{ph} = \left| \frac{n_A n_B}{n_{AB}} \right|_{eq} \qquad (4)$$

is used. The calculations show that the two limiting cases practically coincide for a decaying high-current arc.

Three-body collision reactions have the general form

$$A + B + M \rightleftarrows AB + M + h\upsilon \tag{5}$$

M is the third reaction partner which carries away the energy and momentum deficit. For reactions with negligible radiation the change in concentration of component A is

$$\frac{\partial n_A}{\partial t} = -r_{III} n_A n_B n_M + d_{III} n_{AB} n_M. \tag{6}$$

The rate coefficients r_{III} and d_{III} are related by

$$\frac{d_{III}}{r_{III}} = \left.\frac{n_A n_B}{n_{AB}}\right|_{eq} = \frac{Z_A(T) Z_B(T)}{Z_{AB}(T)} \cdot e^{-E_R/kT} \cdot n_{tot,eq} \tag{7}$$

where E_R is the reaction energy (energy of dissociation, attachment, or ionization), Z_A, Z_B, Z_{AB} the total partition function, i.e. including internal degrees of freedom as well as integration over the phase space, n_{tot} the sum of all particle concentrations and k is Boltzmann's constant.

The reaction rates depend in a complex way on such reaction details as cross sections, excited intermediate states, amount of radiation, activation energy, etc. Calculations starting from first principles are not possible with today's knowledge.

Reactions for which no data could be found were approximated by simple models, the details of which can be found in Brand and Kopainsky[4].

The reactions can be arranged in different groups:

- A) recombination of atoms to molecules
- B) association of atom and positive ion
- C) association of atom and negative ion
- D) mutual neutralization of ions
- E) atomic detachment
- F) molecular attachment
- G) recombination of atom ion and electron
- H) recombination of molecular ion and electron
- I) ion exchange
- J) electron exchange
- K) neutralization of molecular and atomic ion.

The most important reactions, including reaction rates and references, are listed in Table 2 in the above mentioned order of groups.

The conditions in a high-voltage circuit breaker are closely isobaric in the time period of interest. The axial arc temperature decays according to Fig. 2. The concentration of the component i on the arc axis is changed by chemical reactions and by the simultaneous decay in temperature $(-n_i/T \cdot \partial T/\partial t)$. Each reaction k causes a

TABLE 2

Reaction		Rate	rate (r or d) = $aT^b \exp(-c/T)_1$ in cgs units			Reaction type
			a	b	c	
F + F + F	→ F$_2$ + F	r	5.4	10^{-35}	0	0	recombination of atoms to molecules (A)
F + S + F	→ SF + F	r	2.8	10^{-34}	0	0	
S + S + F	→ S$_2$ + F	r	4.5	10^{-34}	0	0	
F + S$^+$ + F	→ SF$^+$ + F	d	3.92	10^{-12}	1/2	43750	association of atom and positive ion (B)
S + S$^+$ + F	→ S$_2^+$ + F	d	381	10^{-12}	1/2	74735	
S + F$^-$ + F	→ SF$^-$ + F	d	3.92	10^{-12}	1/2	24950	association of atom and negative ion (C)
F + S$^-$ + F	→ SF$^-$ + F	d	3.92	10^{-12}	1/2	40849	
S$^+$ + S$^-$ + F	→ S$_2$ + F	r	6.32	10^{-20}	-5/2	0	mutual neutralization of ions (D)
F$^-$ + S$^+$ + F	→ SF + F	r	7.32	10^{-20}	-5/2	0	
F + e + F	→ F$^-$ + F	d	1.00	10^{-11}	0	40000	atomic detachment (E)
S + e + F	→ S$^-$ + F	d	4.23	10^{-12}	1/2	24105	
F$^-$ + F	→ F$_2$ + e	r	1.4	10^{-10}	0	22000	molecular attachment (F)
F$^-$ + S	→ SF + e	r	1.4	10^{-10}	0	4642	

BREAKDOWN IN CIRCUIT BREAKERS

Reaction	Rate	rate (r or d) = $aT^b \exp(-c/T)_1$ in cgs units			Reaction type
		a	b	c	
$S^+ + e + e \rightarrow S + e$	d	5.02 10^{-1}	-3/2	120230	recombination of atomic ion and electron (G)
$S^+ + e \rightarrow S + h\nu$	r	3.1 10^{-12}	0	0	
$SF^+ + e \rightarrow S + F$	r	3.46 10^{-6}	-1/2	0	recombination of molecular ion and electron (H)
$S_2^+ + e \rightarrow S + S$	r	3.46 10^{-6}	-1/2	0	
$S^+ + S_2 \rightarrow S_2^+ + S$	r	1.38 10^{-9}	0	0	
$S^+ + SF \rightarrow SF^+ + S$	r	1.00 10^{-10}	0	0	ion exchange (I)
$S^+ + SF \rightarrow S_2^+ + F$	r	9.90 10^{-10}	0	0	
$S^+ + F_2 \rightarrow SF^+ + F$	r	5.89 10^{-10}	0	0	
$S^- + SF \rightarrow S + SF^-$	r	3.28 10^{-11}	1/2	0	electron exchange (J)
$S_2^+ + F^- + F \rightarrow S_2 + F + F$	r	6.61 10^{-20}	-5/2	0	neutralization of molecular and atomic ion (K)
$S_2^+ + F^- + F \rightarrow SF + F + S$	r	6.61 10^{-20}	-5/2	0	
$S_2^+ + F^- + F \rightarrow S_2 + F_2$	r	6.61 10^{-20}	-5/2	0	
$S_2^+ + F^- + F \rightarrow SF + SF$	r	6.61 10^{-20}	-5/2	0	

net generation rate Γ_{ik} of the component i (see eqs. (2), (6)). The net generation of particles leads to a decrease in the pressure p. For isobaric conditions the net generation rate has therefore to be decreased by a term which describes the pressure deficit. Combining the three effects yields the total concentration change of component i:

$$\frac{\partial n_i}{\partial t} = \sum_k \Gamma_{ik} - \frac{n_i}{p} kT \sum_{i,k} \Gamma_{ik} - \frac{n_i}{T} \frac{\partial T}{\partial t} \qquad (8)$$

The simultaneous solution of all equations (8) yields the particle concentrations shown in Fig. 7.

REFERENCES

1. B. Eliasson and E. Schade, XIII ICPIG, Berlin (1977) 409
2. W. Hermann, U. Kogelschatz, K. Ragaller and E. Schade, J. Phys. D 7 (1974) 607
3. W. Frie, Z. Physik 201 (1967) 269
4. K. P. Brand and J. Kopainsky, submitted for publication in Appl. Phys.
5. C. E. Moore, Atomic Energy Levels, Vol. I, NBS (1948)
6. C. E. Moore, Atomic Energy Levels, Vol. II, NBS (1952)
7. C. E. Moore, Atomic Energy Levels, Vol. III, NBS (1958)
8. G. Herzberg, Molecular Spectra and Molecular Structure, Vol. II, Van Nostrand, New York (1966) 501
9. T. L. Gilbert and A. C. Wahl, J. Chem. Phys. 55 (1971) 5247
10. JANAF, Thermochemical Tables, 2nd ed., Nat. Stand. Ref. Data Ser. 37, NBS (1971)
11. American Institute of Physics Handbook, 3rd ed., McGraw Hill, New York (1972)
12. P. A. G. O'Hare, J. Chem. Phys. 59 (1973) 3842
13. A. Carrington, G. N. Currie, T. A. Miller and D. H. Levy, J. Chem. Phys. 50 (1969) 2726
14. W. A. Chupka, J. Berkowitz and D. Gutmann, J. Chem. Phys. 55 (1971) 2724
15. A. G. Gaydon, Dissociation Energies and Spectra of Diatomic Molecules, London, Chapman and Hall (1968)
16. D. L. Hildenbrand, J. Phys. Chem. 77 (1973) 897
17. R. J. Celotta, R. A. Bennett and J. L. Hall, J. Chem. Phys. 60 (1974) 1970
18. J. T. Herron and V. H. Dibeler, J. Chem. Phys. 32 (1960) 1884
19. P. Bradt, F. L. Mohler and V. H. Dibeler, J. Res. Nat. Bur. Stand. 57 (1956) 223
20. A. J. Banister, L. F. Moore and J. S. Padley in G. Nickless (ed.), Inorganic Sulfur Chemistry, Elsevier, Amsterdam (1968) 137-198

DISCUSSION
(Chairman: W. Bötticher, Technical University, Hannover)

J. Mentel (Ruhr University, Bochum)

I would like to show a measured curve of the dependence on current of the dielectric strength of a double-nozzle circuit-breaker configuration for SF_6 at a pressure of 8 bar (Fig. 1). This is a measurement we did at the Siemens Research Centre. The abscissa is the peak current of a 50 Hz sine wave, and the points are measured breakdown voltages. In these tests we used a high rate of rise of the trv, namely 8 kV/μs. We carried out tests for different positions of the arc-catcher electrodes in the nozzle. As you can see, this has no influence on the breakdown voltage. This measured curve confirms the results given here, at least qualitatively.

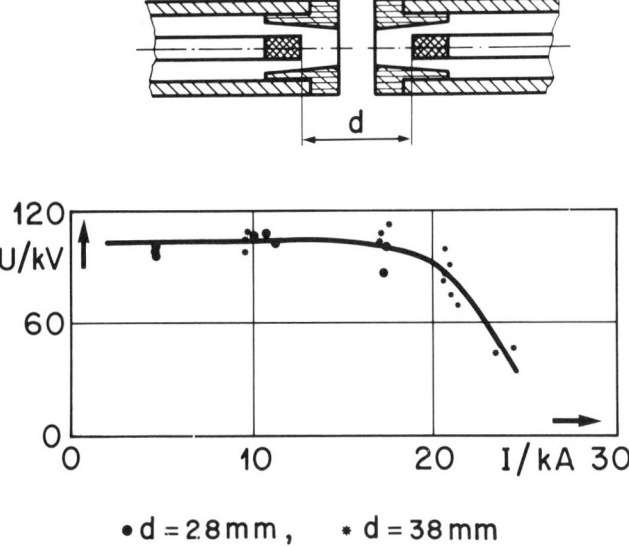

Fig. 1. Breakdown voltage versus current of a double-nozzle circuit-breaker set-up.

J. J. Lowke (University of Sydney)

In a paper by Hertz[1] the conductance decay of a wall-stabilized SF_6 arc was measured to be greater than either he or I could predict theoretically on the basis of equilibrium considerations. He gave the explanation, at that time, that this enhanced decay was due to the fact that the formation of molecules out of S and F was not fast enough and therefore one didn't get the heat of formation of these molecules.

J. Kopainsky

You're talking now about thermal properties. We did not look into that as we were interested in the particle composition with respect to dielectric properties. The only thing I can say is that in the decaying arc channel it is not possible, in the time interval 100-1000 μs after current zero, to build up larger than diatomic molecules.

T. E. Browne (Westinghouse)

I just wonder about that last statement because we find in practice that after arcing almost all of the S and F has gone back to SF_6.

J. Kopainsky

Yes, that's correct: it just takes more time than 1 ms.

W. Hertz (Siemens)
Is it correct that the concentrations of S_2^+ and F^- ions deviate by a factor of more than 100 from the equilibrium values and wouldn't that have to be taken into account in the energy equation as well?

J. Kopainsky
You're right if you refer to earlier equilibrium calculations, such as those by Frie[2] where a number of important species were neglected. What we did, on the other hand, was to calculate a partial-equilibrium-diagram for the complete set of 2-atomic species. Our ion concentrations approach this kind of equilibrium relatively quickly.

J. Mentel
You showed us the influence of the ion column on the electric field and you assumed a constant field along the channel. Wouldn't you expect an axial nonuniformity in the turbulent zones for example? If yes, this wouldn't fit to our experiments (Fig. 1 of discussion) where we couldn't find any differences for different positions of the catcher electrodes.

J. Kopainsky
It's hard to tell what influence turbulence would have on the ion channel. This will be part of our future research.

REFERENCES

1. W. Hertz, Z. Phys. 245 (1971) 105
2. W. Frie, Z. Phys. 201 (1967) 269

AUTHOR INDEX

Benenson, D. M., 27, 94, 116, 202, 327
Bötticher, W., 327
v. Bonin, E., 93, 227
Browne, T. E., 94, 182, 353
Catenacci, G., 94, 201
Cowley, M. D., 179
Diesendorf, J. L., 231, 264, 268
Dubanton, C., 185, 202, 203
Frind, G., 27, 67, 93, 94, 117, 135, 182, 183, 227
Hermann, W., 117, 205, 226
Hertz, W., 28, 182, 354
Humphries, M. B., 29, 64, 65, 203, 265
Jones, G. R., 95, 116, 117
King, L. H. A., 184
Kopainsky, J., 329, 353, 354
Lowe, S. K., 231
Lowke, J. J., 27, 180, 294, 299, 327, 353
Mentel, J., 119, 135, 352, 354
Noeske, H. O., 28, 64, 296
Parrott, P. G., 65, 268
Ragaller, K., 1, 26, 27, 28, 64, 65, 116, 134, 180, 182, 183, 205, 228, 296, 327
Reichert, K., 1
Rieder, W., 182
Ruoss, E., 202, 264
Saunders, L., 231
Swanson, B. W., 137, 182
Teich, T. H., 269, 296
Tuma, D. T., 26, 227
Urbanek, J., 26, 64, 202, 226, 264
Zaengl, W., 269
Zückler, K., 93

SUBJECT INDEX

ablation 93, 320
  protected probes 108
absorptivity, see radiation
air 79, 81, 150, 164, 165, 166, 167, 169, 313, 317
amplitude factor, see peak factor
arc
  decay 106, 111, 113, 114, 183, 210, 330, 331, 338, 343
  flow stabilized arc 99, 107
  high current arc 13, 14, 311, 330, 343
  instability 13, 99, 100
  models 137-179, 210
  properties 11, 137, 179, 299-327
  radius 13, 14, 74, 143, 155, 183, 316, 317
  resistance 52, 68, 157, 160
  root 119-134
  temperature 12, 14, 128, 148, 153, 157, 313, 316, 331
  turbulence, see turbulence
  voltage 154, 157, 235
  wall stabilized arc 103, 105, 106
area of application of circuit breakers, see limiting curve diagrams
attachment 293, 344
  coefficient 19, 271, 344
avalanche 272, 278

breakdown 3, 269, 300, 322, 329
  criteria 19, 21, 285, 288, 295
  voltage 277, 344
bus
  bar 51, 188
  fault, see terminal fault

cables 245, 258
cancellation current 235, 326
capacitor
  banks 258
  parallel capacitor 25, 64, 181, 189, 224
  voltage transformer 45
carbon 72, 122, 135
  jet 126
  plasma 125
Cassie and Mayr equation 137, 159, 180, 182, 299, 318
chambers in series 24, 214
charging power 5

cathodic feedback 278
chemical reactions 335, 348
circuit
  models, see model circuits
  parameters 29-64, 77, 185-201, 231-264
  Weil circuit 69
circuit breaker
  air circuit breaker 11, 138, 212
  area of application of circuit breakers, see limiting curve diagrams
  duties 53
  gas-blast circuit breakers 11, 67, 69, 108, 109, 330
  ideal circuit breaker 4, 186, 207, 235
  oil circuit breakers 11
  physics of circuit breakers 10, 67-93, 95-116, 119-134, 137-179, 269-294, 299-327, 329-352
  SF_6 circuit breakers 11, 70, 134, 138, 180, 205, 269
  testing of circuit breakers 25, 232
clogging 72, 93, 138, 152, 172, 300, 318, 320
coefficients for ionization, see ionization
comparison of theory and experiment 83, 147, 148, 152, 153, 254, 256, 317, 320
conditioning effect 286
conductance decay, see arc decay
conductor
  clashing 29
  geometry 246
constricted arc modes 313
contact
  fingers 135
  separation 11, 13
contamination 120, 133, 287
continuity equation 144, 179
control of slf severity 56
convection 13, 111, 311, 333
copper 72
  tungsten 72, 120
  vapor 120
corona breakdown 291
crest voltage, see peak transient recovery voltage
current
  asymmetry, see dc component
  chopping 181
  deformation 78, 94, 109, 219
  injection method 232

357

SUBJECT INDEX

interruption of small inductive currents 22, 181
keep alive current 70
limit 22, 216, 220, 224
oscillogram 3, 15, 161, 228, 236, 253
post-arc current 158, 163, 182, 209, 211, 227, 228
transformer 45, 203

damping resistor 232
d.c. component 9, 55, 54, 236
decay of charge-carrier concentration 335
decaying axially-blown arc, see arc decay
design
  of the breaker 10, 182
  parameters 76, 80, 119, 138
dielectric
  breakdown, see breakdown
  failure 1, 3, 21, 329
  interruption mode 19, 24, 68, 231-264, 299, 329-352
  limiting curve 22, 27, 345, 353
  strength 121, 269-294, 352
diffusion 342
discharge phenomena 269-294
double-nozzle circuit breaker 10, 28, 134, 330, 341, 352
double-scan technique 183

earth resistivity 42
earthed fault, see fault
eddies, see turbulence
electrical
  conductivity 20, 84, 112, 120, 145, 306
  field strength 14, 17, 21, 109, 110, 148, 182, 183, 227, 341
electrode, see also contacts
  configuration 119
  distance, optimum 76
  graphite, see carbon
  material 72, 119, 122, 132
  vapor 72
electron
  avalanche 274
  density 120, 271, 336
  emission 278
elementary
  circuit responses 5, 7, 248
  networks, see model circuit
emission coefficient 301
energy balance 16, 144, 179, 306, 313, 318, 332
equivalent circuit, see model circuit
erosion rate 126, 128
excursion factor, see peak factor
experimental model, see model interruptor

fault
  current 9, 40, 52, 55, 60, 200, 264
  duty 233
  earthed fault 8, 38, 64, 265, 269
  location 34, 56, 231-264
  out of phase fault 22, 329
  percent current fault 26
  remote fault 235, 236
  short line fault 1, 4, 19, 29-64, 78, 181, 202, 216, 234
    factor 7, 31, 40
    system aspects of the slf 52
  single-phase fault 240
  terminal fault 1, 3, 34, 213, 231-264
  three-phase fault 9, 36, 37, 58, 64, 240, 265
  unearthed fault 9, 64, 240, 265
feedback coefficient 278, 281
field testing 199, 232
figure of merit 169
first-pole-to-clear factor 4, 9, 36, 37, 234, 240
flow field structure 85, 98, 100, 108, 173
gas blast
  arcs, see arcs, flow stabilized
  circuit breakers, see circuit breakers
gas mixtures 84, 171, 293
gas property functions 164, 168
generation of vorticity 14, 98, 141
glow discharge 295
graphite, see carbon
grounded fault, see fault (earthed)

heat-flux potential 332
high-current phase 13, 14, 330, 343
high frequency parameters 32, 33, 264

inception voltage 291
inductance, self and mutual 38
influenced field 282, 341
inherent trv, see transient recovery voltage
inhomogeneous fields 281
initial transient recovery voltage, see transient recovery voltage
interaction arc/network 19, 162, 186, 207, 232
interrupting
  capability 70, 138, 149, 169, 176
  influence of
    current 75, 149
    gas 82, 85, 171
    geometry 76, 80, 86
    pressure 81, 87
    rate of fall of current 78, 79
    see also limiting curve diagrams
interruptors, see circuit breakers
ionization
  coefficient 19, 269, 271, 293, 344
  energy 120

SUBJECT INDEX

ion 21, 130, 271, 272, 330, 335
   column 340, 354
   concentration 336, 338, 342
   currents 130, 273
   drift velocity 273
isolator 50
isothermal approximation 152
Joulean heating 17

kinetics of the reactions 336

Laplace transform 239
Laval nozzle 71, 101, 175
lattice diagram 238, 241
layout of substation, see substation
limiting curve diagrams 17, 23, 24, 64, 67, 76, 88, 150, 181, 211, 216, 223
line
   side oscillation 30, 34
   parameters, see transmission line
   terminal capacitance 29
   traps, see wave traps
linear networks 235
load 247, 256, 268
local thermal equilibrium 301
lumped capacitance 44

Mach number 152, 318
metal vapor 12, 28
mixing length 97, 103, 114
mobility 21
model circuit 4, 5, 50, 187, 207, 213, 238, 243, 248, 255
model interruptor 18, 20, 71, 87, 100, 111, 132, 330, 353
modeling technique 87
molecules, metastable 278
momentum equation 139

network
   model 244, 250, 251, 255, 257, 261
   phenomena 2, 29-64, 185-201, 231-264
   radius of influence 9, 10, 243
nitrogen 149, 166, 304, 330
nominal network voltage, see rated voltage
nozzle
   arc theory, see arc models
   clogging, see clogging
   diameter 76, 318
   expansion angle 86
   pressure ratio 173

Ohm's law 145, 306, 342
oscillograms, see current or voltage oscillograms

partial
   discharge 291
   equilibrium 335
particle densities 125, 270, 336, 338
Paschen relationship 285

peak
   factor 6, 26, 32, 226, 234
   recovery voltage, see transient recovery voltage
   regime 231-264
phase factor, see first-pole-to-clear factor
photoionization 323, 327
plant, see substation
power
   frequency parameters 30, 239, 245, 264
   losses 106, 108
post-arc current, see current
Prandtl number 97, 140
pre-breakdown currents 21, 286, 294
pressure 81, 88, 89, 152
   distributions 11, 151, 176, 320, 340
   ratio 11, 80, 173
   see also flow field
protection 2, 55
protrusions 287, 289

radial intensity profiles 184
radiation 13, 127, 142, 299-327
   continuum radiation 143, 301, 312
   emission coefficient 314
   energy transport 20, 300
   flux density 308
   line radiation 312
   net emission coefficient 310
   spectral absorptivity 304
   spectral lines 143, 302
      broadening 302, 303
      visible region 318
rate of fall of current 69, 77
rated
   breaking current 24, 31, 52, 232
   voltage 24
recovery speed 75, 79, 82, 83, 84, 85, 106
relaminarization 27

sawtooth oscillation 7, 30, 248
sensitivity analysis 256
sequence component method 8, 33, 36, 194, 238, 240
series inductance 191, 243
SF_6 18, 79, 81, 107, 120, 135, 142, 149, 150, 164, 165, 166, 169, 182, 270, 276, 285, 293, 302, 313, 314, 317, 330, 352
shadowgram 101, 102
shock waves 173
short-circuit, see fault
   power 5
short line fault, see fault
shunt capacitance 56, 181, 213, 241
similarity 107, 327
single-flow device 71, 135
single phase
   diagram 7, 186, 244
   fault, see fault
   simulation 4, 193

source side
  capacitance 33, 56, 213
  waveform, see trv
space-charge 282, 296, 330, 341
stagnation point 11
statistical process 134
straining 112, 116
streak photography 15, 101, 102, 183, 282, 331
streamer criterion 280
substation 7, 44, 50, 187, 192, 197
superposition, 8, 235
surge impedance 36, 38, 39, 43, 54, 194, 237, 245
  load 7
switchgear, see circuit breaker
symmetrical components, see sequence components
synthetic circuit 69

temperature, see arc temperature
terminal
  capacitance, see source side capacitance
  fault, see fault
test oscillogram 3, 4
theoretical model, see arc models
thermal
  arc boundary layer 12, 13, 20, 323, 325
  conductivity, equivalent 97, 140, 310
  diffusivity 106, 156, 323, 332
  failure, see thermal interruption mode
  interruption mode 1, 3, 14, 17, 18, 24, 68, 150
  recovery speed, see recovery speed
throat diameter, see nozzle diameter
time
  constant 114, 155
  delay 45, 64, 202, 213
  to peak 198, 218, 234
tower footing resistance 42
Townsend
  breakdown criterion 280
  equation 19
transfer element 241
transformer 243, 247, 268
transient recovery voltage (trv) 2
  calculation 2-10, 29-64, 185-201, 231-264

effect of plant on the trv 44
estimation of the trv 262
frequencies 246
inherent trv 4, 186, 207, 217, 231, 235
initial trv 10, 51, 185-201, 205
line-side trv 6, 266
peak of trv 3, 6, 10, 19, 21, 22, 62, 231, 264
rate of rise of trv 2, 64, 73, 74, 173
source-side trv 6, 29, 33, 254, 266
specification 234
total trv 57, 59, 232
wave shape 26, 265
transmission line 36, 237, 245
  conductor clashing 39
  conductor construction 38
  50 Hz parameters 40
  line height 41
  overhead line 38
  parameters 36, 238
transient network analyzer 235
travelling waves 9, 30, 237, 239
turbulence 1, 14, 16, 27, 95-116, 153, 180, 183, 354
turbulent
  heat transfer 17, 96, 97, 140
  length scale 103
  Prandtl number 140
  shear layers 111
  viscosity 97, 139

unearthed fault, see fault
ungrounded fault, see fault

vacuum ultra violet, see radiation
vapor eruptions 129
vaporization 126
velocity of propagation 31, 43
voltage
  oscillogram 4, 161, 236, 253
  transformer 44, 187
vortex 135

wave
  propagation, see travelling waves
  trap 48, 188, 191, 195